W9-CWW-039

SOCIETY FOR EXPERIMENTAL BIOLOGY

SEMINAR SERIES · 9

BRAIN MECHANISMS OF BEHAVIOUR IN LOWER VERTEBRATES

BRAIN MECHANISMS OF BEHAVIOUR IN LOWER VERTEBRATES

Edited by
P. R. LAMING
Lecturer in Zoology
The Queen's University of Belfast

CAMBRIDGE UNIVERSITY PRESS
Cambridge
London New York New Rochelle
Melbourne Sydney

Published by the Press Syndicate of the University of Cambridge
The Pitt Building, Trumpington Street, Cambridge CB2 1RP
32 East 57th Street, New York, NY 10022, USA
296 Beaconsfield Parade, Middle Park, Melbourne 3206, Australia

First published 1981

Printed in Great Britain by
Fakenham Press Limited, Fakenham

British Library Cataloguing in Publication Data

Brain mechanisms of behaviour in lower vertebrates.
(Society for Experimental Biology.
Seminar series; 9).

1. Vertebrates – Behaviour
2. Vertebrates – Physiology
3. Nervous system – Vertebrates
I. Laming, P. R. II. Series
596′.0188 QL751 80–41368

ISBN 0 521 23702 5 hard covers
ISBN 0 521 28168 7 paperback

CONTENTS

CONTRIBUTORS

Abramson, I. S.
Dept of Physiology, University of Witwatersrand Medical School, Hospital Street, Johannesburg, South Africa.

Aronson, L. R.
Dept of Animal Behavior, American Museum of Natural History, Central Park West at 79th Street, New York 10024, USA.

Belich, A. I.
Laboratory of Comparative Physiology of Sleep, Sechenov Institute of Evolutionary Physiology and Biochemistry, Academy of Sciences of the USSR, Leningrad, USSR.

Borsook, D.
Dept of Physiology, University of Witwatersrand Medical School, Hospital Street, Johannesburg, South Africa.

Davis, R. E.
Neuroscience Laboratory, 103 East Huron, University of Michigan, Ann Arbor, Michigan 48109, USA.

Demski, L. S.
School of Biological Sciences, 101 Thomas Hunt Morgan Building, University of Kentucky, Lexington, Kentucky 40506, USA.

Ebbesson, S. O. E.
Culebra Biological Station, Culebra, Puerto Rico, USA 00645, *and* Dept of Anatomy, Catholic University of Puerto Rico School of Medicine, Ponce, Puerto Rico, USA 00731.

Ewert, J.-P.
Neuroethology and Biocybernetic Laboratories, University of Kassel, Heinrich-Plett Strasse 40, 3500 Kassel, West Germany.

Flood, N. C.
Colorado West Regional Mental Health Center, PO Box 955, Glenwood Springs, Colorado 81601, USA.

Guthrie, D. M.
Dept of Zoology, University of Manchester, Williamson Building, Manchester M13 9PL, UK.

Hara, T. J.
Canada Dept of Fisheries and Oceans, Freshwater Institute, 501 University Crescent, Winnipeg, Canada R3T 2N6.

Karmanova, I. G.
Laboratory of Comparative Physiology of Sleep, Sechenov Institute of Evolutionary Physiology and Biochemistry, Academy of Sciences of the USSR, Leningrad, USSR.

Kassell, J.
Neuroscience Laboratory, 103 East Huron, University of Michigan, Ann Arbor, Michigan 48109, USA.

Laming, P. R.
Dept of Zoology, Queen's University of Belfast, BT7 1NN, Northern Ireland.

Lazarev, S. G.
Laboratory of Comparative Physiology of Sleep, Sechenov Institute of Evolutionary Physiology and Biochemistry, Academy of Sciences of the USSR, Leningrad, USSR.

Martinez, M.
Neuroscience Laboratory, 103 East Huron, University of Michigan, Ann Arbor, Michigan 48109, USA.

Overmier, J. B.
Dept of Psychology, University of Minnesota, Minneapolis, Minnesota, USA.

Savage, G. E.
Dept of Zoology and Comparative Physiology, Queen Mary College, University of London, Mile End Road, London E1 4NS, UK.

*Shapiro, C. M.**
Dept of Physiology, University of Witwatersrand Medical School, Hospital Street, Johannesburg, South Africa.

Vanegas, H.
Centro de Biofísica y Bioquímica, Instituto Venezolana de Investigaciones Científicas, Apartado 1827, Caracas 101, Venezuela.

Vellet, D.
Dept of Physiology, University of Witwatersrand Medical School, Hospital Street, Johannesburg, South Africa.

Woolf, C. J.
Dept of Physiology, University of Witwatersrand Medical School, Hospital Street, Johannesburg, South Africa.

Wright, D. E.†
Dept of Zoology and Comparative Physiology, Queen Mary College, London E1 4NS, UK.

* Present address: Dept of Psychiatry (Royal Edinburgh Hospital), Morningside Park, Edinburgh EH10 5HF, UK.

† Present address: Dept of Biology, Plymouth Polytechnic, Drake Circus, Plymouth, UK.

PREFACE

Since the times of the early philosophers, man has been preoccupied with the study of how the human brain functions. The development in the last 50 years of sophisticated microscopes for fine anatomical study, and high-gain electronic amplifiers for recording neural impulses, has turned this preoccupation into a scientific study of how brain functions control behaviour. In conjunction with the new technology, there have developed numerous techniques for exploring the anatomy and physiology of the central nervous system in relation to behaviour.

With a few notable exceptions most of these neuroethological studies were performed with mammals, though the brains of these animals are in many respects more complex than those of other vertebrates. It is largely the belief that fish, reptiles and amphibians have simpler brains than birds and mammals that has prompted the recent revival of interest in comparative neuroethology. This, and the interest in the control of behaviour in such species in its own right, has caused the emergence of a large number of researchers, from many disciplines, whose prime interests are in understanding the brains and behaviour of fish, amphibians and reptiles.

The contribution such research makes to our overall comprehension of brain mechanisms in vertebrates has become very evident, especially in relation to visual information processing and learning, yet few opportunities are available for the general neuroethologist to gain an insight into current research in non-mammalian vertebrates.

The dearth of literature and research communication in the neuroethology of 'lower' vertebrates prompted the organization of an SEB symposium in July 1979 at Queen's University, Belfast, to present current research in the field to biologists in general. The bias of such research has been largely directed towards fish and amphibians and so this book reflects this interest.

It is evidence of the fruitfulness of research into neuroethology in fish and amphibians that this book merely scratches the surface of our current knowledge. I hope, however, that the major areas of advance are summarized and that readers from diverse fields of biology may capture, in essence, the excitement generated by some of the leading workers in this rapidly expanding field of research.

P. R. LAMING

Introduction

Man has an intrinsic interest in his own behaviour, which has led him to explore the brains of primates, rats and cats, amongst other mammals, intensively over the last few decades. Early research into brain function in relation to behaviour might be compared to the examination of the function of a large computer with a sledge-hammer. More recently, however, two major changes have taken place in neuroethological research. Firstly, a growing number of researchers have begun to look at brain function in non-mammalian vertebrates, which often have smaller and less complex brains, and secondly, new, more sophisticated techniques have become available. Now we are examining a desk-top calculator with a screwdriver.

It is the aim of every book to reach as wide an audience as possible, and this is especially true when an initial attempt is being made to fill a vacuum of information – here in the relationships between the brains and behaviour of lower vertebrates. To this end this book is divided into five parts, the first dealing with general evolutionary considerations of brain structure and function, the other four with specific biological functions like sensation, attention, appetite and learning.

Part I begins with an introductory chapter, which attempts to provide a ground-work of comparative vertebrate neuroethology for those without a classical biological training. Its brief introduction to the evolution of the brain and behaviour of lower vertebrates is, therefore, superfluous to those working in the field. More specialized views of the evolution of vertebrate brain structure and function are given in the chapters by Aronson and Ebbesson. These chapters together, give an evolutionary perspective from which the subsequent detailed analysis of specific functional systems within the brain can be viewed.

Part II examines researches into the ways in which fish and amphibians receive information from their environment. Guthrie considers the relationship between neuronal response in the teleost brain and behaviour, whilst Vanegas looks at the neuronal circuitry of the optic tectum and its responses in relation to behaviour. Another sensory system in teleosts, that of olfaction, is given consideration by Hara, who examines the relationship between the responses of fish both behaviourally and neurophysiologically to the chemical nature of the water in which they live. One of the most outstandingly successful areas of research in comparative neuroethology has been in the examination of visual processes in amphibians. Ewert describes, in his chapter, how toads may detect and respond appropriately to worm-like objects in particular spatial orientations in the environment.

The response of an animal to an environmental change does not, however, depend alone on its ability to detect that change, but also on its general level of responsiveness. In Part III these changes of responsiveness, from sleep to wakefulness to alertness, are considered. Shapiro *et al.* discuss the ontogeny of sleep in a teleost and Karmanova *et al.* its behavioural and physiological correlates in fish and amphibians, especially in relation to the evolution of sleep in vertebrates. Laming then looks at the correlates of aroused or alert behaviour and considers the possible adaptive functions of the physiological changes seen. Although these general changes in responsiveness will affect the way in which an animal behaves, so also will the specific need or appetites discussed in Part IV.

Part IV briefly examines possible brain regions for control of hunger (Demski) and sexual and parental behaviour (Davis *et al.*) in teleosts. This is of especial interest to those who relate their reading of this book to data obtained in mammals. Both the regions identified as being concerned in feeding (hypothalamus) and sexual/parental behaviour (telencephalon) are at least analogous if not homologous to areas performing similar functions in higher vertebrates. Whatever the appetitive or attentional state of a vertebrate, behaviour is not expressed in isolation from previous experience. The mechanisms of learning, considered in the final part of the book, have received considerable interest in recent years, especially in relation to the function of the teleost telencephalon.

In Part V, Flood & Overmier extensively examine the teleost telencephalon's function in learning, especially in relation to experiments in which removal of this area has contributed to our knowledge. Savage & Wright then demonstrate how electrical stimulation of the telencephalon may affect the stability of learnt associations in teleosts. Finally Barsook *et al.* show that the ability to store learnt information as memory may be affected not only by electrical stimulation but also by the presence of pituitary peptides.

The range of the scientific endeavours in neuroethology is vast, but I hope that those reported here will whet the appetite of the reader for the comparative approach to the subject. I should like to thank the SEB and Cambridge University Press, without whom this volume would never have appeared.

My thanks also go to the Zoology Department, Queen's University, especially Professor L. T. Threadgold, and his secretaries, Liz Purdy and Ruth Watt; and to Drs D. R. Bamford, A. Ferguson, R. V. Gotto and R. E. Elwood of the Zoology Department and Professor K. Brown of the Psychology Department for their advice and help during the editing of this volume.

PART I

Evolutionary perspectives in vertebrate neuroethology

P. R. LAMING

An introduction to the functional anatomy of the brains of fish and amphibians

(with comments on the reptiles)

The evolution of research in brains and behaviour (neuroethology) is easy to trace, as it is a recent product of the convergence of many disciplines, and its development is well documented. The same, unfortunately, cannot be said for the evolution of the brain functions and behaviour patterns which are the topics of study. Brains and behaviours leave few fossil records of their evolution; therefore we have to deduce this from rather tenuous evidence. Two major areas of knowledge have contributed to our understanding of the comparative neuroethology of vertebrates: (1) the ontogeny of the brains of vertebrates and (2) the comparative brain anatomy and physiology, and the behaviour, of extant species.

The pattern of embryological development of the brain is remarkably constant throughout the vertebrates. This constancy is most evident in the parts of the brain which develop first and which, from comparative anatomical evidence, we regard as most primitive or unspecialized. The summary recapitulation of evolution which occurs during ontogeny has proved invaluable in understanding brain morphology. Technical and interpretive difficulties make such studies less helpful when physiology and behaviour are being considered, yet such research is vital for the full understanding of brain function, at present largely explored by the methods of comparative neuroethology.

The advantages of comparative study lie in the ability to examine function within the brain, and the associated behaviour in the fully developed animal. The difficulties lie in the extrapolation of results to other vertebrates. Present-day vertebrates are not members of a linear phylogenetic scale, rather they represent the twigs on a tree whose trunk or trunks have long ceased to exist. Each species has undergone specialization during evolution and different parts of the brain may have gained or lost particular functions in the process. Extrapolation of results of neuroethological studies from one present-day species to another may be performed only with caution in the light of this evolutionary history. It is nevertheless unrealistic to neglect comparisons entirely, especially when they relate to basic brain functions like awareness, learning and appetitive behaviour.

It is from an understanding of basic brain structure and function in vertebrates with apparently simple brains, that much can be learnt about the phylogenetically older parts of the highly complex brains of mammals.

The ontogeny of vertebrate brains

The nervous system of vertebrates is one of the earliest groups of tissues to develop embryologically. In most cases the neural tube, formed by the closure of the neural groove in the dorsal mid-line, is developed fully by the time 10–15% of embryonic life has passed. The anterior thickening of this tube which occurs, to form the brain, indicates the cephalization of both sense organs and integrative centres which has occurred in vertebrates. Outgrowths of this early nervous system and connections with nervous tissue outside the neural tube form the spinal and cranial nerves. These are the routes by which the animal receives information from its own tissues and the environment, and relays commands to muscles, glands and sense organs through which a response is mediated.

Fig. 1. Lateral (*a*) and dorsal (*b*) views of a generalized lower vertebrate brain to show the main superficial features. Cer, cerebellum; Di, diencephalon; Ep, epiphysis (pineal); Inf, inferior lobe of hypothalamus; Med, medulla oblongata; OB, olfactory bulb; OT, olfactory tract; Sp, spinal cord; Tec, optic tectum; Vag, vagal lobes; Tel, telencephalon; 1–11, cranial nerves.

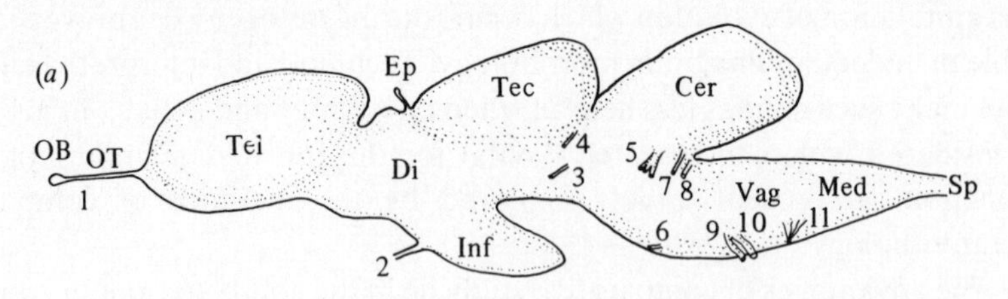

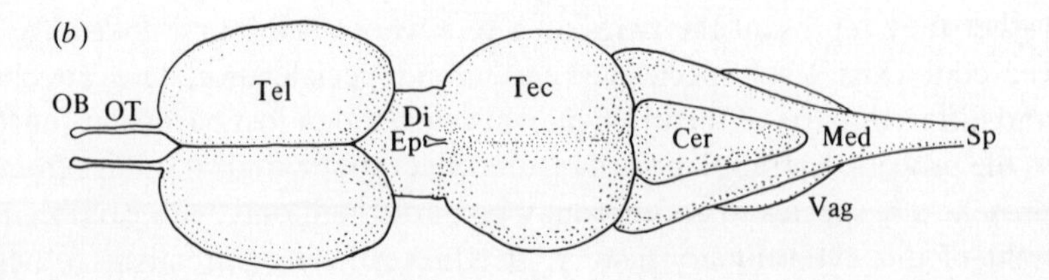

The anterior thickening of the neural tube develops into three distinguishable components: the prosencephalon (forebrain), mesencephalon (midbrain) and rhombencephalon (hindbrain). These three hollow chambers or vesicles are traditionally associated with the three primary senses of olfaction, vision and audition respectively, and together constitute the brainstem

Table 1. *The major divisions of the vertebrate brain*

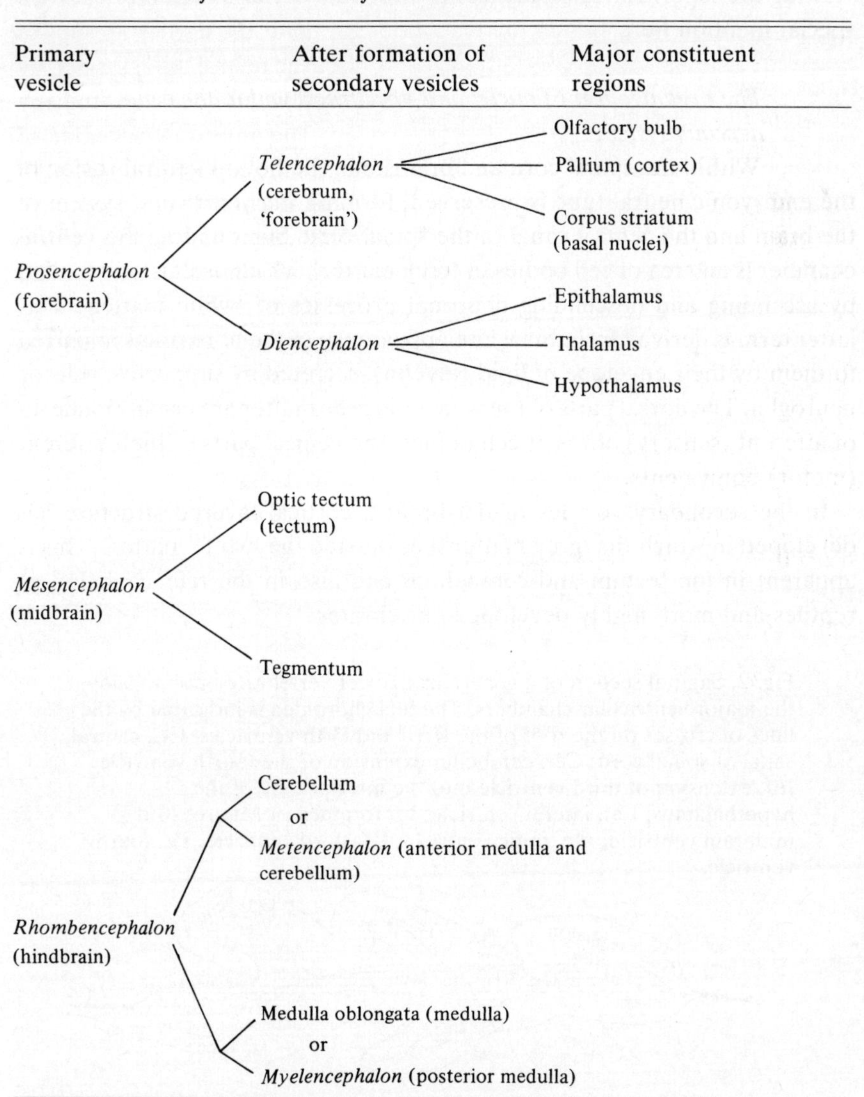

Primary vesicle	After formation of secondary vesicles	Major constituent regions
Prosencephalon (forebrain)	*Telencephalon* (cerebrum, 'forebrain')	Olfactory bulb
		Pallium (cortex)
		Corpus striatum (basal nuclei)
	Diencephalon	Epithalamus
		Thalamus
		Hypothalamus
Mesencephalon (midbrain)	Optic tectum (tectum)	
	Tegmentum	
Rhombencephalon (hindbrain)	Cerebellum or *Metencephalon* (anterior medulla and cerebellum)	
	Medulla oblongata (medulla) or *Myelencephalon* (posterior medulla)	

Embryological divisions are shown in *italic* type.

in the adult. Each develops a secondary outgrowth: the telencephalon (cerebrum), optic tectum and cerebellum respectively.

The external morphology of a generalized adult vertebrate brain thus has six major components (Fig. 1). Some confusion in terminology of these components is apparent in the literature, and is best clarified by a table (Table 1).

Most of the important features of vertebrate brains are not revealed by viewing the superficial anatomy alone, and some of these features deserve a special mention here.

The organization of nuclei and fibre tracts within the central nervous system

Within the spinal cord and brainstem, the hollow central region of the embryonic neural tube is preserved, forming the ventricular system of the brain and the central canal of the spinal cord. Surrounding this central chamber is an area of cell bodies or 'grey matter', which itself is surrounded by ascending and descending neuronal processes or 'white matter'. This latter term is derived from the white appearance of the fibre tracts imparted to them by their envelope of lipid (myelin), secreted by supportive cells or neuroglia. The dorsal parts of the white and grey matter are usually made up of afferent (sensory) fibres or cell bodies, the ventral parts of their efferent (motor) equivalents.

In the secondary vesicles of the brain a cortical layered structure has developed in which the 'grey matter' lies outside the 'white matter'. This is apparent in the tectum and cerebellum and also in the telencephalon of reptiles and more highly developed vertebrates.

Fig. 2. Sagittal section of a generalized lower vertebrate brain to show the major ventricular chambers. The tela choroidea is indicated by the lines of crosses on the roof of the IIIrd and IVth ventricles. CC, central canal of spinal cord; Cer, cerebellar extension of the fourth ventricle; Inf, extension of third ventricle into the inferior lobe of the hypothalamus; Lat, lateral ventricle; M, foramen of Munro; Mid, midbrain ventricle; Op, optic ventricle; III, third ventricle; IV, fourth ventricle.

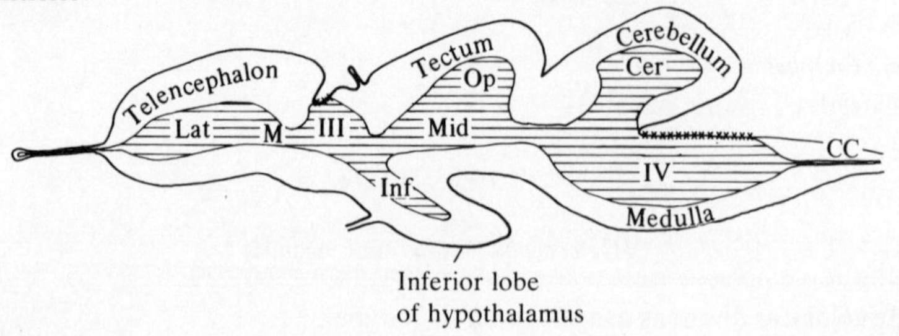

The ventricular system

All the major regions of the brain retain their hollow structure in the adult, and these vesicles are filled with cerebrospinal fluid. This fluid is essentially a filtrate of plasma which has been passed through fine plexuses of blood vessels to reach the ventricular chambers. It has a lower concentration of amino acids and a slightly different ionic constitution to plasma (Prosser, 1973) and is believed to have both nutritive and protective functions. The blood plexuses which provide the fluid, the tela choroidea, lie on the roof of the third ventricle at the telencephalic – diencephalic border and the roof of the fourth ventricle in the medulla (Fig. 2).

In fish, amphibians and some reptiles these two ventricles are connected directly by a midbrain ventricle, which has a dorsal extension (the optic ventricle) separating the tectum from the underlying tegmentum. The fourth ventricle of the medulla, as well as being continuous with the central canal of the spinal cord, extends dorsally in elasmobranchs into the cerebellum. Similarly there is in fish a ventral extension of the third ventricle into the inferior lobe of the hypothalamus (Sarnat & Netsky, 1974). Anteriorly the third ventricle connects to the two lateral ventricles of the telencephalon via the foramen of Munro. The shape of these ventricles and their extensions is related to the degree of development of the telencephalon itself. In holosteans and teleosts the evolution of an 'everted' telencephalon causes the lateral ventricles to lie on the surface of the brain rather than within it (Nieuwenhuys, 1969).

Forebrain development and evolution

The early developmental stage of the forebrain in all vertebrates has the form of a simple vesicle, the dorsal and ventral walls of which are thin and membranous. This arrangement is found in adult cyclostomes (Fig. 3). From both ontogenetic and comparative anatomical evidence an overall view of the evolution of the telencephalon has emerged (Fig. 3). In elasmobranchs, amphibians and reptiles there is a progressively greater evagination or bulging out of the thick side walls of the telencephalic vesicle. This evagination causes the dorsal and ventral parts of the thick lateral walls to lie above and below the lateral extensions of the telencephalic vesicle. The invagination of these newly formed dorsal and ventral walls causes them to meet in the median plane, forming two lateral ventricles.

In holosteans and teleosts the dorsal part of the thick lateral walls of the simple telencephalic vesicle bends outwards or everts – clearly the opposite of its development in other vertebrates. This eversion brings the telencephalic vesicle to lie on the dorsal and medial surfaces of the two telencephalic hemispheres as in thin ependymal sheet (Fig. 3). These differences in forebrain development and evolution mean that comparisons between forebrain functions in vertebrates rarely can be made on the basis of homology, though they may be analogous.

The structure and function of fish and amphibian brains

It is the aim of the neuroethologist to localize functions within the brain, yet often such endeavours have been thwarted by the apparent lack of such localization of functional units (Ebbesson, this volume). This lack of functional separation of brain regions is often even more apparent in 'lower' vertebrates (fish, amphibians and reptiles) than in birds and mammals. The

Fig. 3. The possible lines of evolution of the telencephalon illustrated by transverse sections of primitive, hypothetical and extant species. E, eversion; Evag, evagination; I, inversion; CS, corpus striatum; N, neocortex; P, pyriform; S, septum; H, hippocampus; V, lateral ventricle.

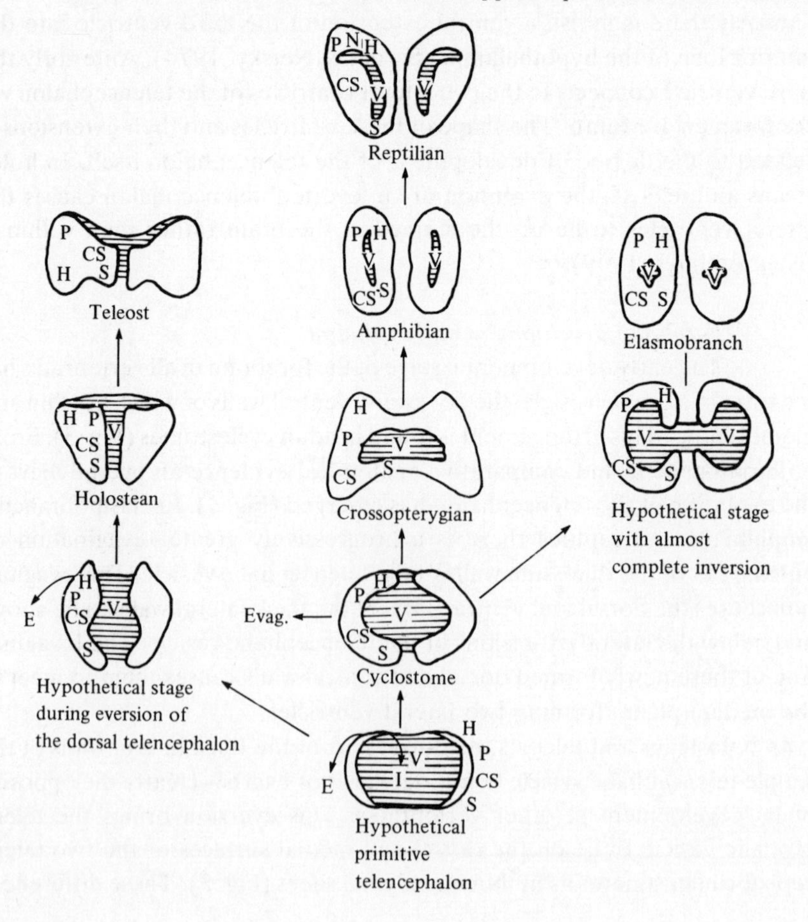

apparently diffuse influence of many regions on behaviours involving attention, learning, appetite, etc., make the traditional region-by-region examination of function of limited use to those concerned with how the brain, as a whole, works.

The variability of behaviour in lower vertebrate species is restricted because of the limitations on the size and complexity of the central nervous system (CNS). Indeed it is this apparent simplicity which makes such species suitable 'models' of more complex systems. However, during the approximately 500 million years since the emergence of the vertebrates, a considerable degree of specialization has occurred in both the behaviour and brains of extant species. The efficiency of the basic plan of the vertebrate CNS is evident from the similarity of the functional systems within the brains of diverse classes of both the aquatic and terrestrial species which have been studied. These systems are largely studied by examining the way in which environmental change affects the behaviour of the animal. Techniques of study will be considered later, but for the moment let us postulate the simplest expression of behaviour as being a result of two processes, the detection of an environmental (or internal) change and an integration of that information with other information available to the animal, thus:

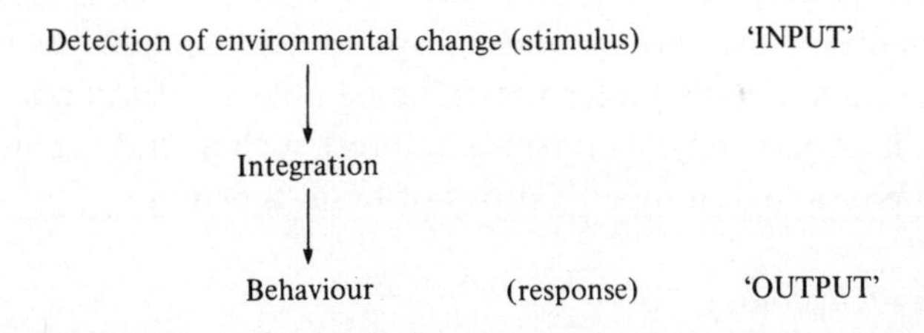

The simplest unit of behavioural response to an environmental change is the reflex arc, which requires little integration and in which a sensory neuron synapses directly with a motor neuron. The mammalian withdrawal reflex is of this type. Such relatively simple mechanisms are few in the behavioural repertoire of vertebrates, and the tremendously complex integrative functions which are prevalent are the ones primarily studied by neuroethologists. The development of interest in the integrative functions of the brain, however, has been dependent on an examination of the input and output systems, so as to narrow the gap in understanding how the stimulus may elicit a response. The development of electronic recording and microscopical techniques in the last 50 years has encouraged researchers of many disciplines to look at the anatomy of the brain from a more functional view than had been possible hitherto.

Gross anatomy

The simplest level of examination of the brain is the study of its gross anatomy. Such an examination, in isolation, gives a rough indication of the animal's capabilities. Thus, large brains in relation to the size of the animal may indicate a potential for complex behaviour. Similarly, relatively expanded regions of the brain may reflect the functional importance of that region in the life of the animal. A large cerebellum may be associated, therefore, with fine postural control, while the size of the vagal or olfactory lobes may reflect the importance of the sensory modalities of vibration and olfaction to cyprinids and elasmobranchs respectively. The danger of such a simplistic interpretation is related to the lack of localization of function described earlier, since many species have evolved to utilize regions for more than the function with which they were primitively concerned. The vagal lobes of cyprinids are concerned with taste and general cutaneous sensation as well as vibration sense.

Similarly the optic tectum of teleosts in general is not only a region for the synthesis of visual information but is also concerned with the integration of visual and auditory sensory inputs. The telencephalon of teleosts is concerned with both olfactory reception and such complex functions as learning and appetitive behaviour.

The gross anatomy of brains, therefore, may only serve as a rough guide to the importance of the various sensory and motor functions to an animal, more precise and detailed information being obtained from microscopical examination. Apart from its intrinsic interest such a study is an essential adjunct to physiological investigations of brain function.

Brain histology

Most of our knowledge of the fine structure of the brain is derived from careful histological examination of thin sections, from which a three-dimensional picture is built. Traditionally, wax-embedded brains, sectioned and stained with basic and acidic dyes such as haematoxylin and eosin, indicate the larger accumulations of neuronal processes (tracts) and cell bodies (nuclei) within the CNS. Anatomical relationships established in this way often also indicate functional relationships which can be confirmed by physiological studies and by the tracing of individual fibre connections using specific staining techniques. These include the staining of lipid material of neuroglial origin with heavy metals, and the use of recently developed histochemical techniques to trace the finest dendritic connections. Of the latter, two techniques have proved most valuable: staining for neuronal degeneration and staining for absorbed horseradish peroxidase (HRP). The first of these techniques relies on the observation that damaged neurons degenerate to the soma, a degeneration which can be detected histologically. As many axonal processes are long, this allows the tracing of neuronal pathways over considerable distances. A more recent technique depends on the free absorption of HRP by neurons and the distribution of this substance throughout the cytoplasm of the cell. Staining for HRP can theoretically provide information about most, if not all, of the interconnections of neurons within the CNS, though interpretation of the functional importance of such interconnections relies on physiological study. The details of the rapid advances being made in understanding the 'wiring' of the vertebrate brain that have been obtained by these techniques is beyond the scope of this chapter, but Fig. 4 presents a summary of the major 'input' and 'output' systems present in fish and amphibians.

CNS input systems

The spinal cord

This least-differentiated part of the CNS receives cutaneous, tactile, temperature and proprioceptive information from segmental nerves which enter the cord via the dorsal root. Many of these inputs contribute to the spinal reflexes mentioned later, but all send information to the brainstem. Cutaneous sensation is primarily transmitted to the medulla and midbrain in the dorsal and lateral white matter, frequently after crossing to the opposite side. As a general rule, fish and amphibians possess few well-defined ascending tracts, most information being conveyed to the brain in the relatively undifferentiated, multisynaptic spinal reticular system. An exception occurs in those animals with a well-developed cerebellum, in which a crossed spinocerebellar tract conveys tactile and proprioceptive information to the brain (Sarnat & Netsky, 1974). In reptiles a greater degree of input differentiation is shown by the appearance of gracile and cuneate tracts to the medulla. These convey information about touch, pressure and kinaesthesia (Sarnat & Netsky, 1974). All this input information concerns the interaction of the animal with the environment, but information from visceral afferents also, is conveyed in the spinal cord.

Cranial nerves

The cranial equivalents of the spinal nerves are derived from the embryonic nerve supply to the branchial arches. In fish, amphibians and reptiles there are usually 11 nerves with a purely cranial origin. These are the terminal nerves (nerve 0) and the cranial nerves 1–10. The 11th cranial nerve has a large spinal component in fish and amphibians and the 12th is represented by the occipito-spinal nerve or is absent. The cranial nerves are considerably more specialized in their function than are their spinal

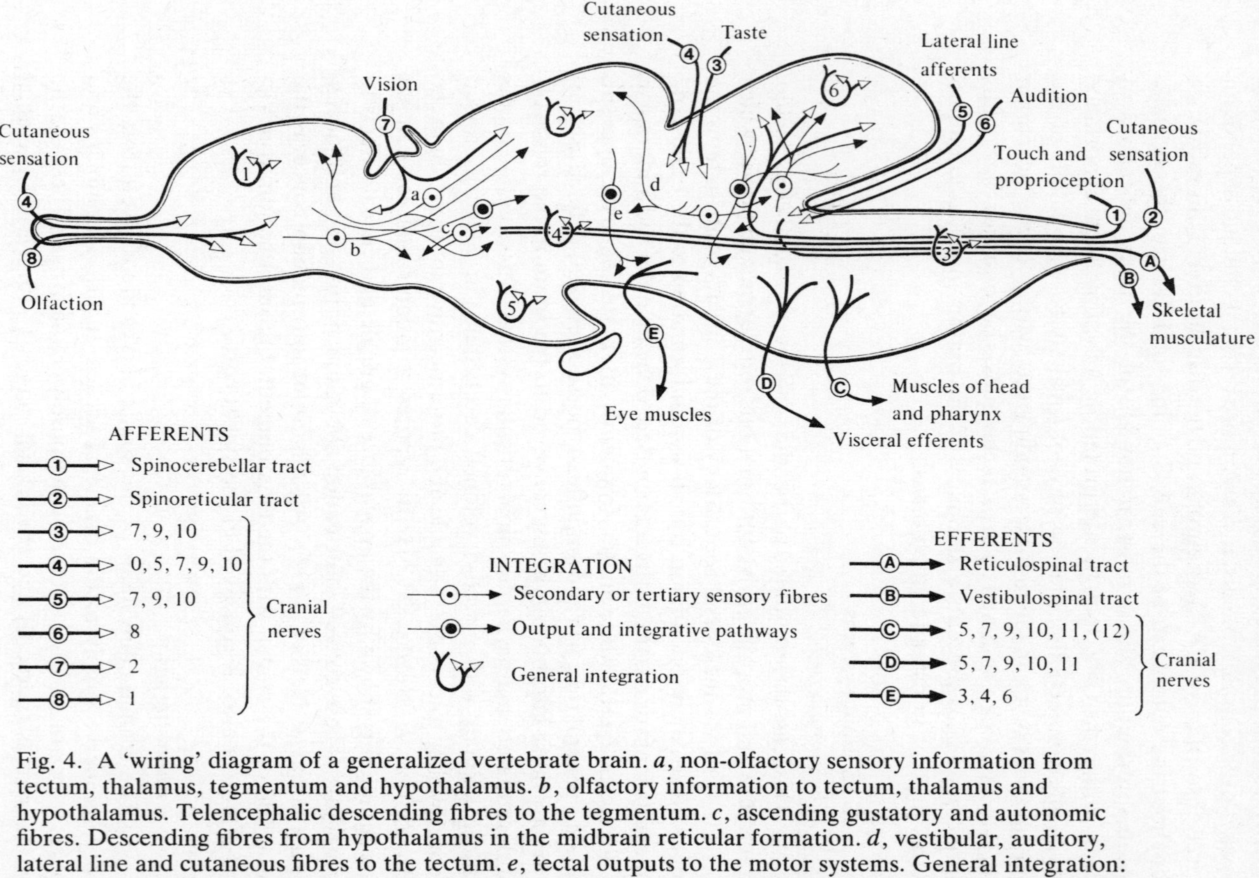

Fig. 4. A 'wiring' diagram of a generalized vertebrate brain. *a*, non-olfactory sensory information from tectum, thalamus, tegmentum and hypothalamus. *b*, olfactory information to tectum, thalamus and hypothalamus. Telencephalic descending fibres to the tegmentum. *c*, ascending gustatory and autonomic fibres. Descending fibres from hypothalamus in the midbrain reticular formation. *d*, vestibular, auditory, lateral line and cutaneous fibres to the tectum. *e*, tectal outputs to the motor systems. General integration: 1, learning, appetitive behaviour, attention; 2, sensory coordination, motor integration; 3, postural control and autonomic regulation; 4, alerting mechanisms; 5, homeostatic and appetitive coordination; 6, motor coordination.

counterparts. Although many of the cranial nerves are associated with sensory input from the head region, many also receive afferents from other regions of the body.

Taste. The 7th, 9th and 10th cranial nerves are important in conveying taste sensation, which in teleosts comes from not only the mouth but also most regions of the body. The primary sensory fibres of these nerves terminate in their nuclei in the medulla, which in teleosts with a strong gustatory sense are enlarged to form the vagal and facial lobes. The secondary gustatory fibres project to more anterior centres in the medulla, to integrate taste sensations with general visceral sensitivity. Taste is probably extremely important in feeding behaviour, in which the hypothalamus appears to have as important a role in fish as it does in mammals (Demski, this volume).

Cutaneous sensation. The trigeminal (5th) nerve is the major cutaneous sensory nerve of the head, with a minor contribution being made by the small terminalis nerve. Cutaneous sensory fibres of the 7th, 9th and 10th cranial nerves join those of the 5th at its nucleus in the anterior medulla.

Orientation, vibration sensitivity and audition. All fish and larval amphibians possess aggregates of specialized neuroepithelial cells which lie in pits or channels in the epidermis. These cells have ciliary projections often covered with gelatinous material. In teleosts the channels containing these cells have developed into canals, the most prominent of which are the lateral line and associated canals of the head region. Changes in the position of the fish, or water vibrations or displacements, distort the hairs of the neuroepithelial cells or neuromasts. The resultant neural impulses are transmitted to the CNS via the 7th, 10th and 11th cranial nerves (Romer & Parsons, 1977). The lateral line is sensitive to low-frequency vibrations (near-field sound) of about 50–150 Hz in *Acerina* sp. for example (Kuiper, 1967). In reptiles and most adult amphibians the lateral line system is absent, but its function as an indicator of change in acceleration and spatial orientation is performed by the more highly developed labyrinth canal systems of the inner ear. The basic arrangement of this system is of three linked semicircular canals, with a chamber at one end, each oriented in a plane at right-angles to the others. The chamber contains neuromasts and a mass of gelatinous material, the cupula. The movement of endolymphatic fluid in the canals, caused by movement of the animal's head, moves the cupula and distorts the hairs of the sensory neuromasts (Romer & Parsons, 1977). Information about the animal's orientation in space is derived from neuromasts in two other endolymphatic chambers, the utriculus and sacculus. The cupula of these chambers contains crystalline calcium carbonate and is called the otolith. The heavy otolith changes position as the animal's orientation changes and stimulates specific regions of hairs in the sacculus and utriculus, thus registering the 'tilt' of the animal.

True acoustic (far-field) information in fish is transferred via cranial bones or cartilage to the semicircular canal system and endolymphatic chambers. In teleosts, greater hearing ability and higher frequency sensitivity (up to 6 kHz in *Phoxinus* sp.) is afforded by transmission of vibrations from the swimbladder to the fluid chambers of the inner ear by a series of bones (Prosser, 1973). These bones, the Weberian ossicles, have developed from detached processes of the anterior vertebrae (Romer & Parsons, 1977). In most amphibians and reptiles sound is transmitted by a membrane, the tympanum, through the air cavity of the middle ear via the stapes bone, and thence via the fenestra ovalis to the ear capsule. In some amphibians, especially urodeles, the tympanum and middle-ear cavity are absent and the stapes may pick up vibrations from the quadrate bone of the skull, with which it articulates. In tetrapods, vibrations reach the inner ear from the fenestra ovalis of the auditory capsule by a sac-like cisterna filled with fluid perilymph. A similar basic arrangement exists in teleosts, except that the Weberian ossicles transmit vibrations directly to the perilymphatic sac.

The neuromast population of the inner ear is sensitive to distortion and low-frequency oscillation and also to vibrations of relatively high frequency (above 100 Hz) often described as far-field sound when fish audition is considered. The primary sensory neurons from the inner ear form the auditory (8th) nerve which projects, along with afferents of the lateral line system, to the acousticolateralis region of the dorsolateral medulla. Many of these primary neurons send collaterals to the cerebellum (Sarnat & Netsky, 1974). It is indicative of the importance of balance, inertial and acoustic information to vertebrates that the secondary neurons have diverse connections with many parts of the brain. Secondary fibres either:

1. enter the ipsilateral medullary reticular formation, or
2. ascend in the lateral longitudinal fasciculus (with or without crossing) to the torus semicircularis of the midbrain, or
3. cross in the acousticolateral commissure, or
4. enter the medial longitudinal fasciculus, and form the well-defined Deiters' nucleus of the acousticolateralis region, or
5. enter the cerebellum (Sarnat & Netsky, 1974).

Ascending fibres to the torus semicircularis provide tertiary projections to the optic tectum in teleosts, for the integration of visual information with that from vestibular, auditory and lateral line systems.

The visual system. The structure of the eye is similar in all vertebrates, though differences exist in the way accommodation is achieved. In fish, and amphibians, accommodation is by movement of the rigid lens forward and backward in the eye so that its position changes relative to the retina, whereas in reptiles it is achieved by the lens changing shape. Most vertebrates control pupillary size by muscles, but in some teleosts change in size is achieved by pressure on the iris by the forwardly positioned lens.

The retinas of vertebrates are similar, consisting of rods and cones (the receptors) which relay the electrical changes, engendered by pigment breakdown, to bipolar cells. These in turn synapse with ganglion cells, the axons of which form the optic nerve. Intra-retinal coding of visual information is probably aided by the horizontal and amacrine cells which interconnect receptor–bipolar cell junctions and bipolar–ganglion cell junctions respectively. The ganglion cell axons decussate (cross over) in the optic chiasma, though incompletely in amphibians and reptiles, providing a limited degree of binocularity in these animals (Ebbesson, 1970). The majority of fibres then project to the optic tectum, though some go to the pretectal region of the thalamus (Ewert, Guthrie, this volume). The fate and importance of these optic projections is discussed in later chapters by Vanegas and by Guthrie for teleosts and by Ewert for amphibians.

The olfactory system. The olfactory receptors lie in a specialized epithelium. In teleosts this forms the lining of paired sacs which communicate to the exterior by openings with the appearance of nares. In elasmobranchs and tetrapods the epithelium lines a channel from nares to the roof of the mouth and is exposed to the ventilatory current be it air or water. The olfactory neurons are also the receptors, with ciliary sensory processes embedded in the olfactory epithelium. Their axons are sometimes associated with axons from the vomeronasal organ, a blind sac opening on to the hard palate and separated from the nasal cavity. In reptiles these sensory regions are important for the analysis of odours constantly carried to the mouth by the flicking of the tongue. Primary chemosensory fibres of both olfactory and vomeronasal origin are frequently short, as the olfactory bulbs are often close to the sensory epithelium. They may be separated from the rest of the telencephalon, as they are in teleosts, by a long olfactory tract composed of secondary olfactory fibres. These project to the lateral and ventrolateral telencephalon in the shark (Ebbesson and Heimer, 1970) and to the basal, lateral and medial areas in other lower vertebrates. The anterior part of the telencephalon is more concerned with olfaction than the posterior. Tertiary olfactory fibres leave the telencephalon for the thalamus and tectum via the olfactohabenular tract, and for the hypothalamus by the olfactohypothalamic tract (Sarnat & Netsky, 1974). Though olfaction is important in lower vertebrates (see Hara, this volume) even in fish, the telencephalon cannot be accurately considered as a 'smell brain', as it performs many other functions (Savage & Wright, Davis *et al.*, Flood & Overmier, Aronson, this volume).

The sensory systems of fish, amphibians and reptiles, summarized here, allow the animals to detect changes in the environment and thus produce the appropriate response. After integration of environmental information, sometimes by a simple reflex link but more often by complex pathways which are not well understood, a response of the animal is mediated through effectors, usually muscles. The nervous control of these effectors is the 'output' system of the brain.

CNS output systems

The spinal cord

Just as afferents to the spinal cord arrive by the dorsal root and usually have their cell bodies in the dorsal horn, so efferents usually have their cell bodies in the ventral horn and leave by the ventral root. In cyclostomes and elasmobranchs the two roots may not join to form a mixed spinal nerve in the way they do in most other vertebrates. Indeed, they may emerge from alternate vertebral segments. In all vertebrates the ventral roots are the primary pathways for efferents to the somatic musculature, and carry visceral efferents more frequently the more phylogenetically 'advanced' the animal (Romer & Parsons, 1977). Efferents to skeletal, post-cranial muscles run largely in the reticulospinal system originating in the tegmental and medullary reticular formation, and in the vestibulospinal system which originates in the acousticolateralis region of the medulla (Ariëns Kappers, Huber & Crosby, 1936). Motor areas of these regions themselves receive information from the tectum (tectospinal tract), cerebelum (cerebellotegmental tract), hypothalamus and forebrain (medial forebrain bundle), and from a variety of forebrain sources via the habenulo-interpeduncular tract to the midbrain.

As well as conveying information to and from the brain the spinal cord performs simple integrative functions, linking more or less directly the appropriate motor response to the incoming afferent signals. An intermediate, decussating neuron in the grey matter of the spinal cord links sensory input and motor response to form the basis of the withdrawal reflex from a noxious stimulus. These reflexes have been most clearly demonstrated in larval urodele amphibians by Coghill (1929) and are probably a general vertebrate feature.

Spinal reflexes are often intersegmental, as are some proprioceptor/motor reflexes which cause the lateral undulations of the body typical in the movement of fish, urodeles and many reptiles.

Cranial efferents

Eye musculature. The 3rd, 4th and 6th cranial nerves supply the external eye muscles. In most fish and amphibians these muscles and their nerve supply are well developed, though in cyclostomes they are not.

Muscles of the head and neck. The muscles of mastication are well developed in jawed vertebrates and are supplied by the motor nucleus of the 5th nerve, fibres of which also descend in the spinal cord. In amphibians these spinal efferents are responsible for jumping reflexes in response to cutaneous stimulation of the head. The superficial muscles of the head are supplied by the 7th nerve, as are the special visceral muscles of the hyoid arch. The 9th, 10th and 11th nerves also supply pharyngeal musculature. The spino-occipital or 12th nerve also has this function in reptiles. This latter nerve supplies the tongue in reptiles.

Autonomic efferents. The musculature and glands of organs concerned with homeostasis are largely under nervous control. Two, often antagonistic, components of the autonomic nervous system are recognized: the sympathetic component, that has a spinal derivation, and the parasympathetic component, that is derived from visceral efferents of the 5th, 7th, 9th, 10th and 11th cranial nerves (the 'vagus'). The sacral spinal nerves in amphibians and reptiles also have a parasympathetic component.

The hearts of all vertebrates contract rhythmically owing to the activity of intrinsic pacemakers and conduction systems, the rate of contraction being inhibited by the vagus. In anuran amphibians and reptiles, which have an established sympathetic innervation to the heart, an increase in rate and force of contraction accompanies sympathetic nerve stimulation. Changes in heart rate often naturally accompany the responses of animals to stimuli. Teleosts have been effectively conditioned to respond to experimental stimuli with a cardiac deceleration. In lower vertebrates, the effects of autonomic stimulation on the gut produce no clear distinction between sympathetic and parasympathetic actions. Both cause contractions of gastric musculature, though muscle tone and peristaltic activity may be more affected by sympathetic stimulation (Nicol, 1952). Peripheral autonomic efferents are known to affect ventilatory rate in higher vertebrates, but their role in the classes considered here is uncertain. Another peripheral autonomic effect of stimulation is the release of adrenaline from chromaffin tissue.

This hormone prepares the animal for immediate action by increasing carbohydrate catabolism, inhibiting gut motility and accelerating the heart (Prosser, 1973). It may also produce rapid colour changes in the animal, though this effect may be attributable to the activity of the autonomic nervous system directly, or to the pituitary gland.

Brain centres for motor coordination. The behaviour shown by lower vertebrates may be initiated in many regions of the brain, but there are two areas which have a well-established importance in coordinating muscular activity, which in turn may form part or all of a behavioural pattern.

(i) *Brainstem reticular and vestibular nuclei.* Few direct descending pathways are evident in the brains of the vertebrates considered here, a more diffuse multisynaptic reticular pathway being pre-eminent. Within the medulla and midbrain reticular formation are groups of large motor neurons comprising the reticular nuclei. In cyclostomes, some of these cells are huge (the cells of Mueller) and mediate locomotion and balance by directing tail musculature. In teleosts giant Mauthner cells occur, which have dendritic arborizations in the vestibular nuclei. The axons of these Mauthner neurons descend the spinal cord to cause contractions of the tail musculature (Retzlaff, 1957). They may be responsible for the 'tail-flip' or startle response of teleosts to strong environmental stimuli, especially those involving pressure changes. Associated also with the medullary reticular formation are localized regions concerned with the control of cardiac and ventilatory function and general muscle tone.

Closely related to the brainstem reticular elements are the vestibular nuclei and the descending vestibulospinal tract. This tract appears to be one of the phylogenetically oldest discrete motor pathways, receiving its input from areas of the brain concerned with balance and posture, such as the acousticolateralis and cerebellum respectively. This tract, with descending influences originating in the tectum (tectospinal tract), probably forms the pathway most concerned with finely coordinated responses of lower vertebrates to environmental changes. Much of the muscular coordination mediated through these tracts is probably due to the close links between vestibular nuclei, tectum and cerebellum.

(ii) *The cerebellum.* The cerebellum is well developed in elasmobranchs and teleosts, and in the latter an anterior extension, the valvula, lies under the tectum. The large cerebellum of these animals is related to the development of the acousticolateralis region, with which it is intimately connected; thus amphibians and reptiles with no lateral line system have correspondingly small cerebelli. The cerebellum has a large medial part, the corpus, concerned with coordination of trunk musculature, and lateral lobes or auricles related to the acousticolateralis of the medulla. The fine cytoarchitecture of the cerebellum will not be considered here; suffice it to say that the layered cerebellar cortex contains cells with large dendritic trees and associated connections. It is thus able to collect information from a variety of sources and coordinate the required motor output. Apart from ascending spinocerebellar tracts, carrying proprioceptive information, the cortical region appears to receive afferents from all the primary sensory regions of the brain. These sensory projections are somatotopically localized within the cerebellum. The multitude of fibre connections within the cerebellar cortex and the effects of ablation in higher vertebrates have implicated the cerebellum in behaviours involving fine muscular coordination. Specific nuclei for these cerebellar efferents are lacking in lower vertebrates, their function being served by reticular and vestibular nuclei of the medulla.

Integration

Reflex responses, like those already considered, are easy to understand in the context of input and output integration leading to a particular behaviour. Most behaviours of vertebrates are not, however, the result of a simple link between sensory input and motor response but are complex, often flexible responses to a wide variety of environmental conditions and previous experience. The wide range of integrative mechanisms which can be studied and the diversity of disciplines from which researchers have emerged, have resulted in a considerable variety of techniques being used in neuroethological research. The basic technique consists of providing a stimulus to an animal and monitoring the response, with or without direct interference in the brain's ability to process the stimulus information into a 'normal' response. By broadly categorizing experimental technique into the areas of: (*a*) recorded responses, (*b*) stimulus types and (*c*) experimental manipulations of the brain, the methods of study and some examples of the types of result obtained can be conveniently described.

Recording techniques

Behavioural observations. The ethologist traditionally observes the behaviour of animals in the natural environment. The nature of neuroethology makes such study difficult, if not impossible, both because of legal restrictions on the release of operated animals and because of the difficulty of tracing and observing them after release. These latter problems may be partially overcome with the use of biotelemetric recordings, whereby the position of the animal and some of its physiological processes can be monitored. Thus Priede & Young (1975) used ultrasonic telemetry of electrocardiograms in fish to monitor both position and cardiac correlates of behaviour in the natural environment. In general, however, the variability of the behaviour of an animal in such circumstances makes it important for the neuroethologist to remove his subjects to a carefully controlled laboratory environment. The common criticism that manipulations of nervous tissue may affect the behaviour of the subject in a manner not specifically monitored by his experiments, makes it necessary for him to have adequate controls, and also to be very sensitive to any deviation of behaviour in his animals from that normally expressed. The types of behaviour commonly observed fall into three broad categories: (i) changes in general activity, (ii) changes in species-specific behaviour patterns and (iii) changes in behaviour resulting from experience.

(i) *Changes in general activity.* The overall activity of an animal is often measured by its locomotory activity. It can also be interpreted in terms of activation of the CNS and measured by behaviourally determined sensory thresholds to stimuli presented by the experimenter. In poikilotherms both activity and sensory threshold are probably affected by temperature, nutritional and reproductive states, season, period of the daily cycle and state of health, to name but a few variables. Any of these may interact with the conditions created by the experimenter to complicate the interpretation of results.

(ii) *Species-specific responses to stimuli.* Many of the appetitive (hunger, thirst, sexual) responses of vertebrates to stimuli are specific to the species being studied. Experimental manipulations which produce changes in such responses as those of courtship (De Bruin, 1977) or parental behaviour (Davis *et al.*, this volume) of teleosts, for example, require observations of all the components of the response in relation to normal behaviour. As species-specific behaviours are often inter-related, affected by experience and the general level of activity, observations of other, apparently unaffected behaviours are essential, though often under-used, as controlling factors.

(iii) *Changes in response as a result of experience.* Representatives of all lower vertebrate classes have been shown to exhibit learning, but most of the research into brain mechanisms of learning in these classes has been performed with teleosts. The telencephalon has been implicated most consistently and its function in learning is amply reviewed by Flood, Overmier & Savage (1976) and Flood & Overmier (this volume). A variety of teleosts have been conditioned to respond classically to an electric shock by gross motor movements (Flood *et al.*, 1976). Fish have also been trained to swim down a trough for a food reward (Flood & Overmier, 1971) or to retreat from an impending shock (Savage, 1969). These experiments, performed with both normal fish and those with damage to the telencephalon, have yielded information on the involvement of this part of the brain in the acquisition, maintenance and extinction of the ability to perform the types of task learnt.

Physiological observations. Probably the most fundamental belief of the neuroethologist is that behaviour is ultimately an overt manifestation of physiological and biochemical changes taking place in the brain of an animal. Thus there is considerable interest in physiological changes which seem to be related to behaviour.

(i) *Peripheral physiological changes.* Many of the physiological responses of animals are, like behaviour itself, peripheral manifestations of changes taking place in the CNS. These include changes in the electrophysiological activity of muscles, monitored in the electromyogram (EMG), and autonomic changes such as heart rate, blood flow and breathing. In the absence of overt behavioural changes these variables can often be used to monitor responses.

Electromyographic activity (EMG). Fine implanted electrodes can be used to record the EMG of skeletal musculature. In the freely moving animal EMGs can be used to indicate phasic electrical changes in muscle. Tonic changes, however, are difficult to detect against the background phasic EMG activity of the non-anaesthetized subject (Laming & Savage, 1980).

Cardiac activity. Most vertebrates show brief changes in heart rate coincident with behavioural responses to environmental change. Electrocardiograms (ECGs) have been monitored using implanted stainless steel electrodes in teleosts (Roberts, Wright & Savage, 1973) and anuran amphibians (Karmanova *et al.,* this volume). Decelerations occur during alerting responses (arousal) in teleosts (Laming & Savage, 1980); accelerations occur in anurans (Karmanova *et al.*, this volume). Heart rate changes have been conditioned classically by a number of workers interested in mechanisms of learning (Overmier & Savage, 1974). The chronic implantation of electrodes in fish is being superseded by the use of electrodes in the water in which the animal is maintained, although EMG artifacts which accompany the behaviour of the animal can be difficult to eliminate (Rommel, 1973).

Breathing. The rate of breathing of most vertebrates can be directly observed. In fish, mechanical lever systems (Otis, Cerf & Thomas, 1957), pressure transducers linked by a catheter to the buccal cavity (Laming & Savage, 1980) and opercular EMGs have been used. In small terrestrial vertebrates direct observation or recording of EMGs of thoracic musculature are probably the most effective methods of measuring breathing rate. In teleosts, alerting responses are accompanied by a decrease in the rate and depth of breathing (Laming & Savage, 1980); few reports are available for other lower vertebrates.

(ii) *CNS changes. The electroencephalogram (EEG)*. Apart from the peripheral variables considered above, there are also changes in the overall pattern of CNS activity. The EEG is considered to be formed by a combination of action potentials and regular membrane potential fluctuations (Klemm, 1969). The action potentials are difficult to interpret, and when prevalent form an overall pattern of apparently random spikes, or a 'desynchronized' EEG. The regular, slow membrane potential fluctuations are considered to be prevalent in a 'synchronized' EEG. Relatively little work has been carried out on EEGs of lower vertebrates in relation to behaviour. Chronically implanted silver/silver chloride, stainless steel or tungsten electrodes have usually been used. Often the EEGs thus obtained have been found to differ according to the behaviour of the animal at the time. Recordings of this type have been performed on fish (Enger, 1957; Laming, 1980), amphibians (Segura & De Juan, 1966) and reptiles (Karmanova, Belekhova & Tchurnosov, 1971). Changes in EEG pattern according to the level of wakefulness or sleep (Karmanova *et al.*, this volume), during arousal (Laming, this volume), or seasonally (Segura & De Juan, 1966; Godet, Bert & Collomb, 1964) have been found in a variety of lower vertebrates.

Some classes of vertebrates produce a desynchronized pattern in the EEG when a stimulus is perceived, which may be the result of action potentials in the pathway of the appropriate sensory modality. If such massed action potentials are averaged over a series of stimulus presentations, data about latency, amplitude and duration of the averaged response may indicate the importance of the recording site in the transmission of the sensory modality concerned.

Unit recording. Gross recordings, like those of the EEG, may act as a rough guide to the activity of parts of the CNS, but more definitive links between behaviour and physiology rely on a determination of the function of individual neurons. Recordings of activity from one, or a very few cells in the CNS can be made intra- or extracellularly using glass or metal microelectrodes. Recordings made in this manner allow neurophysiologists to relate neuronal activity to sensory input or muscle contraction in a very precise manner. At the present time such studies are made at the extremes of the sensory–integration–response continuum, though some notable inroads have already been made into the neurophysiological basis of behaviour (Ewert, this volume).

(iii) *Biochemical changes.* Biochemical as well as electrical changes can be observed in the brain and are especially noticeable during long-term adjustments of behaviour. In teleosts, changes in RNA synthesis occur during learning. Shashoua (1968) demonstrated a clear shift in the base composition of RNA in goldfish trained to swim upright with a ventrally attached float compared with that in controls.

Stimulation techniques

Environmental stimuli. Fish, amphibians and reptiles are sensitive to a large number of physical changes which occur in the environment. In general, smell, taste, vision, audition and tactile and vibration sense are most important, though in some species an enhanced ability to respond to infrared radiation, electric discharge and temperature have evolved. Most stimuli in the natural environment involve the use of a combination of physical features and the concomitant combination of sensory modalities to perceive them. In ethological studies the detection of stimuli by animals can only be indicated by the production of a behavioural response. Through the technique of simultaneous stimulation and physiological recording, however, it is possible to monitor the 'input' of the stimulus given to the animal. Thus the physiology of the input pathway can be studied as well as the more general physiological and behavioural responses of the animal.

Every type of neuroethological research requires the monitoring of a response, and the differentiation of stimulus and response is often a spurious one, as the 'response' of, for instance, retinal photoreceptors, is the stimulus for the first neuronal input (the bipolar cell) to the brain. Dependent upon the parameters recorded, vertebrates seem to detect two main components of environmental change, one being specific to the particular configuration and modality of a stimulus (Ewert, this volume) the other being a general recognition that a change has taken place (Laming, this volume). Thus, stimuli have the characteristics of both a 'releaser' of specific responses and an 'arouser' of brain mechanisms of reception and response.

Appetitive states produced by deprivation or by drugs. Responses of animals to environmental change are often the result of many features of the environment and/or an interaction of the perceived environmental situation with the internal state of the nervous or endocrine system at that time. Thus, changes in the internal state of the animal by, for example, food deprivation or hormone administration will often affect the manner in which animals respond both behaviourally and neuronally to changes in the environment. Many anuran amphibian species can be induced to breed by gonadotrophin injection when water is available. Similarly, the level of motivation of an animal to respond to food can be increased by food deprivation. The effects of these 'internal' stimuli can be monitored by both electro-physiological and behavioural observations.

Electrical stimulation of the nervous system. Many of the effects of stimulating an animal environmentally can be duplicated by electrical stimulation of the nervous system, if the site and characteristics of the input are known. Thus, electrical stimulation, using electrodes similar to those required for recording, is used to confirm inferences derived from recording studies. More crudely, stimulation techniques may be used in an effort to localize areas of the brain concerned with particular behavioural or electrical responses.

The types of stimulation given vary quite considerably, but usually employ trains of square-wave pulses of some 0.1–5.0 ms duration, at frequencies of from 10 to 100 Hz. Voltages used vary from 10 mV to 250 mV, though current is more usually employed as a measure of stimulus strength, being in the range of 1–300 μA.

Experimental manipulations

If the relationship between the stimulus received and the response given by an animal is a simple one, then only minimal interference between the input and output pathway is necessary to disrupt the link and deduce its characteristics. If the integrative system is complex, however, a more drastic interference with the brain, in the first instance, is necessary to broadly localize the areas involved.

Ablation. One of the most commonly used surgical techniques is to completely remove part of the brain. Early work on teleosts (Berwein, 1941; Nolte, 1932) suggested that such ablation techniques had minor effects on behaviour when applied to the telencephalon. The ablation of the teleost telencephalon, performed either by aspiration or incision, has become a popular operation, both because the animal stays alive and because it is suggested that the telencephalon may have a modulatory effect on the rest of the brain, the beginnings of 'higher control' functions epitomized in the cerebrum of mammals. The functions of the telencephalon, as based on experiments of this type, are discussed later in this book (Flood & Overmier, Savage & Wright, Aronson, Davis *et al.*). The operation of telencephalic ablation suffers from several criticisms, in addition to its general crudity. There are few, if any, adequate controls for the ischaemic effects of the operation on the rest of the brain. Thus 'sham' operations do not involve the same degree of tissue or vascular damage, so any impairment of function attributed to loss of the telencephalon may simply be a loss of function of the rest of the brain due to metabolic shock or anoxia. Little research has been performed to assess recovery of function after telencephalic ablation. Similarly, few workers have used olfactory bulbectomy as a control operation for the inevitable loss of specific or general effects of olfactory input. Another criticism of operations involving wholesale ablation is the difficulty of interpreting the results with confidence. Functions attributed to the telencephalon as a result of the operation may not be localized there, but be attributable to a loss of facilitation or inhibition of other brain regions. The difficulties that have been experienced in attempting to discover the functions of the telencephalon may in part be due to the crude nature of this operation. However, the results that have been obtained from this type of experiment have led to a broad picture of some of the behaviours with which the telencephalon is involved (Flood & Overmier, Aronson, Savage & Wright this volume). These behaviours can now be studied with less crude, more localized procedures.

Lesioning. Small areas of brain damage can be created surgically, but it is preferable to destroy the nervous tissue without concomitant damage to the vascular supply. Whilst this is not at present practicable, only limited vascular damage is caused by passing an electric current through the region of nervous tissue to be damaged. Two electrodes are used, and either a direct current of some 200 μA is passed or the more recently developed technique of using an alternating current of radio frequencies is employed. The damage caused by such lesions can be extremely localized and its extent found subsequently by histological examination.

Anaesthesia. A variety of anaesthetics are used in the preparation of animals for surgery, including urethane for fish, MS 222 (methane tricaine sulphonate) for fish and amphibians, and ether for reptiles. A few studies (Enger 1957; Schadé & Weiler, 1959) have looked at changes in CNS activity during recovery from anaesthesia, but the application of this technique has limited uses.

Drugs. One way of 'anaesthetizing' a particular function is to block the transmission of information from neuron to neuron by the use of specific drugs. This technique is commonly used in mammalian studies, but more rarely so in research on lower vertebrates. However, inhibitors of protein synthesis have been used to inhibit the consolidation of memory in fish (Agranoff, Davis & Brink, 1966).

Conclusion

The brains of present-day vertebrates probably bear little superficial resemblance to those of their prehistoric ancestors. Comparative neuroethologists nevertheless hold a well-supported belief that, within the brains of extant species lie basic mechanisms which are common to all vertebrates. It is with these basic mechanisms that we are most concerned in this book. A simple description of the role of the brain can be made in the form of an input–integrative–output continuum. The research into these three areas is multidisciplinary and involves a knowledge of evolution, anatomy, physiology and psychology which even today is rarely obtained formally. This chapter has attempted to review some aspects of relevant vertebrate neuroethology so that the more specialized topics covered later may be better appreciated.

Apologia

In an attempt to present an introduction to neuroethology in the lower vertebrates I have often, inevitably, overgeneralized and strayed outside my own speciality. I hope that any consequential errors are minor and will be forgiven by readers with more knowledge than myself of the various aspects covered.

References

Agranoff, B. W., Davis, R. E. & Brink, J. J. (1966). Protein synthesis and learning. *Brain Research,* **1,** 303–9.

Ariëns Kappers, C. U., Huber, G. C. & Crosby, E. C. (1936). *The Comparative Anatomy of the Nervous System of Vertebrates, Including Man.* Macmillan, New York.

Berwein, M. (1941). Beobachtungen und Versuche über das gesillige Leben von Elritzen. *Zeitschrift für vergleichende Physiologie,* **28,** 402–20.

Coghill, G. E. (1929). *Anatomy and the Problem of Behaviour.* Cambridge, Mass.

De Bruin, J. P. C. (1977). 'Telencephalic functions in the behaviour of the Siamese fighting fish, *Betta splendens* Regan. (Pisces, Anabantidae)'. PhD thesis, Central Brain Research Institute, Amsterdam.

Ebbesson, S. O. E. (1970). On the organisation of central visual pathways in vertebrates. *Brain, Behavior and Evolution,* **3,** 178–94.

Ebbesson, S. O. E. & Heimer, L. (1970). Projections of the olfactory tract fibres in the nurse shark *(Ginglymostoma cirratum). Brain Research,* **17,** 47–55.

Enger, P. S. (1957). The electroencephalogram of the codfish. *Acta Physiologica Scandinavica,* **39,** 55–72.

Flood, N. C. & Overmier, J. B. (1971). Effects of telencephalic and olfactory lesions on appetitive learning in goldfish. *Physiology and Behavior,* **6,** 35–40.

Flood, N. C., Overmier, J. B. & Savage, G. E. (1976). Teleost telencephalon and learning: an interpretive review of data and hypotheses. *Physiology and Behaviour,* **16,** 783–98.

Godet, R., Bert, J. & Collomb, H. (1964). Apparition de la réaction d'eveil telencephalique chez *Protopterus annactens* et cycle biologique. *Comptes rendus des séances de la Société de Biologie,* **158,** 146–90.

Karmanova, I. G., Belekhova, M. G. & Tchurnosov, E. V. (1971). Specifics of behavioural and electrographic patterns of natural sleep and wakefulness in reptiles. *Sechenov Physiological Journal of the USSR,* **57,** 504–11.

Klemm, W. R. (1969). *Animal Electroencephalography.* Academic Press, New York & London.

Kuiper, J. W. (1967). Frequency characteristics of lateral line organs. In *Lateral Line Detectors,* ed. P. Cahn, pp. 105–21. Indiana University Press, Bloomington, Ill.

Laming, P. R. (1980). Electroencephalographic studies on arousal in the goldfish *(Carassius auratus). Journal of Comparative and Physiological Psychology,* **94,** 238–54.

Laming, P. R. & Savage, G. E. (1980). Physiological changes observed in the goldfish *(Carassius auratus)* during behavioural arousal and fright. *Behavioural and Neural Biology,* **29**, 255–75.

Nicol, J. A. C. (1952). Autonomic nervous system in lower chordates. *Biological Reviews,* **27,** 1–49.

Nieuwenhuys, R. (1969). A survey of the structure of the forebrain in higher bony fishes (Osteichthyes). *Annals of the New York Academy of Sciences,* **167,** 31–64.

Nolte, W. (1932). Experimentelle Untersuchungen zum Problem der Lokalisation des Assoziations vermogens im Fischgehirn. *Zeitschrift für vergreichende Physiologie,* **18,** 255–79.

Otis, L. S., Cerf, J. A. & Thomas, G. J. (1957). Conditioned inhibition of respiration and heart rate in the goldfish. *Science,* **126,** 263–4.

Overmier, J. B. & Savage, G. E. (1974). Effects of telencephalic ablation on trace-classical conditioning of heart rate in goldfish. *Experimental Neurology,* **42,** 339–46.

Priede, I. G. & Young, A. Y. (1975). The ultrasonic telemetry of cardiac rhythms of wild brown trout *(Salmo trutta)* as an indicator of bio-energetics and behaviour. *Journal of Fish Biology,* **10,** 299–318.

Prosser, C. Ladd (1973). *Comparative Animal Physiology.* W. B. Saunders, Philadelphia.

Retzlaff, E. (1957). A mechanism for excitation and inhibition of the Mauthner cells in teleosts. A histological and neurophysiological study. *Journal of Comparative Neurology,* **107,** 209–25.

Roberts, M. G., Wright, D. E. & Savage, G. E. (1973). A technique for obtaining the electrocardiogram of fish. *Comparative Biochemistry and Physiology,* **44A,** 665–8.

Romer, A. S. & Parsons, T. S. (1977). *The Vertebrate Body.* W. B. Saunders, Philadelphia.

Rommel, S. A. (1973). A simple method of recording fish heart and operculum beat without the use of implanted electrodes. *Journal of the Fisheries Research Board of Canada,* **19,** 417–22.

Sarnat, H. B. & Netsky, M. G. (1974). *Evolution of the Nervous System.* Oxford University Press, London.

Segura, E. T. & De Juan, A. (1966). Electroencephalographic studies in toads. *Electroencephalography and Clinical Neurophysiology,* **21,** 373–80.

Schadé, J. P. & Weiler, I. J. (1959). EEG patterns of the goldfish. *Journal of Experimental Biology,* **36,** 435–52.

Shashoua, V. E. (1968). RNA changes in goldfish brain during learning. *Nature, London,* **217,** 238.

LESTER R. ARONSON

Evolution of telencephalic function in lower vertebrates

Introduction

During the heyday of comparative anatomy, comparative neurologists were particularly concerned with the structural evolution of the vertebrate telencephalon (Johnston, 1906; Papez, 1929; Ariëns Kappers, Huber & Crosby, 1936; Herrick, 1948). The dominant role of the mammalian cerebrum in the mediation of higher mental processes was undoubtedly a major inspiration for these studies. By the end of this era (approximately 1920 to 1930) the basic neuroanatomical picture had been established, but it has been elaborated substantially by present-day investigators using newly developed techniques for more accurately tracing nerve fibres.

On the other hand there has been a dearth of interest in the functional evolution of the telencephalon, despite the fact that as this structure became larger and more complex, its functional properties obviously changed. Besides myself, only the Soviet neurologist Karamian (1968, 1972) and a few other investigators (e.g. Rensch, 1967) have dealt specifically with this topic. There are obvious reasons for this imbalance, such as difficulties in procuring, maintaining and performing behavioural studies on the variety of species used so successfully by the neuroanatomists. The major obstacle, however, seems to be a theoretical issue. While evolutionary study constitutes a time-honoured endeavour for morphologists, no such tradition exists in physiology, and for behaviour there is still a question in the minds of some investigators (e.g. Skinner, 1966; Lockard, 1971) to what extent this is a heuristic enterprise.

The phylogenetic scale

Concern about the comparative method as applied to behaviour was elaborated a decade ago by two psychologists, William Hodos & C. B. G. Campbell (1969), who published a theoretical article entitled '*Scala naturae*: why there is no theory in comparative psychology'. Actually the title is misleading. Comparative psychology abounds in theory; for example, the nature–nurture dichotomy, the concept of homology as applied to behaviour

(Atz, 1970), the concept of levels, approach–withdrawal theory (Schneirla, 1959), the Thorndike dogma that fundamental learning abilities are similar in all vertebrates (Bitterman, 1975), and so on. What these authors were apparently saying was first, present theories are not pertinent or acceptable, and second, the *Scala naturae*, or phylogenetic scale, is being used incorrectly, particularly by psychologists, as the evolutionary lineage of vertebrates. According to Bitterman (1975) the major Hodos & Campbell thesis is that animal psychologists are unable 'to deal with the "intricacies" of evolutionary history'. My own suspicion is that these authors are not sufficiently cognizant of the objectives and achievements of comparative psychologists in dealing with these 'intricacies'.

It is possible that some psychologists with only minimal training in biology have used the phylogenetic scale incorrectly. Yet, it appears to me that the majority of comparative scientists, whether in morphology, physiology or behaviour, recognize the true nature and value of this conceptual device, and they continue to use it despite the admonitions of Hodos and his associates. Thus, the expression 'higher versus lower vertebrates', as in the title of this volume, is widely used. The sequence fish, amphibian, reptile, bird, mammal, and so on, seems to have important meaning beyond mere convention for a great many life scientists. This use of the phylogenetic scale is particularly widespread among comparative behaviourists, where knowledge of fossil brains and fossil behaviour is limited and where prospects for future knowledge are not encouraging.

Hodos & Campbell (1969) hold that the major task of the comparative scientists is to study lineages, or evolutionary sequences. Because of the special attributes of behaviour, because of the very limited behavioural information in the fossil record, and because of the severe limitations of the concept of homology as applied to behaviour (Atz, 1970), I seriously doubt that we can reasonably expect to establish in the foreseeable future such behavioural lineages in a meaningful way across major phyletic gaps.

In a later article Hodos (1970) states that with respect to behaviour 'The fundamental question is: "Can the study of animals that are alive today tell us anything about animals that lived tens of millions or hundreds of millions of years ago?"' The answer he gives is only a qualified yes. An equally fundamental question is: Can comparisons of the behaviour and behavioural mechanisms among living species and among higher taxa, together with available ontogenetic and fossil evidence, contribute to our understanding of the evolution of behaviour and behavioural processes? The answer as I see it is an unqualified yes.

Rather than attempting to deduce behavioural genealogies, comparative behaviourists are much more interested in comparing species and higher categories analytically, as regards similarities as well as differences at two or more phyletic levels of complexity, with respect to behaviour, associated structures and physiological mechanisms. The comparisons are often made with the confidence and understanding that they are directly pertinent to the evolution of the brain and behaviour.

In a chapter on elasmobranch brain organization, Northcutt (1978*a*) echoes arguments similar to those of Hodos & Campbell, but he stresses the study of adaptive radiations as the main focus of comparative neurobiology. By implication, the overall trends towards increasing complexity in vertebrate evolution, towards telencephalization, towards a progressively more elaborate neocortex, and many similar trends, have for him no biological significance. Thus he emphasizes the specialized features of living forms and ignores the core of commonality that is characteristic of each vertebrate group. A more extensive criticism of Northcutt's position is presented by Karamian, Veselkin & Belekhova (1978), wherein they emphasize the distinction between specialized adaptive radiations and major features characteristic of an entire class.

The concept of levels

The theoretical basis for the comparisons suggested above is the concept of levels of integration and organization as developed by Needham (1929), Novikoff (1945), Schneirla (1949, 1953), Tobach (1976) and others which follows directly from the theory of evolution. Schneirla (1952) emphasized that this is the most effective conceptual device for studying the evolution of behaviour. As summarized by Tobach (1970, 1976):

1. Natural phenomena may be hierarchically ordered with respect to differentiation and organizational complexity, beginning with the most fundamental aspects of matter, namely the elementary particles.
2. Through increasing degrees of organization various levels, such as particles, atoms, molecules, compounds are recognized. Further divisions may be made between inanimate and animate and within the animate we find an array of phyletic levels of increasing complexity ranging from viruses to mammals and man. One may also distinguish transitional states or mesolevels.
3. In addition to phyletic levels, hierarchies can be recognized in numerous life processes such as organismic structure and function, communication (Tavolga, 1970), emotions (Tobach, 1970), social behaviour (Tobach & Schneirla, 1968), etc.

4. Each level has its individual characteristics; its own laws and principles.
5. Information on one level cannot predict phenomena on higher levels without concomitant information at these levels. For example, from the characteristics of sodium and chlorine one cannot predict the nature of sodium chloride. Similarly from the knowledge of the reptilian brain one cannot predict the nature of the bird or mammalian brain. Nor, as proposed by MacLean (1973), can we view the mammalian brain as a reptile brain plus some added mammalian features, because the integration of reptilian and mammalian features produces a new and more complex neural organization.
6. Explanations of phenomena at one level are most effective when coupled with the understanding of previous levels. Thus we will have a better understanding of the mammalian nervous system if we have a good knowledge of lower vertebrate and invertebrate nervous systems.

It is with this concept in mind that I and others view comparisons of living forms as representing increasing levels of complexity, that is, higher levels of organization. This does not mean that I advocate placing every species of animal on a linear scale from lowest to highest. But it is possible to recognize levels of complexity and intricacy for disparate groups. For very widely separated organisms (e.g. viruses compared with mankind), the differences are obvious beyond doubt. I know that the idea of complexity bothers some people who prefer to think only in terms of differences. I cannot deal with this problem at length here, but the description by Karamian (1972) of the phylogenetic increase in complexity of the nervous system provides a concise example. This is the pronounced trend of evolutionary change in both invertebrates and vertebrates – from diffuse, less centralized forms of functioning to specialized forms of neural activity; from the homogeneous to the heterogeneous; from the generalized to the discrete.

Comparing distantly related taxa one has little trouble discerning a hierarchy of organizational levels: thus, virus, protozoan, fish, reptile, mammal, man. But within more closely related groups different systems evolved in different ways, so that man would occupy the highest level with respect to brain organization, plasticity of behaviour and higher mental processes, but not necessarily with respect to, say, the skeletal system. For this reason a hierarchy of organizational levels based on brain and behaviour would not necessarily parallel, except in broad outlines, a phylogenetic scale based essentially on skeletal evidence coupled with other morphological data.

While neural levels do not provide information on lineages they do tell us in what ways and to what extent neural organization has evolved in different species, whether by direct descent (common ancestry), by parallel evolution or by other means. In this way the neural picture at a given level provides the clue to the brain ingredients that were available for the formation of the next higher level.

An important corollary of the levels concept is that the methods – such as phylogenetic trees, cladograms and homologies – used for the study of phenomena at one level (say morphology) are generally not applicable to higher-level phenomena (say behaviour). 'Each level requires its own methods of investigation and yields its own laws and principles' (Tobach, 1970). One should be aware of the pitfalls in trying to build phylogenetic trees of behaviour or to establish behavioural homologies across large phyletic gaps by using morphological methods.

Hodos & Campbell (1969) emphasize the importance of the fossil record for establishing evolutionary sequences, but they are aware of the extremely limited information concerning the structure and function of the brains of fossil forms. Moreover, inferences concerning behaviour of fossil forms are not only scarce but highly speculative as well. To get around this they advocate the comparisons of those extant species that are close descendants of a common evolutionary lineage. The latter is determined largely from fossil skeletal evidence, and the rather tenuous assumption is made that if the skeletons are similar then the brains and behaviour will also be similar.

These investigators also make quite a point about the limitations of research on teleosts for understanding the evolution of behaviour, since bony fishes separated from other vertebrate lines so long ago. Yet as we see from titles of the chapters in this volume, much of our knowledge of brain function and behaviour in lower vertebrates comes from studies on the teleost fishes. So we may ask: Does this growing body of research on teleosts have relevance only for teleosts or does it have broader meaning? With respect to this question I have serious reservations concerning the view of Hodos & Campbell that 'much of the current research in comparative psychology seems to be based on comparisons between animals that have been selected for study according to rather arbitrary considerations and appears to be without goal other than the comparisons of animals for the sake of comparisons'. Further along we read that from the point of view of primate phylogenesis the rat, cat, monkey comparisons are also meaningless. To the contrary I maintain that with respect to the living representatives of various vertebrates, their brains can be arranged in an order of increasing complexity with respect to structure, physiological function and behaviour even though direct lineage cannot be postulated in most cases. Moreover, this ordering is based on evolutionary change.

Practical problems

Leaving theory for the moment, there are some practical problems. Many species that might be useful because of relatively recent common ancestry are scarce or might even be endangered species. For morphological studies where one or two individuals suffice, this may not be too serious. But for physiological and behavioural studies where dozens of individuals are usually needed, this could be very serious. Even among species that are available, a good proportion are just not suitable for laboratory experimentation or field observation. *Latimeria*, for example, may be considered representative of the descendants of ancestral amphibians, but so far no one has been able to observe them in their natural habitat or even bring them to the surface in a condition in which their behaviour can be observed.

I have discussed at some length here the questions raised by Hodos & Campbell because their paper has created much interest, and its influence seems to have spread far beyond its original audience. To the best of my knowledge their position has not been questioned seriously except for brief comments by Bitterman (1975) and Tobach (1976). I feel that the Hodos papers are having, on the one hand, an inhibiting effect on comparative behavioural research and are inadvertently giving support to non-evolutionary systems of behaviour (Tobach, 1976). On the other hand they should provoke continuing interest and discussion into the real nature, value and limitations of phylogenetic scales.

Application of the levels concept

Depending on our needs we can consider an orderly arrangement of integrative levels such as structure, physiology, overt behaviour and higher mental processes, or we can order animals as to phyletic levels based on the sum of all their characteristics, or we can base a phyletic hierarchy on specific characteristics such as nervous system and behaviour. This phyletic arrangement seems most appropriate for studying the functional evolution of the nervous system.

Levels of organization in the lower vertebrate telencephalon

Level one

The first level is represented by the immediate ancestors of vertebrates. The acorn worm *Saccoglossus* possesses a dorsally located, hollow nerve cord in the collar region that is epidermal in origin and resembles the neural tube of vertebrates (Bullock, 1965). In some species it extends into the proboscis. Damage to this anterior extension affects the spontaneity of peristaltic burrowing behaviour, suggesting that the front end of the nerve cord has activating properties (Bullock, 1940; Knight-Jones, 1952). Certain cells at the base of the proboscis seem to be chemoceptive and sensory information from these cells is presumed to reach the neurocord via the nerve net of the proboscis.

Another group of ancestral relatives of the vertebrates is the sessile tunicates. The larval stage is fish-like and possesses a dorsally situated hollow nerve cord. A swelling in some species at the anterior end of the nerve cord has been labelled 'forebrain' (Martini, 1909). Nerves from here extend to the lips, mouth cavity and anterior sensory organ. There is evidence that the anterior sensory organ has olfactory receptors and the lips contain chemoreceptors, possibly gustatory. Thus the anterior part of the brain is related to chemoreception and may serve to integrate information from the lips, pharynx and snout.

Another well-known relative of the vertebrates is amphioxus. Although there is some disagreement (Papez, 1929; Karamian, 1968) the prevalent view is that the front end of the anterior expansion of the neural tube represents the telencephalon (Ariëns Kappers *et al.*, 1936). From a special set of sensory cells thought to be olfactory, a nerve runs to this anterior swelling. This small forebrain may integrate chemical and tactile information from the snout (Aronson, 1970). Karamian (1968), citing the research of Sergeyev, described the formation of highly unstable conditioned responses even within isolated portions of the neural tube.

In summary, the very limited knowledge of these primitive relatives of the vertebrates suggests that at the lowest level that we can recognize, the telencephalon is diffuse and poorly differentiated. It receives information from the anterior end of the animal that is mostly chemical with some possibility of added tactile information in some forms. There is also the suggestion that this primordial telencephalon may have activating properties.

Level two

This level is represented by the jawless agnathans or cyclostomes (lampreys and hagfishes). They have been studied extensively by neuroanatomists and these studies have been supplemented in recent years by electrophysiological experiments. The cerebral hemispheres are well differentiated but are dominated by the large olfactory bulbs. Olfactory fibres extend to all parts of the hemisphere (Nieuwenhuys, 1967). This observation is supported by electrophysiological evidence (Bruckmoser, 1971). According to some authors, the major forerunners of the vertebrate pallium, that is, primordia of the hippocampus and pyriform lobe, can be recognized in these fishes, but they are poorly differentiated.

The dominance of the olfactory system supports traditional theory that this part of the brain is essentially an olfactory system. However, in recent studies (Karamian, 1972; Karamian, Agayan & Veselkin, 1973) it has been reported that evoked potentials have been obtained in the primordium hippocampus following somatic and visual stimulation, thus indicating additional functions other than olfaction.

Although elementary conditioned reflexes could be established rapidly, they were extremely unstable and varied from one experiment to another (Karamian, 1968). These characteristics were substantially unchanged after cerebral extirpation (Karamian, 1972). I am not aware of any other behavioural studies on the telencephalon of cyclostomes. We must, therefore, rely heavily upon the indirect evidence cited above for our understanding of telencephalic function. This seems to show that the telencephalon is dominated by its olfactory input, but we also see the beginning of an integrative system involving somatic, visual and olfactory input. It is possible that the eventual use of newer methods such as the horseradish peroxidase technique may change this picture, as they are doing in elasmobranchs.

Level three

The third level is represented by the amphibians. The telencephalon retains many primitive characters (Nieuwenhuys, 1967), which are especially pronounced in the salamanders (Herrick, 1948; Kokoros & Northcutt, 1977). Most of the bodies of the neurons are crowded close to the ventricle in their embryonic position and show only partial differentiation into discrete areas or nuclei. Axons and dendrites extend outward into the white substance where they form dense synaptic networks or neuropil, and many of the fibre tracts are not much more than diffuse networks of axons. Olfactory fibres from the olfactory and accessory olfactory bulbs and from the anterior olfactory nucleus reach most parts of the cerebrum much as they do in cyclostomes (Herrick, 1948; Nieuwenhuys, 1967; Scalia *et al.*, 1968; Northcutt & Royce, 1975).

The limited behavioural plasticity so characteristic of frogs and salamanders correlates well with the relative simplicity of the brain. Although Chu & McCain (1969), Boice (1970) and others, using traditional learning tests, have claimed some success in modifying the behaviour of anurans, especially in species of *Bufo*, Yaremko, Jette & Utter (1974) emphasized the restricted and transient nature of the learning, which they related to limitations in the capacities of the animals. Van Bergeijk (1967) attributed the difficulties in training anurans to inadequate or unnatural situations, but this, if true, would be a further reflection of their limited ability to adjust behaviourally to changes in the environment. With respect to classical conditioning, Karamian (1966) speaks of the 'imperfection of the conditioned-reflex activity of the amphibian' which remains unchanged after removal of the forebrain.

As early as 1824 Flourens observed a loss of spontaneous movements when the frog telencephalon is extirpated. The animals just sit still unless a strong stimulus is applied; they then behave normally. This loss of spontaneity, which does not occur if just the olfactory bulbs are removed, has since been confirmed many times for frogs, toads and salamanders (Aronson, 1970). The finding by Ewert (1970) that avoidance (withdrawal) responses occur more readily in the toad *Bufo bufo* after total decerebration can be understood as a disruption of the non-specific forebrain modulatory system causing a loss of inhibition (see p. 51).

Sexually active males readily swim to and clasp receptive females. With the forebrain removed, a male will rarely swim to the female; but if he accidentally touches her the male will usually turn to and clasp her in a normal manner and they eventually spawn. This may mean the loss of visual, but not tactile orientation, or it could mean that a stronger stimulus was required to trigger the response (Aronson & Noble, 1945).

Prey-catching behaviour in toads, according to Ewert (1970), is normally activated by visual stimuli to the tectum and is normally inhibited by neural activity in the pretectal area and dorsal thalamus. The telencephalon, in turn, inhibits the pretectal and thalamic areas and thus facilitates prey-catching. Ablation of forebrain abolishes prey-catching, but if the pretectal area is also ablated continuous prey-catching ensues.

Although, as noted above, amphibians are difficult to train, several authors have claimed success and have compared intact and decerebrated subjects. Most of these experiments resulted in major decrements in or complete loss of the conditioned responses (Aronson, 1970).

Because of the partial shift to a terrestrial environment and the concomitant changes in sensory and motor control, amphibians are usually considered more 'advanced' than fishes (Maier & Schneirla, 1964). However, the strongly stereotyped behaviour so characteristic of amphibians, the limited plasticity previously mentioned, the general nature of the deficits following telencephalic extirpation (such as loss of spontaneity) and the anatomically primitive nature of the telencephalon lead me to conclude that we are dealing with a structure that is functionally at a considerably lower level than that of the teleost fishes, and more on a par with cyclostomes. On the other hand, the extensive anatomical and electrophysiological evidence points to a varied sensory representation in the telencephalon consisting of olfactory, vomeronasal, dual visual, somesthetic, gustatory and possibly auditory modalities. This arrangement represents a higher level of organization than that of cyclostomes. Obviously this conclusion requires considerably more behavioural study.

Level four

Elasmobranchs. Until a few years ago (Aronson, 1963) I had the impression, based on the classical neurological literature on elasmobranchs, that the telencephalon in this group is sufficiently similar in structure and function to that of cyclostomes to be included in the second level. However, recent anatomical, electrophysiological and behavioural studies have altered this picture considerably.

First, the olfactory tracts, as now seen, extend to only a relatively small portion of each hemisphere on the lateral and ventral side (Ebbesson & Heimer, 1970). Only in these areas were secondary evoked potentials obtained following electrical stimulation of the olfactory mucosa and olfactory bulbs (Bruckmoser & Dieringer, 1973).

Second, the dorsal and ventral thalamic nuclei receive retinal fibres as well as tactile, cerebellar and spinal pathways, thus providing a variety of sensory inputs and probable separation of sensory modalities. These nuclei give rise to ascending telencephalic pathways that terminate largely in the central nucleus of the telencephalon (Schroeder & Ebbesson, 1974; Northcutt, 1978*a*). This region of the telencephalon may thus be viewed as a multisensory correlation area that has become largely freed from direct olfactory domination and may be especially important for vision. These conclusions are supported by electrophysiological studies on sharks and rays. Electrical stimulation of the optic nerve evoked potentials in the telencephalon (Vesselkin, 1964; Vesselkin & Kovaćević, 1973) which were localized by Cohen, Duff & Ebbesson (1973) in the posterior part of the ipsilateral central nucleus. Among the many species of elasmobranchs the telencephalon varies considerably in size and in ventricular volume. It is small in the widely studied squalids and becomes much larger in the lesser-known galeomorphs and batoids. Of particular importance is the bulge on the dorsal side caused by an enlarged central nucleus.

Third, one juvenile nurse shark (*Ginglymostoma*) was clearly able to perform visual discrimination tasks (black versus white, and horizontal versus vertical black and white stripes) after extensive bilateral lesions in the optic tectum (Graeber & Ebbesson, 1972). Following lesions of varying sizes in the central nucleus of the telencephalon, the ability to perform visual discriminations was completely lost in two out of 11 nurse sharks. The remaining subjects showed decrements of varying magnitudes which correlated with the amount of damage to the central nucleus (Graeber *et al.*,

1978). From the anatomical, physiological and behavioural experiments just described these investigators have postulated partially independent retinotectal and retinothalamotelencephalic visual systems which possibly represent primordia of their counterparts in mammals.

Classical and instrumental conditioning has been established in several species of sharks and rays, including simple problems of visual and auditory discriminations (reviewed by Gruber & Myrberg, 1977). Learning curves obtained when nurse sharks were trained to discriminate light from dark paddles did not differ from those obtained from teleosts and white mice given comparable problems (Aronson, Aronson & Clark, 1967). Learning studies have been too few for broad generalizations but I suspect that we have by no means approached the limits of elasmobranch learning capacities. I am aware of only one study of learning after telencephalic injury, where, as noted above, decrements in visual discrimination were attributed to lesions in the central nucleus (Graeber *et al.*, 1978). On the other hand we must keep in mind the conclusions reached by Karamian (1966) that in fishes the connections in conditioned responses occur in the cerebellum. More studies on the contribution of the telencephalon to learning in this group are urgently needed.

Teleosts. The bony fishes are unique among vertebrates in that the telencephalon develops ontogenetically by *eversion* of the lateral walls of the neural tube whereas in all other vertebrates it develops by lateral evagination of the median ventricle and *inversion* of the dorsal portion of the side walls (Herrick, 1922). The process of eversion is not just a simple turning out instead of turning in. Rather, it is a process of remoulding, so that identification of component parts, particularly in the upper or pallial division, is neither simple nor certain and is the source of much controversy. Nevertheless, the major fibre connections, to the extent that they are known, are roughly similar to those in other lower vertebrates (Nieuwenhuys, 1962; Schnitzlein, 1964), suggesting that the basic functions of the telencephalon may be similar to those of the elasmobranchs despite the topographical rearrangement. Eversion in teleosts results in a very compact brain with marked reduction of the ventricles. This, we have suggested, is an adaptation for streamlining, rather than a functional change, which may be critical in some species during the fry and juvenile stages when the brain is packed tightly into the cranium (Aronson & Kaplan, 1968).

Because of the uniqueness of the teleost forebrain it is often treated briefly in evolutionary studies (e.g. Papez, 1929; Karamian, 1968). However, teleosts are easy and economical to experiment with and the telencephalon of this group has been by far the most studied in terms of physiology and behaviour of all the lower vertebrates. It is important, therefore, to place the telencephalon of these fishes in proper evolutionary perspective.

Anatomically, the telencephalon is marked by many discrete fibre tracts and specialized nuclei. Secondary olfactory fibres terminate primarily in the subpallium but apparently reach the caudal portion of the pallium in some species (Scalia & Ebbesson, 1971). As in the elasmobranchs the major part of the pallium is free of secondary olfactory fibres (Aronson, 1963; Nieuwenhuys, 1967) and appears to serve other sensory systems. The topography of the telencephalon varies considerably in different families of fishes and presumably this reflects specialized behavioural adaptations. It is very much enlarged in many reef-inhabiting species such as *Holocentrus*, where the enlargement is mainly a major expansion of the pars dorsalis (or dorsolateralis) of the pallial area. There is indirect evidence that in this visually dominant species this is an enlarged visual area, comparable with, if not homologous to, the central nucleus of elasmobranchs. Northcutt (1977) discusses the possibility that these two areas developed similar structure and sensory functions as a result of parallel evolution from a common genome in organisms facing similar environmental problems.

There is abundant evidence that most or all areas of the telencephalon are connected with the ventral thalamus and hypothalamus by large bundles of afferent and efferent fibres, but evidence for the projection of sensory fibres from the dorsal thalamus to the telencephalon is much weaker. There are also reports of EEG changes or the recording of evoked potentials in the dorsal pallium following photic stimulation, but this too is controversial (reviewed by Kaplan & Aronson, 1969). Thus, there are still questions concerning the extent of the sensory projections to the telencephalon and whether the telencephalon exerts sensory control by modulating sensory processes in the diencephalon, midbrain and cerebellum as suggested by the experiments of Zagorul'ko (1965) and Timkina (1965).

A large array of species-typical behaviour patterns have been described in teleosts, some of which are remarkably elaborate. Most of the species that have been studied intensively are highly adaptable. Many are rapid learners. One-trial learning can easily be demonstrated (Aronson, unpublished observation), and many species seem able to master relatively difficult or complex problems.

In a large number of experiments the behavioural consequences of partial or total extirpation of the telencephalon have been examined. Several papers in this volume deal with the subject, which has been reviewed many times in recent years (Aronson, 1963; Segaar, 1965; Aronson & Kaplan, 1968; Aronson, 1970; Flood, Overmier & Savage, 1976; Hollis & Overmier, 1978). While a further summary seems unnecessary the following conclusions drawn from this research are pertinent to the present discussions.

1. After a brief period of post-operative recovery one sees no obvious changes in the typical aquarium behaviour – e.g. swimming, feeding, approach, withdrawal, simple learning tasks – of numerous teleost species.
2. Classical conditioning and simple instrumental tasks are not affected by extirpation of the telencephalon, which led Karamian (1965) to conclude that the connections are established in the midbrain and cerebellum in teleosts. More complex learning tasks are almost always affected adversely. Only in a few cases have no change or improvement in some measures of learning been reported (reviewed by Bernstein, 1970; Hollis & Overmier, 1978).
3. Extirpation of the telencephalon in various teleostean species results in marked decrements in almost every kind of behaviour studied, as, for example, in reproductive behaviour, parental activities, aggressive behaviour, schooling, optomotor responses, feeding, general activity and a variety of learning procedures other than classical conditioning as noted in point 2. Only occasionally have behavioural patterns been completely eliminated (Aronson, 1970).

Any general hypothesis concerning the function of the telencephalon in teleosts must take the above factors into account. Recent attempts to develop hypotheses based on certain factors such as learning alone have obvious limitations.

The scanty neurological evidence concerning the telencephalon of elasmobranchs is not at variance with that of teleosts. Similarly, from what is known about their behavioural repertoire (Gruber & Myrberg, 1977), there is nothing to suggest a level of organization that is substantially different from that of teleosts. Tentatively, then, I am suggesting that telencephalic structure and function in elasmobranchs and teleosts have reached the same level of organizational complexity, even though their evolutionary lines separated so long ago and, superficially at least, their forebrain structures appear so different.

Mesolevels

There is an array of primitive fishes – the chimaerans related to ancestral sharks and rays, the polypterids, sturgeons, paddle fishes, bowfins and gars related to ancestral teleosts, the lung fishes and coelacanths related to ancestral amphibians – which, for many years, have had considerable interest for the neuroanatomists. The descriptions of the telencephalon in these forms (Nieuwenhuys, 1962, 1967) suggest that they should be considered as transitional or mesolevels (see p. 35). Unfortunately, they have not attracted the attention of the neurobehaviourists because most species seem poorly suited for functional studies.

Level five

Reptiles are the first truly terrestrial vertebrates and the complex adaptations and adjustments to this new mode of existence correlate with a much more complex brain. The dorsal thalamus is much enlarged and contains a number of well-differentiated nuclei which receive visual and auditory impulses as well as somatic information from the brainstem and spinal cord. They have a clearly defined visual system which consists of a retinogeniculate pathway that extends to the dorsal cortex and a retinotecto-rotundus pathway that projects to the lateral part of the dorsal ventricular ridge. Recent evidence shows that this ridge is part of the general or dorsal cortex rather than part of the striatum as previously described. An auditory pathway from the torus semicircularis runs to the dorsal thalamus and then to the striatum and to the medial part of the dorsal ventricular ridge. A somesthetic projection from the dorsal thalamus runs to the central portion of the dorsal ventricular ridge (reviewed by Northcutt, 1978*b*; Butler, 1978). The cortex is much larger and more complexly organized than in amphibians and has a laminar configuration (Ulinski, 1974).

A variety of motor responses have been obtained by electrically stimulating the cortex, dorsal ventricular ridge, striatum and olfactory areas (Schapiro & Goodman, 1969; Distel, 1978; and review by Aronson, 1970), but the cortical responses may be an experimental artifact (Orrego, 1961). These motor responses seem to stem mostly from striatal projections to the tegmentum which may be a precursor of one of the mammalian extrapyramidal systems. A pyramidal (corticospinal) system is not evident at this level.

Electrophysiological studies correlate rather well with the anatomical descriptions, showing that olfactory, visual, auditory, somesthetic and visceral information reaches the major telencephalic structures. Intracellular recordings provide evidence of a spatial arrangement of the different sensory inputs, but there is also considerable overlap. Several studies (reviewed by Aronson, 1970) point to the cortical associations between the visual, auditory and somatic systems. With respect to the optic system, Mazurskaya (1971) has demonstrated a topical arrangement of retinal projections in the dorsal cortex of turtles. In a further study Mazurskaya (1973) analysed the interactions of the visual projections and concluded that they have many features of mammalian associative cortex, namely wide receptive fields, complex interactions between local areas of the receptive field and high sensitivity of the neurons to movement of visual objects.

One would expect that the abundant research depicting the advances in the structure and physiology of the telencephalon would correlate with studies showing a considerably more extensive and complex behavioural repertoire, but this expectation is only partly realized. In the past scientists tended vastly to underestimate the learning capacities of reptiles. We now know from the extensive review of Burghardt (1977) that learning abilities have been demonstrated in almost all orders of reptiles, and that in some species tasks of some complexity have been mastered. Nevertheless, one gains the impression from this review that many reptiles, like the amphibians, are bound to some extent by environmental constraints and do not adapt as readily to strange situations and unusual problems as is often the case in mammals. For the most part their learning abilities seem to be on a par with teleosts, although Bitterman (1975) showed that turtles are superior in some tasks such as reversal learning.

Sikharulidze (1969) reported difficulties in establishing new conditioned reflexes in turtles after complete removal of the telencephalon, but when only the cortex was destroyed conditioned reflexes were unaffected. Similarly, Morlock (1972) found no changes in spatial discrimination learning after cortical ablation, and Bass, Pritz & Northcutt (1973) reported similar results for the side-necked turtle, in a problem involving discrimination of horizontal from vertical lines after bilateral removal of the dorsal cortex or dorsal cortex plus dorsal ventricular ridge.

As in amphibians, removal of the telencephalon in a variety of reptiles causes marked decrements in spontaneous movements without interfering with motor performance. The operation also causes major decrements in feeding, optokinetic responses, climbing and visual cliff responses. Older studies of these phenomena (see Aronson, 1970) have been supplemented by recent observation on the loss of spontaneity (Tarr, 1977) and decrements in feeding behaviour (Morlock, 1972). Sikharulidze (1969) also reported decrements in orientation.

Terrestrial habitats are more complex than their typical aquatic counterparts and require more involved locomotory processes and sensory refinements, especially in those reptiles with well-developed limbs. The advances in telencephalic organization considered above may very well provide the substrate for this behaviour. Many reptiles are predators and active foragers. Regal (1978) has emphasized the complexity of foraging behaviour, the skill and agility needed, and the selective advantage to be derived from the development of an enlarged memory and advanced information processing capacity.

Reptiles have also developed rather elaborate patterns of social organization and behaviour (Greenberg & MacLean, 1978) and the telencephalon seems to be importantly involved in the mediation of these behaviours. Keating, Kormann & Horel (1970) observed decrements in attack and retreat responses following amygdala (striatal) lesions and Tarr (1977) reported loss of aggressive behaviour with similar lesions. Greenberg (1977) found that lesions in the paleostriatum caused deficits in challenge displays without affecting levels of activity or impairing vision. Although the evidence is still very sketchy, it seems that for most of the behaviours considered in this section, the subpallial structures are more critically involved. The major anatomical advance in the reptilian cerebrum is the enlargement and specialization of the cortex and its extension, the dorsal ventricular ridge. Yet, despite the efforts of a number of investigators, neurobehavioural studies have not involved the cortex in the mediation of any behaviour patterns or processes. Perhaps new approaches and new techniques are needed to understand the function of the reptilian pallium.

Hypotheses concerning telencephalic function

In this section I shall describe briefly the best-known hypotheses concerning telencephalic function in lower vertebrates. Some are relevant to most of the lower vertebrate groups; others refer to a specific group such as teleosts. Some are mutually exclusive; others are not.

The 'no function' hypothesis. During the last century a number of investigators were unable to find any differences between normal and telencephalon-ablated fishes in a number of items such as locomotion, equilibrium, spontaneity, feeding, schooling, simple discriminations of colour and general aquarium behaviour. Steiner (1888), apparently ignoring the obvious relation of the anterior part of the brain to olfaction, concluded that the teleost telencephalon has no function. This hypothesis, which now has only historical interest, nevertheless serves to remind us that this division of the teleost brain is not critically involved in the life-supporting processes on either physiological or behavioural levels.

The olfactory hypothesis. This traditional theory of neuroanatomists holds that the telencephalon serves exclusively as an olfactory mechanism or for the correlation of olfaction and taste. It has been applied to all vertebrates up to reptiles. Until recently the hemispheres were often labelled 'olfactory lobes', with associated tracts labelled accordingly. This hypothesis has been pretty well refuted on anatomical, physiological and behavioural grounds in most vertebrates, but it dies slowly and is still seen occasionally in the current literature. On the other hand, while olfaction may not be the sole function of the telencephalon in any vertebrate, it is certainly one of the important functions in practically all vertebrates. Recent anatomical studies in elasmobranchs, teleosts and reptiles have shown that the terminations of secondary olfactory fibres are largely confined to limited areas on the ventral side of the cerebrum. The possibility that tertiary fibres bearing olfactory information are more broadly distributed is considered only occasionally (Ebbesson & Northcutt, 1976), and studies of the telencephalon using lesions or ablations seldom include olfactory controls. Often overlooked are efferent fibres from the cerebrum to olfactory bulbs which may regulate olfactory acuity (Döving & Gemne, 1966).

The teleost telencephalon as a striatum. This old hypothesis holds that the entire cerebrum is the corpus striatum while the pallium is just the non-nervous membranous tela which covers the cerebrum and has not yet developed nervous tissue or function. Suggested by Rabl-Ruckhard (1883), it was shown to be wrong by Gage (1893) on embryological grounds, but it nevertheless became a very popular conception until almost recent times and has encouraged investigators to look for striatal-like functions in the cerebrum. Until one or two decades ago it was not unusual to find the telencephalon of teleosts labelled 'corpus striatum'.

Functional localizations. According to this hypothesis, the telencephalon has localized areas that organize specific behaviours – such as an area that organizes spawning patterns, another for aggressive behaviour, another for parental behaviour, and so on. Evidence supporting this hypothesis comes mainly from experiments on teleosts (Noble, 1936, 1937; Segaar, 1961; Segaar & Nieuwenhuys, 1963) and reptiles (Greenberg, 1977). It is not clear how this accords with the more widely accepted limbic hypothesis (p. 50) or the fact that species-typical behaviour usually survives total telencephalic ablation to some extent.

Sensory integration. Herrick was first to suggest that in lower vertebrates the telencephalon integrates olfaction and taste. This idea has been expanded by many investigators to also include olfactory, visual, auditory and somatosensory and visceral information. Karamian (1968, 1972) has developed a rather impressive picture, based heavily on electrophysiological evidence, of how sensory integration becomes more and more complex in orderly fashion as one compares amphioxus, lampreys, elasmobranchs, amphibians, reptiles and mammals. This picture is much in accord with the levels concept, although Karamian never refers to it as such. Some degree of sensory integration seems to be one function of the telencephalon in all vertebrates, but it is not the only function in any vertebrate.

The 'forebrain–cerebellum shift'. Karamian (1966) proposed that in conditioning, the connections between the conditioned stimulus, unconditioned stimulus and response take place in the cerebellum in the lower vertebrates up to and including teleost fishes and shift in reptiles, birds and mammals to the forebrain. Amphibians represent a transitional state. More recently he has proposed that while in elasmobranchs and teleosts the telencephalon does not participate directly in the formation of conditioned links, it already has non-specific effects on the functions of those regions of the brain which carry out analytical and synthetic activity (Karamian, Malyukova & Sergeev, 1967). This hypothesis, which has received little attention outside of the USSR, emphasizes a role for the cerebellum of lower vertebrates in non-motor functions, an area that has been largely neglected elsewhere.

The limbic hypothesis. The view that, except for olfactory functions, the forebrain of lower vertebrates is largely the forerunner of the mammalian limbic system (Aronson, 1963; Segaar, 1965) has gained many adherents in recent years. It has even been proposed that this was the primary function in ancestral vertebrates and that the olfactory system evolved from the limbic system (Riss, Halpern & Scalia, 1969). Although this seems doubtful, it is important to keep in mind that limbic structures have increased enormously in vertebrate evolution in both size and differentiation and we should anticipate parallel changes in limbic function.

Learning hypotheses. Neural mechanisms of learning in lower vertebrates have attracted by far the most attention among those interested in the role of the telencephalon in behaviour. From a large number of studies, mainly on teleost fishes, it is evident that partial or complete ablation of the telencephalon causes learning decrements of varying dimensions except in classical conditioning and in simple instrumental tasks (Aronson & Kaplan, 1968; Flood *et al.*, 1976; Hollis & Overmier, 1978), and this is the subject of other chapters in this volume. Whether the telencephalon plays a direct and specific role in some stage of the learning process, as suggested by these hypotheses, or whether the observed decrements are the result of a more general function of the telencephalon is still an open question (see next hypothesis). However, it should be apparent that a specific effect on a learning mechanism cannot be the sole function in any vertebrate.

Forebrain as a non-specific modulating mechanism. This hypothesis stems from the ideas of Herrick (1948), who called it non-specific arousal. Since at that time the telencephalon was considered primarily an olfactory structure, he viewed 'arousal' as a major attribute of olfaction. That is, through olfaction the animal becomes responsive to the environment. In my early papers, following the lead of Herrick, I had proposed that the telencephalon of fishes and anurans had arousal or facilitative functions that had become largely independent of olfaction. More recently I have adopted the terms modulatory or regulatory functions since these imply homeostatic-like mechanisms which are either excitatory or inhibitory. The thought is that behaviour organized and integrated in lower centres is modulated by forebrain activity. This modulation is most effective with respect to complex behaviour and least important with respect to simple behaviour. I favour this hypothesis because it seems to explain many things, as for example, why the effects of telencephalic ablation are so widespread, why this operation usually causes deficits in behaviour while complete loss of behaviour is uncommon, and why classical conditioning and very simple avoidance problems are unaffected. Moreover, it conforms with the observation that the loss of spontaneity is so widespread after removal of the lower vertebrate telencephalon. It is also in accord with the concept of a primitive limbic system, for the most inclusive statement that one can make about the limbic system is that it is a master regulatory system which predisposes the individual to behave in a given way. The limbic system in cyclostomes regulates relatively simple behavioural activities. In succeeding levels as the behaviour becomes more complex so does the limbic regulation. We can follow this all the way to man where a most intricate limbic system seems to be needed to regulate the most complex behaviours known to man.

It is not possible within the confines of this article to evaluate adequately the relative merits of the foregoing hypotheses. With respect to teleost fishes Hollis & Overmier (1978) argue that the primary function of the telencephalon is to provide a component of the learning process, namely utilization of secondary reinforcers. Hence they interpret the deficits in species-typical behaviour following telencephalic ablation as stemming from impairment of the secondary reinforcement mechanism. How this relates to olfactory function is not discussed. Alternatively, I see three basic functions for the telencephalon: (1) mediation of olfaction; (2) multisensory integration; (3) limbic modulation. The third function accounts for the disruption of learning processes as well as the changes in species-typical behaviour that follow telencephalic ablation.

Recapitulation

Using the concept of organizational levels we have developed the following composite picture of the functional evolution of the telencephalon in lower vertebrates. At the first level we recognize a chemical sensory input to the anterior end of the neural tube and, associated with this, the possible beginning of a chemically related modulatory function. In level two, the agnathans, a specific olfactory input is evident. Associated with this we postulate a non-specific modulatory system and the elements, at least, of a multisensory integrative mechanism. In level three, the amphibians, the three systems are discrete entities although considerable diffuseness remains. By level four, elasmobranchs and teleosts, the modulatory system becomes independent of olfaction and is greatly enlarged in some species. This is associated with the elaboration of specific sensory inputs especially visual. Finally in reptiles, level five, along with the modulatory system there is considerable elaboration of the multisensory integrative mechanism as a general cortex. This is associated with the increased sensory and motor requirements of a terrestrial habitat. Also at this level the motor mechanism of the telencephalon provides a discrete and direct pathway to the lower motor centres which undoubtedly influences the other telencephalic functions in ways that we are just beginning to glean from current reptilian research. The picture as outlined is based essentially on those living representatives of the vertebrate classes that have been studied most extensively. Although we are dealing with species having specialized adaptive traits they form groups that have cores of commonality with respect to telencephalic structure and function. Thus it is possible to apply the levels concept without making assumptions about evolutionary lineages. In essence, the levels approach provides the basis – in fact the only realistic basis at the present time – for reconstructing the course of the progressive evolution of vertebrate telencephalic function.

References

Ariëns Kappers, C. U., Huber, G. C. & Crosby, E. C. (1936). *The Comparative Anatomy of the Nervous System of Vertebrates Including Man.* Hafner, New York.

Aronson, L. R. (1963). The central nervous system of sharks and bony fishes with special reference to sensory and integrative mechanisms. In *Sharks and Survival*, ed. P. W. Gilbert, pp. 165–241. D. C. Heath, Boston.

Aronson, L. R. (1970). Functional evolution of the forebrain in lower vertebrates. In *Development and Evolution of Behavior: Essays in*

Memory of T. C. Schneirla, ed. L. R. Aronson, E. Tobach, D. S. Lehrman & J. S. Rosenblatt, pp. 75–107. W. H. Freeman, San Francisco.

Aronson, L. R., Aronson, F. R. & Clark, E. (1967). Instrumental conditioning and light–dark discrimination in young nurse sharks. *Bulletin of Marine Science*, **17,** 249–56.

Aronson, L. R. & Kaplan, H. (1968). Function of the teleostean forebrain. In *The Central Nervous System and Fish Behavior*, ed. D. Ingle, pp. 107–25. University of Chicago Press, Chicago.

Aronson, L. R. & Noble, G. K. (1945). The sexual behavior of Anura. II. Neural mechanisms controlling mating in the male leopard frog, *Rana pipiens. Bulletin of the American Museum of Natural History*, **86,** 87–139.

Atz, J. W. (1970). The application of the idea of homology to behavior. In *Development and Evolution of Behavior: Essays in Memory of T. C. Schneirla*, ed. L. R. Aronson, E. Tobach, D. S. Lehrman & J. S. Rosenblatt, pp. 53–74. W. H. Freeman, San Francisco.

Bass, A. H., Pritz, M. B. & Northcutt, R. G. (1973). Effects of telencephalic and tectal ablations on visual behavior in the side-necked turtle, *Podocnemis unifilis. Brain Research*, **55,** 455–60.

Bernstein, J. I. (1970). Anatomy and physiology of the central nervous system. In *Fish Physiology*, vol. 4, pp. 1–90. Academic Press, New York & London.

Bitterman, M. E. (1975). The comparative analysis of learning. *Science*, **188,** 699–709.

Boice, R. (1970). Avoidance learning in active and passive frogs and toads. *Journal of Comparative and Physiological Psychology*, **70,** 154–6.

Bruckmoser, P. (1971). Elektrische Antworten im Vorderhirn von *Lampetra fluviatilis* (L.) bei Reizung des Nervus olfactorius. *Zeitschrift für vergleichende Physiologie*, **75,** 69–85.

Bruckmoser, P. & Dieringer, N. (1973). Evoked potentials in the primary and secondary olfactory projection areas of the forebrain in Elasmobranchia. *Journal of Comparative Physiology*, **87,** 65–74.

Bullock, T. H. (1940). The functional organization of the nervous system of the Enteropneusta. *Biological Bulletin*, **79,** 91–113.

Bullock, T. H. (1965). Chaetognatha, Pogonophora, Hemichordata, and Tunicata. In *Structure and Function in the Nervous Systems of Invertebrates*, ed. T. H. Bullock & G. A. Horridge, chapt. 27. W. H. Freeman, San Francisco.

Burghardt, G. M. (1977). Learning processes in reptiles. In *Biology of the Reptilia*, ed. C. Gans & D. W. Tinkle, pp. 555–681. London: Academic Press, New York & London.

Butler, A. B. (1978). Forebrain connections in lizards and the evolution of sensory systems. In *Behavior and Neurology of Lizards*, ed. N. Greenberg & P. D. MacLean, pp. 65–78. National Institute of Mental Health, Rockville, Maryland.

Chu, P. K. & McCain, G. (1969). Discrimination learning and extinction in toads. *Psychonomic Science,* **14,** 14–15.

Cohen, D. H., Duff, T. A. & Ebbesson, S. O. E. (197?). Electrophysiological identification of a visual area in shark telencephalon. *Science*, **182,** 492–4.

Distel, H. (1978). Behavioral responses to the electrical stimulation of the brain in the green iguana. In *Behavior and Neurology of Lizards*, ed. N. Greenberg & P. D. MacLean, pp. 135–47. National Institute of Mental Health, Rockville, Maryland.

Döving, K. B. & Gemne, G. (1966). An electrophysiological study of the efferent olfactory system in the burbot. *Journal of Neurophysiology*, **29,** 665–74.

Ebbesson, S. O. E. & Heimer, L. (1970). Projections of the olfactory tract in the nurse shark (*Ginglymostoma cirratum*). *Brain Research*, **17,** 47–55.

Ebbesson, S. O. E. & Northcutt, R. G. (1976). Neurology of anamniotic vertebrates. In *Evolution of Brain and Behavior in Vertebrates*, ed. R. B. Masterton, M. E. Bitterman, C. B. G. Campbell & N. Hotton, pp. 115–46. Wiley, New York.

Ewert, J.-P. (1970). Neural mechanisms of prey-catching and avoidance behavior in the toad (*Bufo bufo L.*). *Brain, Behavior and Evolution*, **3,** 36–56.

Flood, N. C., Overmier, J. B. & Savage, G. E. (1976). Teleost telencephalon and learning: an interpretative review of data and hypotheses. *Physiology and Behavior*, **16,** 783–98.

Flourens, J. P. M. (1824). *Recherches expérimentales sur les propriétés et les fonctions du système nerveux dans les animaux vertébrés.* Paris.

Gage, S. (1893). The brain of *Diemyctylus viridescens* from larval to adult life and comparisons with the brain of *Amia* and *Petromyzon.* In *The Wilder Quarter-Century Book*, pp. 259–315. Comstock, Ithaca.

Graeber, R. C. & Ebbesson, S. O. E. (1972). Visual discrimination learning in normal and tectal-ablated nurse sharks *(Ginglymostoma cirratum). Comparative Biochemistry and Physiology*, **42,** 131–9.

Graeber, R. C., Schroeder, D. M., Jane, J. A. & Ebbesson, S. O. E. (1978). Visual discrimination following partial telencephalic ablations in nurse sharks *(Ginglymostoma cirratum). Journal of Comparative Neurology*, **180,** 325–44.

Greenberg, N. (1977). A neuroethological study of display behavior in the lizard *Anolis carolinensis* (Reptilia, Lacertilia, Iguanidae). *American Zoologist*, **17,** 191–201.

Greenberg, N. & MacLean, P. D. (eds.) (1978). *Behavior and Neurology of Lizards.* National Institute of Mental Health, Rockville, Maryland.

Gruber, S. H. & Myrberg, A. A. Jr (1977). Approaches to the study of the behavior of sharks. *American Zoologist*, **17,** 471–86.

Herrick, C. J. (1922). Functional factors in the morphology of the forebrain of fishes. In *Libro en honor de D. S. Ramón y Cajal*, vol. 1, pp. 143–204. Jiminez & Molina, Madrid.

Herrick, C. J. (1948). *The Brain of the Tiger Salamander, Ambystoma tigrinum.* University of Chicago Press, Chicago.

Hodos, W. (1970). Evolutionary interpretation of neural and behavioral studies of living vertebrates. In *The Neurosciences: Second Study Program*, ed. F. O. Schmidt, pp. 26–39. Rockefeller University Press, New York.

Hodos, W. & Campbell, C. B. G. (1969). *Scala naturae*: why there is no theory in comparative psychology. *Psychological Review*, **76,** 337–50.

Hollis, K. L. & Overmier, J. B. (1978). The function of the teleost telencephalon in behavior: a reinforcement mediator. In *The Behavior of Fish and Other Aquatic Animals*, pp. 137–95. Academic Press, New York & London.

Johnston, J. B. (1906). *The Nervous System of Vertebrates.* P. Blakiston's Sons & Co., Philadelphia.
Kaplan, H. & Aronson, L. R. (1969). Function of forebrain and cerebellum in learning in the teleost, *Tilapia heudelotii macrocephala. Bulletin of the American Museum of Natural History*, **142,** 141–208.
Karamian, A. I. (1962). *Evolution of the Function of the Cerebellum and Cerebral Hemispheres.* The Israel Program for Scientific Translations, Jerusalem. (Trans. from the Russian.)
Karamian, A. I. (1965). Evolution of functions in the higher divisions of the central nervous system and of their regulating mechanisms. In *Essays on Physiological Evolution*, ed. J. W. S. Pringle, pp. 149–65. Pergamon Press, New York.
Karamian, A. I. (1968). On the evolution of the integrative activity of the central nervous system in the phylogeny of vertebrates. *Progress in Brain Research*, **22,** 427–47.
Karamian, A. I. (1972). Formation of the structural and functional organization of the paleo-, archi-, and neocortex in premammalian phylogeny. *Zhurnal Evolyutsionnoi Biokhimii i Fiziologii*, **6,** 324–32. (English trans. in *Neuroscience and Behavioral Physiology* (1973), **6,** 109–18.)
Karamian, A. I., Agayan, A. L. & Veselkin, N. P. (1973). Evoked potentials in various regions of the lamprey brain following stimulation of dorsal parts of the spinal cord. *Biologicheskii Zhurnal Armenii*, **26,** 56–63. (English abstr. in *Biological Abstracts* (1974), **58,** 32805.)
Karamian, A. I., Malyukova, I. V. & Sergeev, B. F. (1967). Participation of the telencephalon of bony fish in the accomplishment of complex conditioned-reflex and general-behavior reactions. In *Povedenie I Retseptsii Ryb*, pp. 109–14. Akademiya Nauk SSSR (English trans. (1969) by Bureau of Sport Fisheries and Wildlife, Washington, DC.)
Karamian, A. I., Veselkin, N. & Belekhova, M. (1978). Recent advances in the biology of sharks. *Zhurnal Evolyutsionnoi Biokhimii i Fiziologii*, **14,** 415–17. (English trans. in *Journal of Evolutionary Biochemistry and Physiology* (1979), **14,** 341–3.)
Keating, E. G., Kormann, L. A. & Horel, J. A. (1970). The behavioral effects of stimulating and ablating the reptilian amygdala *(Caiman sklerops). Physiology and Behavior*, **5,** 55–9.
Knight-Jones, E. W. (1952). On the nervous system of *Saccoglossus cambrensis* (Enteropneusta). *Philosophical Transactions of the Royal Society of London*, **236B,** 315–54.
Kokoros, J. J. & Northcutt, R. G. (1977). Telencephalic efferents of the tiger salamander *Ambystoma tigrinum tigrinum* (Green). *Journal of Comparative Neurology*, **173,** 613–27.
Lockard, R. B. (1971). Reflections on the fall of comparative psychology: is there a message for us all? *American Psychologist*, **26,** 168–79.
MacLean, P. D. (1973). *A Triune Concept of the Brain and Behavior. The Hinks Memorial Lectures*, pp. 6–66. University of Toronto Press, Toronto.
Maier, N. R. F. & Schneirla, T. C. (1964). *Principles of Animal Psychology*. Dover Publications, New York.
Martini, E. (1909). Studien über die Konstanz histologischer Elemente. I. *Oekopleura longicauda. Zeitschrift für wissenschaftliche Zoologie*, **92**, 563–626.

Mazurskaya, P. Z. (1971). Retinal projection in the forebrain of *Emys orbicularis*. *Zhurnal Evolyutsionnoi Biokhimii i Fiziologii*, **7**, 624–31. (English trans. in *Neuroscience and Behavioral Physiology* (1973), **6**, 75–82.)

Mazurskaya, P. Z. (1973). Organization of receptive fields in the forebrain of *Emys orbicularis*. *Zhurnal Evolyutsionnoi Biokhimii i Fiziologii*, **8**, 617–24. (English trans. in *Neuroscience and Behavioral Physiology* (1973), **6**, 311–18.)

Morlock, H. C. (1972). Behavior following ablation of the dorsal cortex of turtles. *Brain, Behavior and Evolution*, **5**, 256–63.

Needham, J. (1929). *The Sceptical Biologist*. Chatto & Windus, London.

Nieuwenhuys, R. (1962). Trends in the evolution of the actinopterygian forebrain. *Journal of Morphology*, **111**, 69–88.

Nieuwenhuys, R. (1967). Comparative anatomy of olfactory centres and tracts. In *Progress in Brain Research*, ed. Y. Zotterman, vol. 23, pp. 1–64. Elsevier, Amsterdam.

Noble, G. K. (1936). The function of the corpus striatum in the social behavior of fishes. *Anatomical Record*, **64**, 34.

Noble, G. K. (1937). Effect of lesions of the corpus striatum on the brooding behavior of cichlid fishes. *Anatomical Record*, **70**, 58.

Northcutt, R. G. (1977). Elasmobranch central nervous system organization and its possible evolutionary significance. *American Zoologist*, **17**, 411–29.

Northcutt, R. G. (1978*a*). Brain organization in the cartilaginous fishes. In *Sensory Biology of Sharks, Skates and Rays*, ed. B. Hodgson & R. F. Mathewson, pp. 117–93. Office of Naval Research, Washington, DC.

Northcutt, R. G. (1978*b*). Forebrain and midbrain organization in lizards and its phylogenetic significance. In *Behaviour and Neurology of Lizards*, ed. N. Greenberg & P. D. MacLean, pp. 11–64. National Institute of Mental Health, Rockville, Maryland.

Northcutt, R. G. & Royce, G. J. (1975). Olfactory bulb projections in the bullfrog *Rana catesbeiana*. *Journal of Morphology*, **145**, 251–68.

Novikoff, A. (1945). The concept of integrative levels and biology. *Science*, **101**, 209–15.

Orrego, F. (1961). The reptilian forebrain. I. The olfactory pathways and cortical areas in the turtle. *Archivio italiano di biologia*, **99**, 425–45.

Papez, J. W. (1929). *Comparative Neurology*. Thomas Y. Crowell, New York.

Rabl-Ruckhard, H. (1883). Das Grosshirn der Knockenfische und seine Anhangsgebilde. *Archiv für Anatomie und Physiologie, Anatomisches Abteilung*, 279–322.

Regal, P. J. (1978). Behavioral differences between reptiles and mammals: an analysis of activity and mental capabilities. In *Behavior and Neurology of Lizards*, ed. N. Greenberg & P. D. MacLean, pp. 183–202. National Institute of Mental Health, Rockville, Maryland.

Rensch, B. (1967). The evolution of brain achievements. In *Evolutionary Biology*, ed. Th. Dobzhansky, M. K. Hecht & W. C. Steele, pp. 26–68. Appleton-Century-Crofts, New York.

Riss, W., Halpern, M. & Scalia, F. (1969). Anatomical aspects of the evolution of the limbic system and olfactory systems and their potential significance for behavior. *Annals of the New York Academy of Sciences*, **159**, 1096–111.

Scalia, R. & Ebbesson, S. O. E. (1971). The central projections of the olfactory bulb in a teleost (*Gymnothorax funebris*). *Brain, Behavior and Evolution*, **4**, 376–99.

Scalia, R., Halpern, M., Knapp, H. & Riss, W. (1968). The efferent connections of the olfactory bulb in the frog: a study of degenerating unmyelinated fibres. *Journal of Anatomy*, **103**, 245–62.

Schapiro, H. & Goodman, D. C. (1969). Motor functions and their anatomical basis in the forebrain and tectum of the alligator. *Experimental Neurology*, **24**, 187–95.

Schneirla, T. C. (1949). Levels in the psychological capacities of animals. In *Philosophy for the Future*, ed. R. W. Sellers, V. J. McGill & M. Farber, pp. 243–86. Macmillan, New York.

Schneirla, T. C. (1952). A consideration of some conceptual trends in comparative psychology. *Psychological Bulletin*, **49**, 559–97.

Schneirla, T. C. (1953). The concept of levels in the study of social phenomena. In *Groups in Harmony and Tension*, ed. M. Sherif & C. Sherif, pp. 54–75. Harper, New York.

Schneirla, T. C. (1959). An evolutionary and developmental theory of biphasic processes underlying approach and withdrawal. In *Nebraska Symposium on Motivation*, ed. M. R. Jones, vol. 7, pp. 1–42. University of Nebraska Press, Lincoln.

Schnitzlein, N. H. (1964). Correlation of habitat and structure in the fish brain. *American Zoologist*, **4**, 21–32.

Schroeder, D. M. & Ebbesson, S. O. E. (1974). Nonolfactory telencephalic afferents in the nurse shark (*Ginglymostoma cirratum*). *Brain, Behavior and Evolution*, **9**, 121–55.

Segaar, J. (1961). Telencephalon and behavior in *Gasterosteus aculeatus*. *Behaviour*, **18**, 256–87.

Segaar, J. (1965). Behavioural aspects of degeneration and regeneration in fish brain: a comparison with higher vertebrates. In *Progress in Brain Research*, vol. 14, *Degeneration Patterns in the Nervous System*, ed. M. Singer & J. P. Shade, pp. 144–231. Elsevier, Amsterdam.

Segaar, J. & Nieuwenhuys, R. (1963). New electrophysiological experiments with male *Gasterosteus aculeatus*. *Animal Behaviour*, **11**, 331–44.

Sikharulidze, N. I. (1969). On the function of the forebrain in the behaviour of tortoises, *Emys orbicularis* and *Clemmys caspica*. *Bulletin of the Academy of Sciences of the Georgian SSR*, **55**, 169–72. (In Russian with English summary.)

Skinner, B. F. (1966). The phylogeny and ontogeny of behavior. *Science*, **153**, 1205–13.

Steiner, J. (1888). *Die Functionen des Zentralnervensystems und ihre Phylogenese*, part 2, *Die Fische*. Vieweg, Braunschweig.

Tarr, R. S. (1977). Role of the amygdala in the intraspecies aggressive behavior of the iguanid lizard, *Sceloporus occidentalis*. *Physiology and Behavior*, **18**, 1153–8.

Tavolga, W. N. (1970). Levels of interaction in animal communication. In *Development and Evolution of Behavior: Essays in Memory of T. C. Schneirla*, ed. L. R. Aronson, E. Tobach, D. S. Lehrman & J. S. Rosenblatt, pp. 281–302. W. H. Freeman, San Francisco.

Timkina, M. I. (1965). Relationships between different sensory systems in bony fish. *Zhurnal Vysshei Nervnoi imeni I. P. Pavlova*, **15**, 927. (English trans. in *Federation Proceedings Translation Supplement* (1966), **25**, 750–2.)

Tobach, E. (1970). Some guidelines to the study of the evolution and development of emotion. In *Development and Evolution of Behavior: Essays in Memory of T. C. Schneirla*, ed. L. R. Aronson, E. Tobach, D. S. Lehrman & J. S. Rosenblatt, pp. 238–53. W. H. Freeman, San Francisco.

Tobach, E. (1976). Evolution of behavior and the comparative method. *International Journal of Psychology*, **11**, 185–201.

Tobach, E. & Schneirla, T. C. (1968). The biopsychology of social behavior of animals. In *The Biologic Basis of Pediatric Practice*, ed. R. E. Cooke, pp. 68–82. McGraw-Hill, New York.

Ulinski, P. S. (1974). Cytoarchitecture of cerebral cortex in snakes. *Journal of Comparative Neurology*, **158**, 243–66.

Van Bergeijk, A. W. (1967). Anticipatory feeding behavior in the bullfrog (*Rana catesbeiana*). *Animal Behaviour*, **15**, 231–8.

Vesselkin, N. P. (1964). Electrical responses in skate brain to photic stimulation. *Fiziologicheskii Zhurnal SSSR imeni I. M. Sechenova*, **50**, 268. (English trans. in *Federation Proceedings Translation Supplement* (1965), **24**, 368–70.)

Vesselkin, N. P. & Kovaćević, N. (1973). Non-olfactory telencephalic afferent projections in elasmobranch fishes. *Zhurnal Evolyutsionnoi Biokhimii i Fiziologii*, **9**, 585–92. (English trans. in *Biological Abstracts* (1975), **58**, 56437.)

Yaremko, R. M., Jette, J. & Utter, W. (1974). Further study of avoidance conditioning in toads. *Bulletin of the Psychonomic Society*, **3**, 340–2.

Zagorul'ko, T. M. (1965). Interaction between the forebrain and visual centers of the midbrain in teleosts and amphibians. *Zhurnal Evolyutsionnoi Biokhimii i Fiziologii*, **1**, 449–58. (English trans. of abstr. in *Biological Abstracts* (1965), **47**, 67267.)

SVEN O. E. EBBESSON

Interspecific variability in brain organization and its possible relation to evolutionary mechanisms

Summary

The purpose of the present paper is to provide explanations for the variability of some neural structures in the forebrain of vertebrates and to suggest several evolutionary mechanisms that are operating to produce this.

An analysis of recently obtained data leads us to propose that:

1. Neocortical equivalents (NE) have existed in the forebrain since the beginning of vertebrate evolution.
2. Neocortex of mammals, and NE of non-mammalian forms, evolved from a single aggregate in the telencephalon, and that the greater differentiation and multiplication of identifiable systems and aggregates found in some vertebrates is formed by a mechanism of parcellation (the parcellation hypothesis).
3. Existing variability in organization of neocortex and NE is the result of varying selective pressures.
4. The ontogenesis of a given neural structure provides insights into its evolution, since the structure could not either have appeared *de novo* or attained a certain structural organization without going through certain specific steps.
5. At least some of the so-called abnormal projections induced in sprouting experiments represent the ancestral connections of a given system that have been lost during evolution involving parcellation. It is suggested that 'plasticity experiments' can be designed to obtain information about ancestral arrangements.
6. Hints about primordial organizations can be obtained from deprivation experiments, since the *last* evolutionary sequence of development of a given system is more likely to have a component involving current experience, i.e. current selective pressure.
7. Neurons of a given primordial cluster have a greater variety of inputs than the more recent (but less specific), and as differentiation and parcellation of neural systems evolves, some inputs may be lost, i.e. differentiation and parcellation involve losing, *not* gaining, a variety of inputs. This is related to the proposed rejection of the

invasion hypothesis. The result is also that the more parcellated a system is, the more likely it is that some parcels will lack some of the original connections.

8. Parcellation of nuclear groups (as in the thalamus of some reptiles) and layers (as in the tectum of some teleost species) produces parcels that are all homologous to earlier, less parcellated structures. Some of the new parcels are likely to have lost some of their primordial connections (but may regain them in sprouting experiments).

It is suggested that the evidence collected during the last decade *does not* support the following older concepts:

1. that the telencephalon in extant anamniotes is concerned totally, or almost totally, with olfactory functions; and
2. that, therefore, the primordial telencephalon was primarily or totally concerned with olfaction;
3. that the non-olfactory modalities became represented in the thalamus and telencephalon by a mechanism of invasion from lower centres;
4. that one neural system will invade another unrelated system and that this is a mechanism in evolution (or plasticity);
5. the generalization that most non-olfactory modalities in anamniotes are integrated in the mesencephalic tectum of all non-mammalian forms and that the forebrain has little part in the processing of non-olfactory sensory information;
6. that modern anamniotes have a multimodal diencephalic nucleus sensitivus and that modality-specific systems and nuclei evolved from it;
7. that neocortex and NE evolved from two separated primordial sites.

Introduction

One of the major problems confronting comparative neurologists for the last century has been to deal with the interspecific variability in brain structures and behaviours. Bridging the gap between brain organization in non-mammalian versus mammalian forms is particularly challenging, and has received considerable attention lately with the development of many new and powerful research tools.

The diversity of thalamic and telencephalic organization in all vertebrate groups is perplexing and poorly understood. Relatively little has been said in the literature about how this diversity has come about. It is my purpose here to deal with this issue. Although I will address the problem of diversity in neocortical equivalents (NE), the proposed hypotheses about evolution of neural structures probably apply to any neural system.

A theoretical framework to explain the guiding principles of nature's way of evolving brain structures in response to the need for new behaviours is lacking. Playing the devil's advocate, I would like to suggest certain alternative explanations for some currently accepted concepts. Future researchers will be able to accept or refute these hypotheses.

One of the motivations for writing the present paper is that I believe it important for every scientist to, every 10 years or so, put down on paper a summary of his or her recent work and insights gained. For what if something should happen to the investigator and the ideas never be published? This, then, is a review of some current thoughts about brain evolution in the light of the newly available evidence, and my own interpretation of them.

The concept that the telencephalon is primarily an olfactory structure in anamniotes and that the telencephalon in the earliest vertebrates was an olfactory structure

The available evidence until 10 years ago was that most, if not all, of the telencephalon of anamniote vertebrates is under olfactory dominance. This viewpoint was given impetus by the early work of Herrick (1910, 1922, 1924, 1948), Johnston (1911) and Bäckström (1924), supported by descriptive studies of normal material. The conclusion was reached to a large extent from examinations of shark brains that were considered particularly 'primitive'.

Herrick (1922) subdivided the telencephalon into 'area olfactoria dorsalis', 'area olfactoria medialis', 'area olfactoria lateralis', and 'tuberculum olfactorium'. His terminology stemmed from the belief that the entire telencephalon (with the exception of an 'area somatica' in the telencephalon medium) received primary olfactory fibres (see also Bäckström, 1924). In a similar study of the selachian forebrain, Johnston (1911) described the forebrain of these vertebrates as consisting of 'area olfactoria medialis', 'area olfactoria lateralis', and a 'primordium hippocampi' (corresponding to the area olfactoria dorsalis of Herrick). He considered the roof of the selachian forebrain as an olfactogustatory correlation centre. Furthermore, it was suggested that the brains of 'higher' vertebrates have evolved from 'primitive olfactory brains' (Herrick, 1924).

Amphibian forebrains were also essentially regarded as 'smell-brains' (Herrick, 1910, 1933, 1948). It was thought that almost the entire telencephalon was the target of secondary or tertiary olfactory fibres, while non-olfactory information only reached subpallial striatal areas after a relay in the dorsal thalamus.

The revolutionary new anatomical methods have been responsible for a complete revision of this concept. It has now been shown in sharks (Ebbesson & Heimer, 1968, 1970), teleosts (Scalia & Ebbesson, 1971) and amphibians (Scalia *et al*., 1968; Royce & Northcutt, 1969; Scalia, 1972; Northcutt & Royce, 1975) that the telencephalic regions receiving olfactory bulb fibres are well defined, relatively small portions of the lateral hemispheric walls. The tertiary olfactory fibres in the nurse shark are also limited to the ventrolateral quadrant of the hemisphere (Ebbesson, 1972*b*). These discoveries led us later to postulate that some of the non-olfactory regions of the telencephalon might have other modalities represented, and in 1970 projections from the visual thalamus to a well-defined dorsolateral region of the telencephalic hemisphere were discovered (Ebbesson & Schroeder, 1971; Schroeder & Ebbesson, 1971, 1974; Ebbesson, 1972*b*). This finding was later confirmed by neurophysiological methods (Cohen, Duff & Ebbesson, 1973).

Thalamotelencephalic pathways have now also been demonstrated in teleosts (Ebbesson, 1980) and amphibians (Vesselkin, Agayan & Nomokonova, 1971; Northcutt, 1972), but have not as yet been shown in cyclostomes. I would guess that non-olfactory information has always provided an important input to such structures as the hippocampus and that integration of all sensory modalities for most behaviours requires the proximity of these systems.

The concept that in non-mammalian vertebrates the principal integration centre for non-olfactory modalities is located in the mesencephalic tectum

The optic tectum is an exceptionally prominent landmark in most non-mammalian vertebrates and it has over the last century gained the reputation of being the principal brain region concerned with analysis of sensory information and 'correlation' with motor systems. Herrick (1948, p. 24) referred to the tectum of lower vertebrates as the 'supreme center of regulation of motor responses to the exteroceptive system of sense organs'. Based partly on this belief and related neuroanatomical data, the theory of encephalization asserts that there is a gradual shift of most visual function from the optic tectum to the visual cortex as one ascends 'the phylogenetic scale' (Marquis, 1935). Furthermore, the shift in locus of function is generally thought to reflect the evolutionary development of the human visual system.

The recent discoveries of large thalamotelencephalic systems in sharks (Ebbesson & Schroeder, 1971), teleosts (Ebbesson, 1980) and amphibians (Kicliter & Northcutt, 1975) have made earlier concepts of tectal dominance suspect. Our findings, demonstrating that the optic tectum of the nurse shark is not necessary for some types of visual discriminations, cast a further shadow on the old concept (Graeber, Ebbesson & Jane, 1973). The findings of severe visual dysfunction after bilateral lesions of the posterior telencephalon in the nurse sharks, on the other hand, suggest a telencephalic function more similar to mammals (Graeber & Ebbesson, 1972; Graeber *et al*., 1978).

The results of these studies in the nurse shark will no doubt be different from others in other anamniotes, as the degrees of differentiation of the tectum and telencephalon vary greatly between species. One would predict that the more elaborate the structure is, the more important it is to the organism, and the more sensitive it is to damage. The nurse shark tectum is poorly differentiated when compared with, for example, a frog's tectum.

Nature no doubt has many options available in the evolution of neural systems; sometimes the choice is for greater tectal development, while at other times the choice is for telencephalic elaboration. The reasons for one instead of the other are not known at this time.

The invasion hypothesis

The basic premise of this hypothesis is that a given neural system can invade another (presumably unrelated) region and either establish itself on its own or make functional contact with the invaded cell population.

Many have used this theory to account for their observations. Two examples might be mentioned here: (1) the invasion of non-olfactory modalities into thalamus and telencephalon (Ariëns Kappers, Huber & Crosby, 1936; Herrick, 1948; Diamond & Hall, 1969) and (2) the suggestion by Karten (1963) that evolution of neocortex may have involved an invasion of layer IV elements from a site of origin different from that of the remainder of the cortex. These are examples of two kinds of invasion; the first does not necessarily involve establishing connections with the new neighbouring aggregates, whereas the second does.

The evidence *against* this hypothesis comes mainly from three new pieces of evidence: (1) the demonstration of NE located in the telencephalon of teleosts, elasmobranchs, amphibians, reptiles and birds obviates the need for an explanation involving a migration of new systems into the telencephalon to form neocortex; (2) the variability of the NE systems strongly suggests a common origin (see below); (3) our systematic analyses of many systems, including the somatosensory and visual systems, in a broad range of vertebrates of every class reveal not a single unexpected 'new' connection with an unrelated system.

However, one finds, on occasion, overlap of certain modalities, e.g. in the optic tectum, but that is always in the case of less differentiation and poorer development. The trend of greater differentiation, enlargement and parcellation is probably always accompanied by greater segregation of inputs and specialization of neurons (Ebbesson, Schroeder & Butler, 1975). If evolution of the nervous system can involve the invasion of one system by another, one would expect to see evidence of such in the most highly differentiated systems. We have found no example of this, and in fact the entire literature suggests that the opposite occurs, namely that some inputs and outputs are lost in cases of extreme development and differentiation. This is explained further by the 'parcellation hypothesis' (see p. 67).

The new view of the telencephalon

The telencephalons of vertebrates take at least four basic forms: (1) the 'typical' laminated hemispheres, as seen in amphibians; (2) the basically non-laminar organization seen in birds and elasmobranchs; (3) the laminar/non-laminar mixture seen in reptiles; and (4) the everted, non-laminar type seen in teleosts.

Although the configurations vary tremendously, the basic inputs and outputs appear to be basically the same. There is furthermore no strong evidence to suggest that mammalian neocortex evolved from a structure comparable to one or more of these patterns. I personally believe that the neocortex evolved from something comparable to the urodele lateral pallium or the embryonic elasmobranch pallium without ever passing through a phase comparable to that seen in extant reptiles or birds. The reasons for that will be explained in a later section (p. 69).

The new studies show clearly that the olfactory bulb projects to very limited portions of the telencephalon of all vertebrates and that other modalities are also represented in characteristic, limited regions. Although the telencephalic organization of non-olfactory modalities is not known in amphibians (Kicliter & Ebbesson, 1976), it is clear in the nurse shark that the visual area is well defined, dorsally, laterally and caudally in the central telencephalic nucleus, and that the trigeminal region does not overlap with it and is located ventral, medial and rostral to the visual area (Cohen, Leonard, Jane & Ebbesson, unpublished observations). The visual input to the telencephalon in *Holocentrus* is also to the caudal half (Ebbesson, 1980).

The new view of the dorsal thalamus

Solid evidence for the presence of a neothalamus in non-mammalian vertebrates has only appeared during thc last 12 years. It began with several discoveries about the same time. One was the identification of somatosensory regions of the thalamus ventral to the lateral geniculate nucleus (Karten, 1963; Ebbesson, 1966, 1967*a*, 1969), which was then considered homologous to the mammalian pars ventralis (Ariëns Kappers *et al*., 1936). These ascending spinal projections in birds and reptiles, indicate the presence of a ventrobasal complex and intralaminar nuclei, suggesting that at least part of the lateral geniculate nucleus might be homologous to the mammalian pars dorsalis. Later studies have confirmed the presence of spinothalamic fibres in amphibians (Ebbesson, Jane & Schroeder, 1972; Ebbesson, 1976) and nurse shark (Ebbesson *et al*., 1972; Ebbesson & Hodde, 1981). The medial lemniscus has also been traced to the ventrobasal complex in a lizard (Ebbesson, 1978), and cerebellar fibres have been traced to a region of the thalamus of the nurse shark more or less overlapping with the spinal input (Ebbesson *et al*., 1972; Ebbesson & Campbell, 1973).

The big breakthrough in understanding the non-mammalian thalamus came when Karten & Nauta (1968) and Karten *et al*. (1973) described visual and auditory thalamotelencephalic systems. Hall & Ebner (1970) followed with a description of thalamotelencephalic projections in the turtle. Vesselkin *et al*. (1971) and Northcutt (1972) and Kicliter & Northcutt (1975) have traced the thalamotelencephalic systems in amphibians, while Ebbesson & Schroeder (1971) described comparable pathways in the nurse shark. Thalamotelencephalic pathways have also recently been identified in a teleost (Ebbesson, 1980).

These projections vary considerably between the various taxa; for example, the visual projection in birds is approximately 75% ipsilateral and 25% contralateral whereas in the shark almost all, if not all, fibres are crossed. This feature has no satisfactory explanation at this time.

The recipient cell aggregates in the telencephalon vary tremendously, ranging from laminar structures in some turtles to non-laminar structures in sharks and birds.

One of the main reasons why I think that all of these projections are NE is that the telencephalic structures generally project back to the thalamus, pretectum, tectum, red nucleus, reticular formation and dorsal column nuclei in the same manner as neocortex does in mammals (Ebbesson & Schroeder, 1971; Ebbesson, 1972*b*, and unpublished observations; Ebbesson *et al*., 1975).

The concept of a primordial nucleus sensitivus thalami

Herrick (1948) had a particularly good feel for the evolution of the nervous system. Many of his insights came from his studies of the tiger salamander with the rapid Golgi method. One of his conclusions from these studies was that there exists in sharks and amphibians a multimodal dorsal thalamic nucleus that projects to the telencephalon medium. It was thought that unimodal regions had evolved from this.

We have not been able to confirm the essence of his observations, although there is ample evidence for multimodal units in restricted portions of the thalamus (Belekhova & Kosareva, 1971) of some reptiles. Our findings suggest little overlap between visual and somatosensory regions of the thalamus in lamprey (Ebbesson & Northcutt, 1975), sharks (Ebbesson, 1967*b*, 1970*a*, Ebbesson & Ramsey, 1968; Ebbesson *et al*., 1972), frog (Ebbesson, 1976) and lizard (Ebbesson, 1966, 1967*a*, 1969; Ebbesson *et al*., 1972). In all of these species visual nuclei are dorsal and lateral to the 'ventrobasal complex' (except in lamprey where no spinal fibres were traced as far rostrally as the thalamus). These studies, carried out using the Nauta–Fink–Heimer methods (Ebbesson, 1970*b*), support the notions that (1) a dorsal thalamus comparable to that in mammals exists in all vertebrates; (2) modality-specific nuclei exist in the species examined; and (3) that modality-specific as well as multimodal cell groups probably existed from the beginning of vertebrate evolution.

The new view of the optic tectum

The optic tectum, although always clearly identifiable, shows a broad range of development in various vertebrates; but the basic organization is always present:

1. Retinal afferents terminate chiefly in the superficial layers, whereas other modalities, telencephalic afferents and commissural tectal fibres terminate deeply (Ebbesson *et al*., 1975; Ebbesson & Northcutt, 1975).
2. Efferent pathways are of three types: (*a*) those ascending to pretectal and thalamic nuclei (often bilaterally), (*b*) a tectotectal pathway, and (*c*) bilateral descending pathways that terminate primarily in the reticular formation and spinal cord (Ebbesson, 1970*a*).

The variation in the tectum lies basically in the intratectal organization. As stated above, enlargement and differentiation are apparently always accompanied by a greater separation of inputs and the presence of a greater variety of cell types (Ebbesson *et al*., 1975; Ebbesson & Northcutt, 1975).

These observations relate to the very reason for lamination and may account for the various degrees of parcellation of cell aggregates. There is at this time no good explanation for lamination, but it is likely that the organization of functional columns (Hubel & Wiesel, 1963) is one of the key reasons for it. The development of columns is poorly understood, but the impression one gets from reviewing the literature is that a certain 'set' of neurons, of different types and inputs, arrange themselves, during development, into tighter and tighter 'columns' with less and less anatomical 'overlap' with neighbouring columns (see Hubel, Wiesel & LeVay, 1977). My interpretation is that one of the net effects of more and more segregation (lateral parcellation) of columns is greater functional isolation, i.e. less opportunity for disturbing cross-talk between columns. One would expect that this trend would lead not only to greater efficiency and clarity of function, but also to an increase in the potential for processing information in general.

The parcellation hypothesis

The parcellation hypothesis (Ebbesson, 1972*a*) was first suggested to explain the interspecific variability of the two visual systems (Schneider, 1969). Since our publications, additional evidence in its support has been obtained in several laboratories and the generality of the hypothesis can now be broadened to include evolution of neural systems in general.

The hypothesis can be stated as follows. Neural systems evolve not by invasion of other systems, but by differentiation and parcellation resulting from increased selective pressures for the modification of behaviour. The range of arrangements observed in extant vertebrates reflects nature's response to given selective pressures; the outcome of the parcellation is dependent on the state of the organization when the new selective pressure appeared and the strength of that selective stimulus. The parcellation of a given system involves competition between inputs and may involve the loss of connections of certain cells (but never the acquisition of an unrelated input).

The key evidence for the hypothesis rests with the new neuroanatomical information that indicates:

1. Thalamotelencephalic sensory systems probably exist in all vertebrates and therefore probably existed from the beginning of vertebrate evolution. This is a powerful argument against the 'invasion hypothesis'.
2. The various numbers and arrangements of thalamotelencephalic visual systems encountered in modern vertebrates probably evolved from a single system. Apparently only one thalamotelencephalic visual system is present in the nurse shark (Ebbesson & Schroeder, 1971; Cohen *et al*., 1973), the tiger salamander (Ebbesson *et al*., 1972) and the banded water snake (Ulinsky, 1979). In these forms retinal and tectal fibres project to the same dorsal thalamic aggregate, which in turn projects to a single telencephalic target. This, I believe, is the ancestral arrangement of the two visual systems of

Schneider (1969). (The situation is actually much more complicated than two systems, since the pretectum and other structures are also involved; however, this only increases the number of potential combinations of arrangements that have evolved in various groups.)

3. The varieties of interconnections seen among the more or less parcellated visual systems in various vertebrates reflect ancestral arrangements (e.g. the small dorsal geniculate projections to peristriate cortex in the cat).
4. The variety of interconnections and overlap seen among the more or less parcellated somatosensory–motor cortex (Donoghue, Kerman & Ebner, 1979) of mammals reflects ancestral arrangements and suggests that not only did the ventrolateral thalamic and the ventrobasal complex evolve from a common ancestor, but that somatosensory and motor cortex also evolved from one cortical zone.

There are apparently at least two types of parcellation occurring at the same time. One we may label the *horizontal* variety, described above, which explains the development of new nuclei and cortical regions. The other type can be labelled *vertical* and has to do with the development of more stratification in a system, e.g. cortex or optic tectum. Here parcellation appears to involve greater segregation of inputs from different sources, specialization of neuronal types and the loss of certain connections as the neurons become more specialized (e.g. some cells lose a given input and become dominated by the remaining inputs). This is discussed further in the following sections.

The concept of a dual origin of neocortex

Karten & Nauta (1968) and Karten *et al*. (1973) have recently reported that the two visual systems of the thalamus in birds project onto two widely separated telencephalic fields – the dorsal geniculate homologue on to the Wulst dorsally, and the nucleus rotundus upon ectostriatum. Karten makes the interesting suggestion, based on his findings and the embryological studies of Källen (1951, 1962), that the reptilian and avian NE as well as mammalian neocortex have a dual origin: 'an *in situ* development from the ependyma, forming cortex, or a cortex-like structure, and typified by the Wulst in birds and perhaps striate and somatosensory cortex in mammals, and a second type of formation which, in birds, is the ectostriatum, etc., and in mammals, the laminae of cortex with their origin in the lateral ganglionic eminence, migrating into the pallium and receiving projections from the phylogenetically older structures such as the colliculi. In birds, these structures may subsequently project on to the pallial mantle.'

Karten also makes the suggestion that in the development of neocortex the derivatives of the dorsal ventricular ridge enter the pallium, perhaps mingling with cells of local, pallial, ependymal origin to form neocortex. This would indicate that layer IV of neocortex has a different origin to the other layers.

Northcutt (1969, 1974) supports Karten's suggestion on the basis of his own analysis. He suggests that the Wulst is derived from his PII component and that the neostriatal complex evolves from his PIIIa component. This evolutionary line is referred to as the sauropsid line and includes birds and all reptiles except turtles. The therapsid line (turtles and mammals) is also thought to have a dual origin (PIIb and PIIIa) for NE and neocortex. Northcutt, at this time, also came to the important conclusion that all tetrapods possess structures homologous to mammalian neocortex.

My interpretation of the recently available evidence suggests a new interpretation: a unitary origin of neocortex. Although it appears likely that in some sauropsids, and birds, there may very well be a dual origin of NE, I believe this to be a specialization in this line of evolution that has little to do with evolution of neocortex.

The unitary origin of neocortex is supported, I believe, mainly by the facts supporting the parcellation hypothesis. One argument against the dual origin hypothesis is the existence of incredibly strong interdependent connections and functions of striate and peristriate cortex. The boundary between the two cortical belts is of particular functional importance, and suggests, to me, that these belts have always been contiguous. The unitary origin and the parcellation hypothesis are furthermore supported by the variability of afferent and efferent connections seen in various vertebrates. If this hypothesis is correct, we have no immediate explanation for the split of the NE seen in birds, except that it occurred early in saurian evolution and probably resulted in a very different functional organization from that seen in mammals.

The ontogenetic evidence for evolutionary trends

If a given structure or feature does not appear *de novo*, and *de novo* appearance is unlikely, it must have evolved from something else. It is therefore logical that at least some evolutionary steps should be identifiable during development of the structure, especially the more recent ones. To test, in the nervous system, the hypothesis that ontogeny recapitulates phylogeny has not been an easy task, but with the development of more sophisticated neuroanatomical and neurophysiological techniques the day has arrived for meaningful testing.

An example of this approach is the description of the development of ocular dominance columns in monkeys by Rakic (1976*a*, *b*) and Hubel *et al*. (1977). They found no hint of columns 6 weeks before term in layer IVc in the striate cortex, but by 6 weeks post-partum complete segregation (parcellation) of columns had taken place. Hubel *et al*. (1977) suggest that 'the process presumably occurs by a retraction of the two sets of terminals. One may imagine that in any areas occupied by both sets of terminals there is a competitive mechanism in which the weaker set at any point tends to regress. Given such an unstable equilibrium, the normal end result of any initial inequalities would be a complete segregation of terminals'.

Such a model, based on competition, may account for the evolution of the ocular dominance columns as well as parcellation of any neural system. I believe that the developmental sequence recorded by Rakic (1976*a*, *b*) and Hubel *et al*. (1977) reflects how these columns evolved in the monkey. Similar examples of ontogenetic and evolutionary parallels should be present in almost every neuronal system and should provide us with clues as to which selective pressure resulted in which structural and functional modification.

It is worth noting, in this context, that in the fetal eye-injection experiments of Rakic (1976*a*, *b*) the optic afferents, both to the geniculates and to the superior colliculi, occupied their entire targets for an extended period between their first arrival and their eventual segregation into layers (geniculates) or clumps (colliculi). These observations also are likely to reflect ancestral organization.

One can predict, on the basis of the parcellation hypothesis and the observation by Rakic (1967*a*, *b*) and Hubel *et al*. (1977), that unsegregated, or absent, ocular dominance columns and undifferentiated geniculate nuclei and colliculi will be found in some adult mammals not subjected to the same selective pressures as the monkey.

What experimentally induced sprouting tells us

When an animal suffers brain injury during its development there may result alterations in the final pattern of connections (Devor & Schneider, 1975). I would like to propose that one, hitherto not mentioned, factor for the establishment of certain 'improper' connections with certain specific cell groups and not with others depends on the evolutionary history of the system studied. Specifically I would like to suggest that at least some of the 'aberrant' projections represent *ancestral projections* lost during evolution.

The simplest example of this hypothesis comes from the work of Schneider (1973), who induced the sprouting of retinal fibres into the nucleus lateralis posterior thalami (LP) after ablation of the superior colliculus in fetal hamsters. This can be explained if the parcellation hypothesis is correct, which suggests that the lateralis posterior and the dorsal geniculate nuclei evolved from the same nucleus with both retinal and tectal inputs. I believe the retinal–LP connection was lost during parcellation due to increased competition with the tectal input as a result of an, as yet unknown, selective pressure for the isolation of the tecto-LP-cortical system.

The optic tectum can also be manipulated in such a way that the strict, normal lamination is lost (Lund & Lund, 1971; Chung, Gaze & Sterling, 1973; Schneider & Jhaveri, 1974). These new rearrangements, I believe, may mimic ancestral, less parcellated organization.

The sprouting observed in the lateral geniculate of the cat (Guillery, 1972; Kalil, 1972; Hickey, 1975) also is likely to involve the re-establishment of connections lost during evolutionary and ontogenetic parcellation and the related competition by the two eyes.

On considering the various plasticity experiments, I find very few examples that I cannot account for on the basis of other models of brain organization seen in other vertebrates. One that I find no explanation for is the commonly seen recrossing of retinotectal fibres at mesencephalic levels (see Lund, 1978).

If ancestral organization plays a role in the course and termination of 'abnormally' sprouting fibres, it would be possible to use plasticity studies not only to learn about ancestral brain organization but also to manipulate sprouting for specific needs. Devor & Schneider (1975) point out the possibility that more than one system is likely to be stimulated to sprout into a given area, and that each system may be stimulated at different stages of development. I propose that all of the new afferent sprouts were at one time related to the new innervation site and that the temporal differences may some day be used to tell us the sequence of parcellations (i.e. the loss of the connections).

What sensory deprivation experiments may tell us

There are, to my knowledge, no clear-cut examples of deprivation experiments that can elucidate whether an ancestral type of organization can be shown by such an experiment. However, it does seem logical that the most recent evolutionary development would be the most susceptible to deprivation, and that a primordial organization may reveal itself in such an experiment.

Conclusion

Nature has dealt with the evolution of brain structures in many varying and confounding ways and the end results, the brains of extant vertebrates, show such a diversity that one wonders whether we are ever going to understand this evolution. One of the problems facing the comparative neurologists of today and tomorrow is the ever-present question of homology. Can homologous cellular fields be recognized in living vertebrate brains? The more commonly used criteria are that: (1) similar fields are limited by similar ventricular sulci, (2) similar fields possess similar cell types and similar topographical relationships, and (3) similar fields possess similar fibre connections. These criteria have been evaluated by Nieuwenhuys & Bodenheimer (1966), Northcutt (1969), Campbell & Hodos (1970) and others. The general consensus is that any one character alone is not adequate and that all three must be taken into consideration.

The purpose of the present contribution has been to suggest some less traditional avenues for obtaining knowledge about homologies and brain evolution. I have done so at the risk of being presumptuous and I have clearly covered only some of the publications that led to these ideas.

References

Ariëns Kappers, C. U., Huber, G. C. & Crosby, E. C. (1936). *The Comparative Anatomy of the Nervous System of Vertebrates Including Man*. Macmillan, New York.

Bäckström, K. (1924). Contributions to the forebrain morphology in selachians. *Acta Zoologica*, **5**, 123–240.

Belekhova, M. G. & Kosareva, A. A. (1971). Organization of the turtle thalamus: visual, somatic and tectal zones. *Brain, Behavior and Evolution*, **4**, 337–75.

Campbell, C. B. G. & Hodos, W. (1970). The concept of homology and the evolution of the nervous system. *Brain, Behavior and Evolution*, **3**, 353–67.

Chung, S. H., Gaze, R. M. & Stirling, R. V. (1973). Abnormal visual function in *Xenopus* following stroboscopic illumination. *Nature New Biology*, **246**, 186–9.

Cohen, D. H., Duff, T. A. & Ebbesson, S. O. E. (1973). Electrophysiological identification of a visual area in shark telencephalon. *Science*, **182**, 492–4.

Devor, M. & Schneider, G. E. (1975). Neuroanatomical plasticity: the principle of conservation of total axonal aborization. In *Aspects of Neural Plasticity/Plasticité nerveuse*, ed. F. Vital-Durand & M. Jeannerod, vol. 43, pp. 191–200.

Diamond, I. T. & Hall, W. C. (1969). Evolutions of neocortex. *Science*, **164**, 251–68.

Donoghue, J. P., Kerman, K. L. & Ebner, F. F. (1979). Evidence for two organizational plans within the somatic sensory-motor cortex of the rat. *Journal of Comparative Neurology*, **183**, 647–64.

Ebbesson, S. O. E. (1966). Ascending fiber projections from the spinal cord in the Tegu lizard (*Tupinambis nigropunctatus*). *Anatomical Record*, **154**, 341–2.

Ebbesson, S. O. E. (1967*a*). Ascending axon degeneration following

hemisection of the spinal cord in the Tegu lizard (*Tupinambis nigropunctatus*). *Brain Research*, **5**, 178–206.
Ebbesson, S. O. E. (1967*b*). Retinal projections in two species of sharks. *Anatomical Record*, **157**, 238.
Ebbesson, S. O. E. (1969). Brainstem afferents from the spinal cord in a sample of reptilian and amphibian species. *Annals of the New York Academy of Sciences*, **167**, 30–101.
Ebbesson, S. O. E. (1970*a*). On the organization of central visual pathways in vertebrates. *Brain, Behavior and Evolution*, **3**, 178–94.
Ebbesson, S. O. E. (1970*b*). The selective silver-impregnation of degenerating axons and their synaptic endings in non-mammalian species. In *Contemporary Research Methods in Neuroanatomy*, ed. W. J. H. Nauta & S. O. E. Ebbesson, pp. 132–61. Springer-Verlag, New York.
Ebbesson, S. O. E. (1972*a*). A proposal for a common nomenclature for some optic nuclei in vertebrates and the evidence for a common origin of two such cell groups. *Brain, Behavior and Evolution*, **6**, 75–91.
Ebbesson, S. O. E. (1972*b*). New insights into the organization of the shark brain. *Comparative Biochemistry and Physiology*, **42A**, 121–9.
Ebbesson, S. O. E. (1976). Morphology of the spinal cord. In *Frog Neurobiology*, ed. R. Llinas & W. Precht, pp. 679–706. Springer-Verlag, New York.
Ebbesson, S. O. E. (1978). Somatosensory pathways in lizards: the identification of the medial lemniscus and related structures. In *Lizard Neurology and Behavior*, ed. P. D. MacLean & N. Greenberg, DHEW Publication No. (ADM) 77–491, pp. 91–104. Dept. of Health, Education and Welfare, Washington, DC.
Ebbesson, S. O. E. (1980). Thalamotelencephalic pathways in the squirrel fish (*Holocentrus* sp.). *Cell and Tissue Research*, in press.
Ebbesson, S. O. E. & Campbell, C. B. G. (1973). The organization of cerebellar efferents in the nurse shark (*Ginglymostoma cirratum*). *Journal of Comparative Neurology*, **152**, 233–55.
Ebbesson, S. O. E. & Heimer, L. (1968). Olfactory bulb projections in two species of sharks. *Anatomical Record*, **160**, 469.
Ebbesson, S. O. E. & Heimer, L. (1970). Projections of the olfactory tract fibers in the nurse shark (*Ginglymostoma cirratum*). *Brain Research*, **17**, 47–55.
Ebbesson, S. O. E. & Hodde, K. (1981). Ascending spinal systems in the nurse shark (*Ginglymostoma cirratum*). *Journal of Comparative Neurology*, in press.
Ebbesson, S. O. E., Jane, J. A. & Schroeder, D. M. (1972). An overview of major interspecific variations in thalamic organization. *Brain, Behavior and Evolution*, **6**, 92–130.
Ebbesson, S. O. E. & Northcutt, R. G. (1975). Neurology of anamniotic vertebrates. In *Evolution of Brain and Behavior in Vertebrates*, ed. R. B. Masterson *et al*. Halsted Press, New York.
Ebbesson, S. O. E. & Ramsey, J. S. (1968). The optic tracts in two species of sharks. *Brain Research*, **8**, 36–53.
Ebbesson, S. O. E. & Schroeder, D. M. (1971). Connections of the nurse shark's telencephalon. *Science*, **173**, 254–6.
Ebbesson, S. O. E., Schroeder, D. M. & Butler, A. B. (1975). The Golgi method and the revival of comparative neurology. In *Golgi Centennial Symposium*, ed. M. Santini. Raven Press, New York.

Graeber, R. C. & Ebbesson, S. O. E. (1972). Visual discrimination learning in normal and tectal ablated nurse sharks (*Ginglymostoma cirratum*). *Comparative Biochemistry and Physiology*, **42**, 131–9.

Graeber, R. C., Ebbesson, S. O. E. & Jane, J. A. (1973). Visual discrimination in sharks without optic tectum. *Science*, **180**, 413–15.

Graeber, R. C., Schroeder, D. M., Jane, J. A. & Ebbesson, S. O. E. (1978). Visual discriminations following parietal ablations in nurse sharks (*Ginglymostoma cirratum*). *Journal of Comparative Neurology*, **180**, 325–44.

Guillery, R. W. (1972). Experiments to determine whether retinogeniculate axons can form translaminar collateral sprouts in the dorsal lateral geniculate nucleus of the cat. *Journal of Comparative Neurology*, **146**, 407–19.

Hall, W. C. & Ebner, F. F. (1970). Parallels in the visual afferent projections of the thalamus in the hedgehog (*Paraechimus hypomelas*) and the turtle (*Pseudemus scripta*). *Brain, Behavior and Evolution*, **3**, 135–54.

Herrick, C. J. (1910). The morphology of the forebrain in Amphibia and Reptilia. *Journal of Comparative Neurology*, **20**, 413–547.

Herrick, C. J. (1922). Functional factors in the morphology of the forebrain of fishes. In *Libro en honor de D. Santiago Ramón y Cajal,* vol. 1, pp. 143–204. Jimenez & Molina, Madrid.

Herrick, C. J. (1924). *An Introduction to Neurology*, 3rd ed. W. B. Saunders, Philadelphia.

Herrick, C. J. (1933). The amphibian forebrain. VI. *Necturus. Journal of Comparative Neurology*, **58**, 1–288.

Herrick, C. J. (1948). *The Brain of the Tiger Salamander*. University of Chicago Press, Chicago.

Hickey, T. L. (1975). Translaminar growth of axons in the kitten dorsal lateral geniculate nucleus following removal of one eye. *Journal of Comparative Neurology*, **161**, 359–82.

Hubel, D. H. & Wiesel, T. N. (1963). Shape and arrangement of columns in cat's striate cortex. *Journal of Physiology*, **165**, 559–658.

Hubel, D. H., Wiesel, T. N. & LeVay, S. (1977). Plasticity of ocular dominance columns in monkey striate cortex. *Philosophical Transactions of the Royal Society of London*, **278B**, 377–409.

Johnston, J. B. (1911). The telencephalon of selachians. *Journal of Comparative Neurology*, **21**, 1–113.

Kalil, R. E. (1972). Formation of new retino-geniculate connections in kittens after removal of one eye. *Anatomical Record*, **172**, 339–40.

Källén, B. (1951). On the ontogeny of the reptilian forebrain. Nuclear structures and ventricular sulci. *Journal of Comparative Neurology*, **95**, 397–447.

Källén, B. (1962). Embryogenesis of brain nuclei in the chick telencephalon. *Ergebnisse des Anatomie und Entwicklungsgeschichte*, **36**, 62–82.

Karten, H. J. (1963). Ascending pathways from the spinal cord in the pigeon (*Columbia livia*). *Proceedings of the XVI International Congress on Zoology*, **2**, 23.

Karten, H. J., Hodos, W., Nauta, W. J. H. & Revzin, A. M. (1973). Neural connections of the 'visual Wulst' of the avian telencephalon. Experimental studies in the pigeon (*Columbia livia*) and owl (*Speotyto cunicularia*). *Journal of Comparative Neurology*, **150**, 253–78.

Karten, H. J. & Nauta, W. J. H. (1968). Organization of retinothalamic projections in the pigeon and owl. *Anatomical Record*, **160**, 373.
Kicliter, E. & Ebbesson, S. O. E. (1976). Organization of the 'non-olfactory' telencephalon. In *Frog Neurobiology*, ed. R. Llinás & W. Precht, pp. 946–72. Springer-Verlag, New York.
Kicliter, E. & Northcutt, R. G. (1975). Ascending afferents to the telencephalon of ranid frogs: an anterograde degeneration study. *Journal of Comparative Neurology*, **161**, 239–54.
Lund, R. D. (1978). *Development and Plasticity of the Brain*. Oxford University Press, Oxford.
Lund, R. D. & Lund, J. S. (1971). Synaptic adjustment after deafferation of the superior colliculus of the rat. *Science*, **171**, 804–7.
Marquis, D. G. (1935). A phylogenetic interpretation of the functions of the visual cortex. *Psychiatry*, **33**, 807.
Nieuwenhuys, R. & Bodenheimer, T. S. (1966). The diencephalon of the primitive bony fish *Polypterus* in the light of the problem of homology. *Journal of Morphology*, **118**, 415–50.
Northcutt, R. G. (1969). Discussion of the preceding paper. *Annals of the New York Academy of Sciences*, **167**, 180–5.
Northcutt, R. G. (1972). Afferent projections of the telencephalon of the bull frog (*R. catesbeiana*). *Anatomical Record*, **172**, 374.
Northcutt, R. G. (1974). Some histochemical observations on the telencephalon of the bull frog, (*Rana catesbeiana* Shaw). *Journal of Comparative Neurology*, **157**, 379–90.
Northcutt, R. G. & Royce, G. J. (1975). Olfactory bulb projections in the bull frog (*Rana catesbeiana*). *Journal of Morphology*, **145**, 251–68.
Rakic, P. (1976*a*). Prenatal genesis of connections subserving ocular dominance in the rhesus monkey. *Nature, London*, **261**, 467–71.
Rakic, P. (1976*b*). Prenatal development of the visual system in the rhesus monkey. *Philosophical Transactions of the Royal Society of London*, **278B**, 245–60.
Royce, G. J. & Northcutt, R. G. (1969). Olfactory bulb projections in the tiger salamander (*Ambystoma tigrinum*) and the bull frog (*Rana catesbeiana*). *Anatomical Record*, **163**, 254.
Scalia, F. (1972). The projections of the accessory olfactory bulb in the frog. *Brain Research*, **36**, 409–11.
Scalia, F. & Ebbesson, S. O. E. (1971). The central projection of the olfactory bulb in a teleost (*Gymnothorax funebris*). *Brain, Behavior and Evolution*, **4**, 376–99.
Scalia, F., Halpern, M., Knapp, H. & Riss, W. (1968). The efferent connexions of the olfactory bulb in the frog: a study of degenerating fibers. *Journal of Anatomy*, **103**, 245–62.
Schneider, G. E. (1969). Two visual systems. *Science*, **163**, 895.
Schneider, G. E. (1973). Early lesions of superior colliculus: factors affecting the formation of abnormal retinal projections. *Brain, Behavior and Evolution*, **8**, 73–109.
Schneider, G. E. & Jhaveri, S. R. (1974). Neuroanatomical correlates of spared or altered function after early brain lesions in the newborn hamster. In *Plasticity and Recovery of Function in the Central Nervous System*, ed. D. G. Stein, J. J. Rosen & N. Butters, pp. 65–109. Adacemic Press, New York & London.
Schroeder, D. M. & Ebbesson, S. O. E. (1971). Diencephalic projections

to the telencephalon of the nurse shark. *Anatomical Record*, **169**, 421.

Schroeder, D. M. & Ebbesson, S. O. E. (1974). Nonolfactory telencephalic afferents in the nurse shark (*Ginglymostoma cirratum*). *Brain, Behavior and Evolution*, **9**, 121–55.

Ulinsky, P. S. (1979). Tectal efferents in the banded water snake *Natrix sipedon*. *Journal of Comparative Neurology*, **173**, 251–74.

Vesselkin, N. P., Agayan, A. L. & Nomokonova, L. M. (1971). A study of thalamo-telencephalic afferent systems in frogs. *Brain, Behavior and Evolution*, **4**, 295–306.

PART II

Sensory mechanisms and behaviour

D. M. GUTHRIE

The properties of the visual pathway of a common freshwater fish (*Perca fluviatilis* L.) in relation to its visual behaviour

Introduction

A great deal of progress has been made in recent years towards an understanding of the way in which visual sensory systems process information, especially in the case of mammals. At the same time questions have emerged related to the significance of these processing functions, and the exact nature of optimal stimuli. In the search for key features the experimenter is faced with a virtually limitless choice of stimuli to present to the animal, yet for practical reasons he must utilize a relatively small range, which may not include the most effective ones. As far as the broader significance of visual processes is concerned one must attempt to discern their adaptive nature in relation to the peculiarities of individual species. Both these considerations lead us towards combining physiological and behavioural observations in a single study, and emphasizing the importance of considering visual processes in the free-living animal in its natural environment.

Fish offer some special advantages for research of this type. The visual pathway is relatively simple compared with that of higher vertebrates, and the physiology of the retina is probably better understood than in any other animal group. There are a number of suitable indigenous species which can be obtained for laboratory work, and under favourable conditions these can be observed in the field. Unfortunately, perhaps, a great deal of the information gathered so far about the visual pathway of fishes is derived from studies on the goldfish, an ornamental species originating from Far East Asia, which on account of its hardiness and availability has become 'the laboratory fish' but which in the UK cannot be studied in its natural environment.

Our choice has fallen instead on the Eurasian perch (*Perca fluviatilis* L.), a common freshwater species belonging to the great group of advanced spiny-finned fishes – the Acanthopterygii. This group includes many of the most successful marine fishes – groupers, swordfish, trigger fish and a host of brightly coloured reef fish which appear to be very dependent on vision – as well as some freshwater fish distinguished by complex types of behaviour such as the cichlids and sunfishes. Other more specific reasons for the choice of species that seemed important were:

1. It is a shoaling fish, and shoaling behaviour appears to depend on visual recognition between conspecifics (Keenleyside, 1955).
2. The perch is a strikingly marked fish (Fig. 1), with dark vertical bars, orange ventral fins and an eyespot on the dorsal fin – elements likely to be involved in intraspecific recognition.
3. The species is generally distributed throughout the British Isles, and can easily be trapped or netted. It is quite often found in shallow, clear waters where it can be observed under natural conditions.
4. The perch is a visual predator, dependent on effective estimation of range and velocity for the capture of small fish, as well as on an ability to identify the smaller planktonic organisms on which it also feeds.
5. There is a fairly extensive literature on the biology of perch in natural waters, both on *P. fluviatilis* in Europe and Asia and the very closely related *P. flavescens* in North America.

The first part of this account consists of a description of the visual pathway, and this is followed by an attempt to relate this to the visual behaviour of the perch.

Fig. 1. The Eurasian perch, *Perca fluviatilis* L. It is suggested that the contour and contrast features of the perch are important in interactions between shoal-mates. Six of these features are illustrated here. Some are also likely to be involved in interactions with predators.

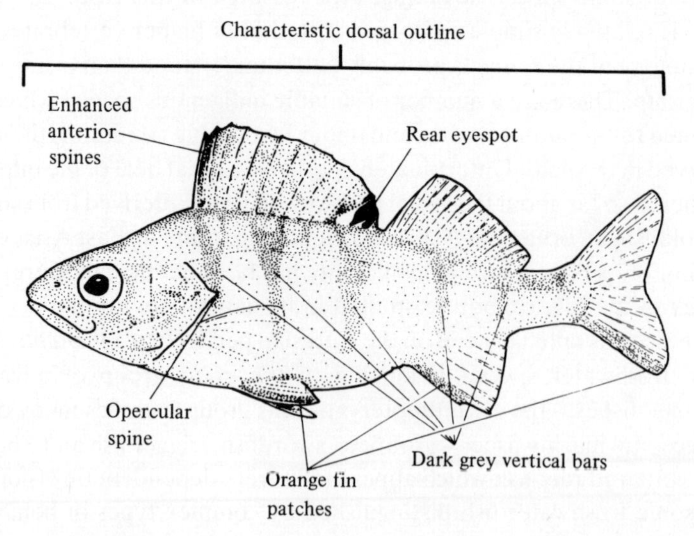

The eye

The structure of the eye differs from the basic teleostean type in a number of particulars.

The cornea contains a yellow pigment absorbing maximally at about 460 nm (470 nm: Muntz, 1974; 460 nm: Cameron, 1977). Absorption is minimal above 550 nm, but remains at over 40% of maximum at 400 nm, so that it tends to screen the retina from the shorter wavelengths (blue–violet). The pupil is pear-shaped with an anterior extension indicating that the major binocular viewing axis is directly forwards. This is characteristic of other predatory perch-like species (Tamura, 1957), and accords with a mid-depth habit.

Accommodation for binocular and lateral focussing is produced in the perch by the action of a complex system of four muscle bundles, in contrast to the situation in trout, goldfish and pike where only a single muscle is present. The perch system is illustrated in Fig. 2. Muscle 1, the thickest, pulls the lens posteriorly, muscle 2 moves it ventrally and slightly inwards, and muscles 3 and 4 pull the lens directly inwards at differing angles. This accommodatory apparatus appears to allow subtle adjustments of focus, and is shared with one or two other advanced fish species such as blennies (Munk, 1971), and kelp bass (Schwassman & Meyer, 1971), which live in well-lighted marine habitats.

The extent of accommodation in the perch (*P. flavescens*) and a number of

Fig. 2. Intra-ocular structures of the perch. Note that there are several accommodatory muscles arranged along several axes (numbers 1 to 4). The inner lenticular muscle (4) is inserted on the lenticular tendon by a separate tendinous thread.

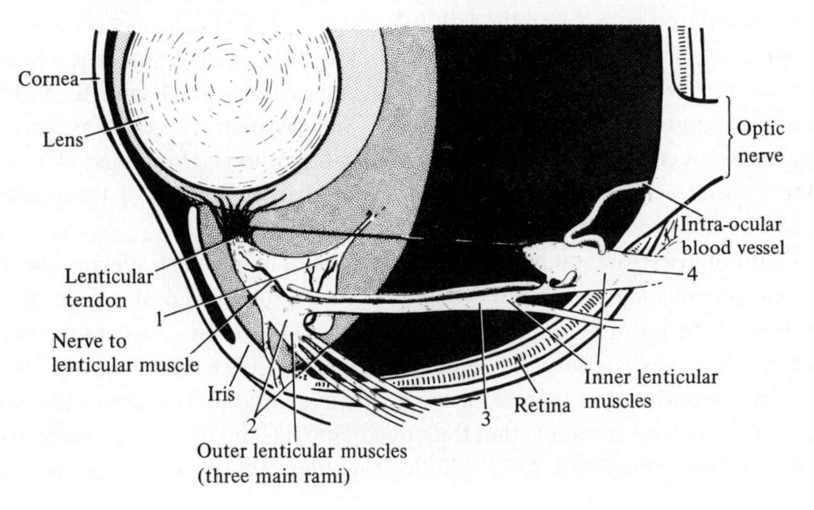

other freshwater species was examined by Sivak (1973) by measuring lens movements. He found that lateral excursions of the lens were about the same for most species (10–12 dioptres), but that the rostrocaudal movements in the binocular focussing axis in perch were the largest observed (40 dioptres) in a sample that included rainbow trout. They were much closer to values found in higher vertebrates and emphasize that image sharpness must be important in perch visual behaviour.

The cone mosaic is of the very regular 'rosette' type regarded by Engström (1963) as characteristic of the visually more active species. Each rosette consists of four pairs of similar double (twin) cones, and a single smaller central cone. There appear to be only two cone pigments, one absorbing maximally in the red ($P615_2$) and one absorbing most strongly in the green ($P535_2$). The first of these is found in the twin cones, the second in the singles (Loew & Lithgoe, 1978), and since the area occupied by double-cone receptors measured in tangential sections is about 12 times that of single cones in the same part of the retina, the system would appear to be very strongly biassed towards discriminating longer wavelengths. On the other hand, measurements of total retinal absorbence show that while the major peak is between 500 nm and 565 nm (mainly from the rod pigments: λ max = 540), there is a quite high absorbence below 425 nm (blue). Tamura & Niwa (1967) found that a number of clearwater marine fishes, like the brightly coloured butterfly fish (Centropomidae), were dichromats (i.e. have only two types of cone pigment), so that the absence of a blue cone does not appear necessarily to entail poor colour vision.

Using the method of Tamura (1957) it is possible to calculate the resolving power of the eye, in terms of minimum visual angle, from the lens diameter and the cone separations. For small perch this works out at 14.4′ (minutes of arc) compared with 24′ in the goldfish (Hester, 1968). This is rather a high value compared with those given by Tamura for 27 marine species, as some of the percoids he examined provided values between 4′ and 6′. Values of this kind were also obtained by Schwassman (1974) for several percoid species using electrophysiological methods (*Lepomis*: 4′), and their visual acuity was again shown to be superior to that of the goldfish (15′).

Although a foveal pit is absent in the perch, as in most fish species, there is a conspicuous aggregation of ganglion cells in the temporal region of the retina, at the point of major binocular focus. The overlap between the fields of the two eyes is about 30°, and ensures an effective binocular function.

Another advanced feature of the perch eye and one it shares with more recently evolved species is that the optic nerve has the form of a compressed ribbon rather than of a solid cylinder (Scholes, 1979). This appears to be related to the way in which new fibres are added to the visual pathway as the eye grows. Visual acuity is related to the size of the eye, so that as the fish grow throughout life this property should increase with age.

The eye of the perch is clearly of an advanced type, specialized for relatively fine-grain image resolution. Although it is a dichromat eye, and there is no blue cone, retinal absorbence indicates that some short-wave discrimination process may be present.

The central visual pathway

In fishes the major destination of nerve fibres from the eye is the optic tectum, This forms the roof of the midbrain, and is a complex laminated structure composed of the terminals of fibres from the eye and from other regions of the brain, and the processes of several million intrinsic tectal cells (described more fully in the chapter by Vanegas, this volume). In addition to the tectal formation, there are also a number of discrete pretectal nuclei which receive a direct input from the retina. These are shown in Fig. 3(*a*) together with the tectum, but as no clear idea has yet emerged as to their function they are not described further.

Although some axons from the tectum return to the eye the great majority of tectal efferents sweep ventrally and posteriorly to the motor columns of the midbrain and hindbrain, which exert a control over movement via the motor neurons of the spinal cord. One of the nuclei which contributes an important element to the descending motor column is the nucleus reticularis mesencephali (Fig. 3*a*). Some indications as to the function of this structure have been obtained from work in our laboratory, and these will be described below. At the point where the major ventral projection of tectal efferents begins to curve medially there is a large nucleus – the nucleus isthmi (Fig. 3*a*). Many of its cells receive powerful inputs from the tectum and their properties are currently under investigation; however, at the present time most of what we know about central visual processing in the perch and other fishes concerns the optic tectum.

Our own studies show that the perch tectum resembles that of other fishes in possessing at least six clearly recognizable strata or layers. These are illustrated in Fig. 3(*b*). There is also a stratum visible in sections between the stratum album and the stratum periventriculare in the perch (Al-Akell, unpublished observation) which appears to be weak or absent in the tectum of some other species (Leghissa, 1955; Vanegas, Laufer & Amat, 1974; Meek & Schellart, 1978). Compared with the large-eyed tropical percoids *Eugerres* and *Holocentrus* the marginal layer, which has connections with the tectal pyramidal cells (Vanegas, Williams & Freeman, 1979) and with the cerebellum (Ito & Kishida, 1978), is comparatively thin in the perch,

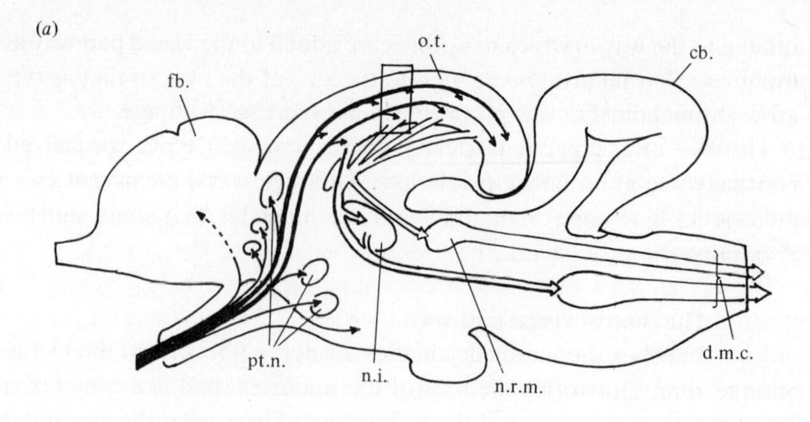

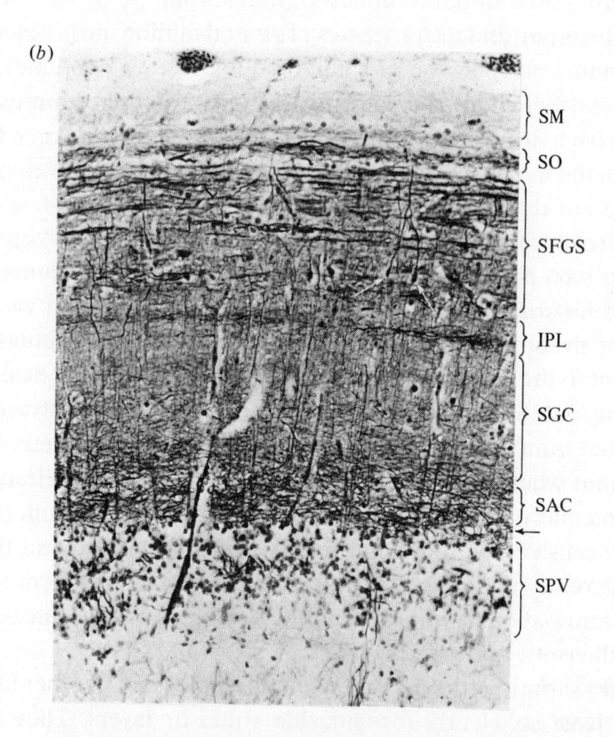

Fig. 3. The anatomy of the visual centres in the perch. (*a*) Diagrammatic longitudinal section of the brain. Besides its main radiation to the optic tectum (o.t.), the optic nerve projects to a number of pretectal nuclei (pt.n.) and to the ventral thalamus and the forebrain (fb.). A major output path sweeps ventrally across the nucleus isthmi (n.i.) to the ventral component of the descending motor column (d.m.c.). The n.i. is shown displaced anteriorly for clarity. Another tectal efferent pathway is believed to project to the nucleus reticularis mesencephali (n.r.m.) which gives rise to part of the d.m.c. running below the cerebellum (cb.). (*b*) Part of a section of the optic tectum similar to the area outlined in (*a*), to

although roughly equivalent to those of goldfish, carp and cod (Guthrie, unpublished observations).

Histological analyses of the tectum in various species reveal an array of 11 to 15 cell types, depending on how these are defined (Leghissa, 1955; Vanegas *et al*., 1974; Meek & Schellart, 1978; Romeskie & Sharma, 1979), and most of these are readily identified in the perch (Al-Akell, unpublished observation).

The tectum is a relatively conservative structure in lower vertebrates generally and this is compatible with the view that it mediates the important and basic function of visually guided orientation (Akert, 1949). This is perhaps most thoroughly demonstrated in amphibians (Ewert, 1976, and this volume).

A cell type that is worth a special mention is the pyramidal cell. This has been subjected to particular scrutiny by Vanegas *et al*. (1974, 1979) in the tropical percoids mentioned above, because of its specific connections with the marginal layer. Their illustrations do suggest that this cell is larger and more highly differentiated than we have found it to be in the perch. At the same time goldfish pyramidal cells seem to be even less well developed, going by published figures (Meek & Schellart, 1978; Romeskie & Sharma, 1979). The difference in habit and habitat here is between clearwater, littoral, marine predators, freshwater predators (perch) in waters of varied turbidity, and bottom-dwelling species that are partly herbivorous (goldfish: Maitland, 1977), so there may be an association between degree of pyramidal cell development and the importance of visually guided predation.

In a previous paper (Guthrie & Banks, 1974) a general theory of tectal function was advanced based on its structure, which will be briefly described here. The majority of intrinsic cells have long dendritic rami arranged at right-angles to the tectal surface, and this includes the most numerous group of cells, small neurons with cell bodies in the periventricular layer (Fig. 3*b*). Functions dependent on the conservation of fine detail, especially as regards position, must be based on cells of this type (System 1). Generalized functions such as the response to a moving object irrespective of identity, anywhere in the visual field of the eye, are more likely to be related to the other system of cells, which have processes spread out in the same plane as that of the tectal surface (System 2).

---

show the layers as follows: SM, stratum marginale; SO, stratum opticum; SFGS, stratum fibrosum et griseum superficiale; IPL, inner plexiform layer; SGC, stratum griseum centrale; SAC, stratum album centrale; SPV, stratum periventriculare. Two pyramidal cells within the SFGS are indicated by arrows.

The other region of the midbrain that we have examined is the nucleus reticularis mesencephali and adjacent neuron groups. A brief study has been made of the stimulus preferences of the neurons lying near the mid-line and beneath the valvula cerebelli (an extension of the cerebellum under the tectum). One of the most striking nuclei in this subtectal region is the nucleus reticularis, illustrated in Fig. 4, which appears to consist of a dorsal large-celled part and a more ventral zone of much smaller cells. The dorsal cells seem to give rise to a distinct dorsal fillet of the longitudinal median fasciculus, and these cells should therefore be closely involved in the origin of motor commands, as the fasciculus projects to spinal motor centres. How-

Fig. 4. Structure and function of the nucleus reticularis mesencephali, and oculomotor complex. (*a*) and (*b*) show responses of two binocular cells as spike frequency histograms. s, spontaneous (unstimulated level); i, response to stimulation of ipsilateral eye; c, response to stimulation of contralateral eye. Arrow indicates start of stimulus. (*c*) and (*d*) are sagittal and transverse sections through the region respectively to show dorsal large-celled area and ventral small-celled area (arrows), and the median longitudinal fasciculus (m.l.f.).

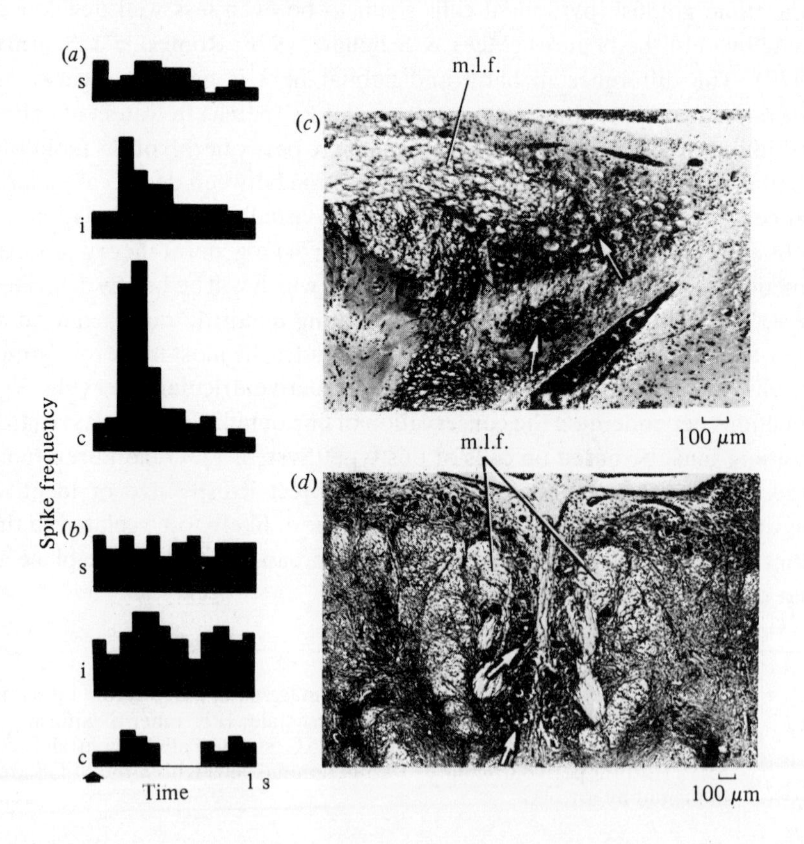

ever this region of the midbrain is a complex one and cranial nuclei, in particular the oculomotor nucleus, lie close to it (according to Hermann, 1971, working with goldfish). The oculomotor nucleus, concerned with the control of eye movements, will be expected to have a largely visual input, but of a precise and localized kind.

Receptive-field properties of intrinsic tectal cells

Each cell which is sensitive to visual stimuli can only be stimulated by light originating from a circumscribed part of the visual field seen by the eyes, its receptive field (RF). The analysis of the properties of the RFs of individual neurons in the visual pathway allows the specific items of information abstracted by the cells to be identified. Since the animal has to act on the basis of significant but restricted stimuli, rather than on a broad range, single-cell properties should tell us something about how the recognition of significant features takes place.

Our most detailed results have come from the analysis of single-unit activity within the optic tectum using the method of raster scanning. Impulses recorded from a neuron appear as dots on the oscilloscope screen in register with the stimulus light spot being swept across a screen in front of the fish's eye. In this way a map 90° × 70° of the spatial dimension of the excitability of each cell can be obtained. Most cells have some kind of spontaneous discharge, so that an area within which the stimulus produces inhibition appears as a zone with few or no impulse dots, while excitatory areas appear densely dotted. For a more detailed account of the methods used the reader should refer to Guthrie & Banks (1978).

Receptive-field types

These fall naturally into two groups: highly structured RFs with well-defined boundaries (type 1 cells), and irregular or poorly defined RFs belonging to types 2 to 5. The former are likely to be involved in visual functions requiring precise positional information (location and contour); the latter have more generalized responses in relation to brightness, contrast, velocity and degree of novelty. Scanning with white light usually allows the general form of a RF to be demonstrated quite clearly. Substitution of a variety of spectral absorbance filters often reveals a more complex structure, but normally fails to alter the gross dimensions or shape characteristics of the RF.

Type 1. These are localized RFs with defined boundaries. Type 1A are unitary RFs with a single centre. It may be difficult to make sure that there are no other major foci. The simplest of these are of the *single-patch* or centre type (Fig. 5), which are either 'off' or 'on' and are without any obvious peripheral annuli (although an antagonistic annulus can sometimes be demonstrated by special methods: Schellart, Riemslag & Spekreijse, 1979). Most of them are relatively small, 5°–20° in diameter. There are also *concentric field cells*, in which, as in the goldfish (Daw, 1968), a distinct centre is surrounded by some kind of antagonistic annulus. In the perch RFs with several annuli (type 1B) are not uncommon, and a class of multiannulate RFs with more than two clearly distinguishable annuli can be demonstrated. There is one group of RF types which is rather striking, and these fall into the category of fields with few annuli. In our experience the centre is near the principal optic axis of the eye, and for this reason they are referred to as *centre eye field cells*, and will be described separately below (Fig. 6).

Centre eye field cells. The same type of cell was identified precisely in 13 different fish. When the light spot is swept horizontally the 'on' centre and the horizontal parts of the 'on' annulus appear as vertical bars. This is due to the extremely powerful directional properties of the field, which respond to centrifugal stimulus movements in a strongly selective way. In one direction part of the centre and one annulus wing respond; in the other the rest of the centre and the other wing appear. The RF is between 30° and 40° in diameter (inner 'on' annulus). The form of the RF together with a functional model of it is shown in Fig. 6. The unit which runs down through the tectum to at least the middle levels (SFGS) responds down to light levels equivalent to

Fig. 5. Examples of small-patch receptive fields recorded from intrinsic tectal cells in response to horizontal raster scanner with a 5° light spot. (*a*) Centre 'on' type, (*b*) centre 'off' type. Each dot represents a spike. The 'off' field was consistent over 10 scans. Each scan includes about 37 horizontal lines.

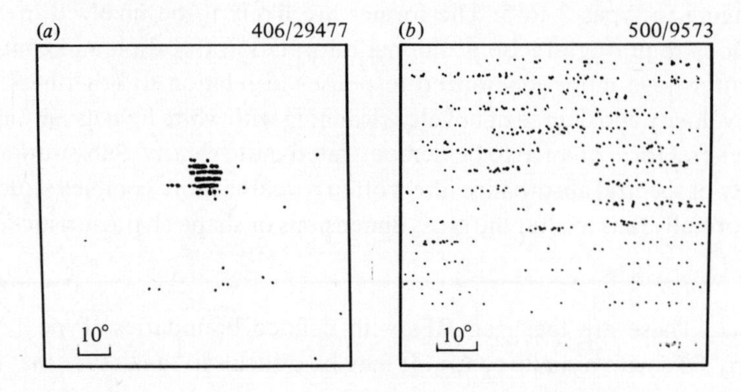

3×10^{-3} cd m^{-2}, so is not amongst the most sensitive to luminosity changes; yet chromatic responses are rather variable, so that it is a little difficult to classify as either a luminosity or a chromaticity cell.

Multiannulate fields. RFs with five or more annuli have been observed by us with centres in different parts of the eye field. The diameter of these fields is usually so large that only a part of them can be accommodated on the tangent screen, but their regularity makes it easy to deduce their structure (shown in Fig. 7). Estimated field diameters range between 70° and 160°, that is to say the largest appear to occupy quite a large part of the eye field. Often the strongest features are not the centres but thick annuli equivalent in diameter to about half the total RF. Directionality is usually weaker than in

Fig. 6. Responses to raster scanning from a centre eye field tectal cell (triple bar unit). (*a*) Horizontal, (*b*) vertical scan. Each part of the 'on' zone only responds to the light spot when it is moving away from the field centre, as indicated in the model shown in (*c*) (upper plan view). In the lower sectional model the 'on' zones are represented as 'downhill' profiles in relation to the scanning spot. In unidirectional scans only half the 'on' centre is excited at each stimulus sweep (see legend to Fig. 4).

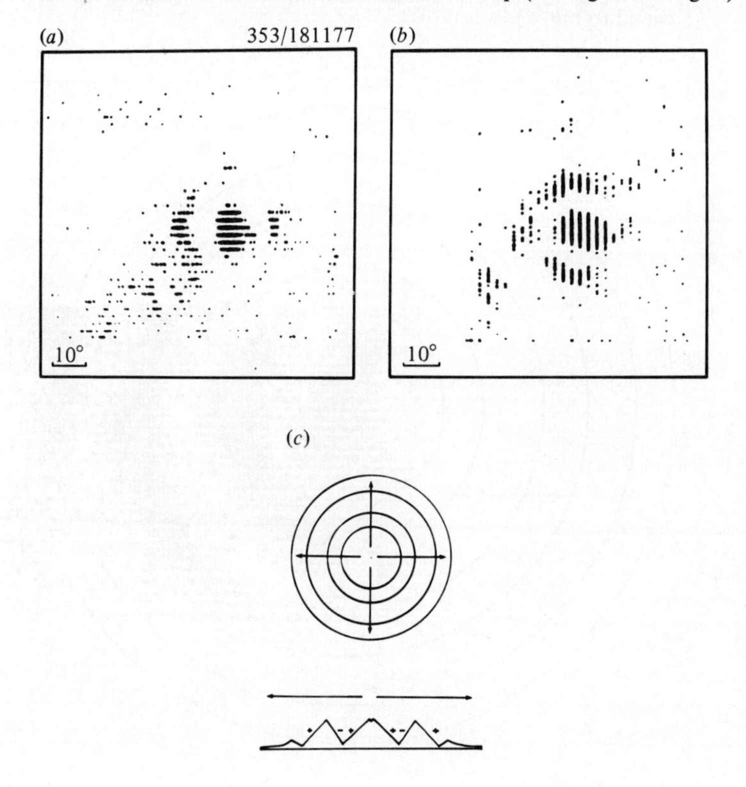

centre field cells so that annuli appear much more complete when scanned along one axis only.

Both the centre field cells and those with large multiannulate fields were previously recognized as belonging to type 1C cells (fields with bar-shaped features: Guthrie & Banks, 1978), but can now be seen as essentially type 1B cells with complex RFs and distinct centres. Type 1C cells do occur, but they are rare.

Multicentre receptive fields (type 1D cells). These appear to be composed of several type 1B RFs either superimposed or arranged edge to edge. In the example illustrated in Fig. 8 a large and a small RF lie adjacent to one another. According to Schellart & Spekreijse (1976) RFs of this type are characteristic of tectal cells, rather than of optic nerve fibres, and the results of our studies on perch are in broad agreement with this. It is difficult, however, to see what function such a complex and irregular pattern of inputs

Fig. 7. A reconstruction of the proposed complete receptive field of a large multiannulate receptive field with rather weak 'on' centre, surrounded by three 'on' annuli and three 'off' or neutral annuli (see legend to Fig. 4).

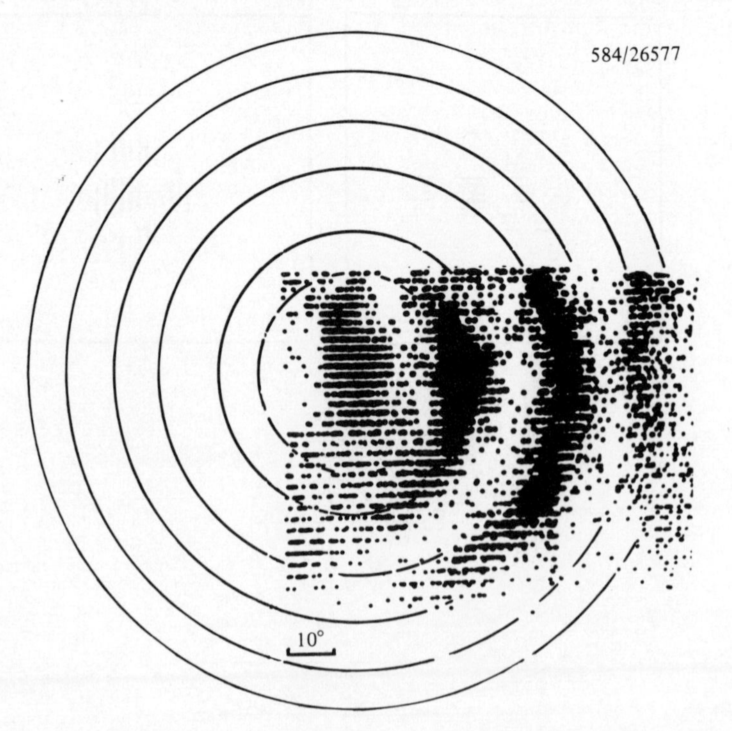

can perform in visual discrimination, unless it corresponds to some particular arrangement of external objects significant for individual fish, an innate or acquired visual template.

As a result of overlapping 'on' and 'off' zones some of these multicentre fields may be much more complex than the one illustrated here.

Receptive fields belonging to types 2, 3, 4 and 5 (field maxima 80°–160°). A wide variety of RFs exist which are weakly structured and possess irregular and poorly defined boundaries. The groups which we have identified so far, are as follows:

Type 2. Cells with a vertical rather than a horizontal directional preference when stimulated by a scanning light spot. These appear to be rarer than those with a horizontal preference.

Type 3. Phasic cells which respond to a moving stimulus with a short latency response irrespective of position, but adapt very rapidly. These may correspond to the novelty units observed by researchers working with amphibians, which draw the attention of the animal to small unexpected movements.

Type 4. Cells which respond to fairly rapid changes in light intensity over a large area – wide-field transient luminosity units. Cells often respond to both light on and light off, but more strongly to the latter.

Fig. 8. Multicentre receptive field recorded from an intrinsic tectal cell. Large field to left with weak 'off' centre, surrounded by orientation-selective 'on' annulus (responds most strongly along radii at right-angles to stimulus axis). Rather weak receptive field to right with marginal 'off' annulus.

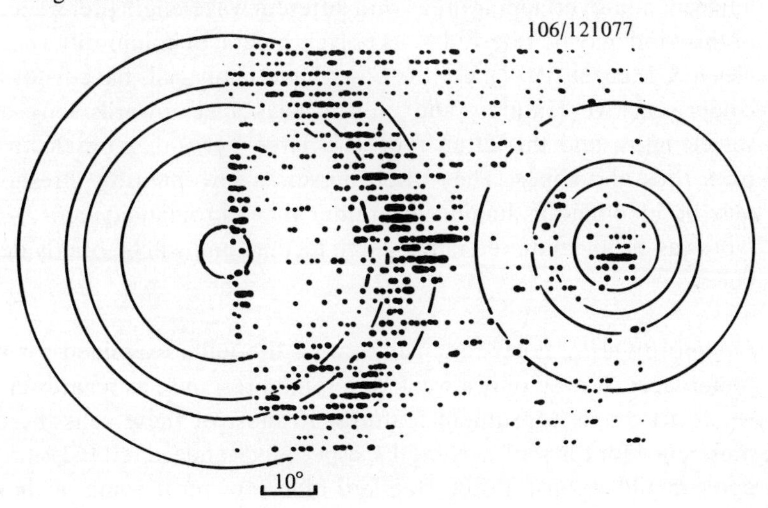

Type 5. A number of units which respond to the local dimming of a scanned light spot have been observed in the absence of a structured RF. These cells are probably allied to type 4 units. Under experimental conditions they can be made to produce impulse plots which outline static figures placed on the tangent screen that intercepts the scanning spot, but this does not mean that they are directly involved in contour detection. It may be significant, however, that one type responds better to upper rather than lower edges, a characteristic of shape discrimination in the goldfish noted by Sutherland (1968).

Columnar organization

It was found that adjacent tectal cells could have RFs with shared features. Multi-unit recordings from the optic nerve also tend to support the idea of RF-related bundles of fibres, connected to the tectum. Our anatomical studies indicate a columnar structure in the deep tectal layers, vertical cell groups in close association in the more superficial layers (SO, SFGS: see Fig. 3), as well as axon columns in the outer fibrous layers. These observations emphasize the way in which the tectum is divided up into areal systems with different 'mesh sizes'. Superimposed on this are wide-field elements with generalized functions like those of types 2 to 5.

Chromatic responses

Hue discrimination in animals generally is believed to depend primarily on cells with colour-opponent RFs, that is to say fields in which within defined areas light of different wavelengths produces opposite effects, i.e. inhibition or excitation (Daw, 1973). Another type of RF may consist of adjacent non-overlapping areas with different wavelength preferences. Cells of this kind may be referred to as polychromatic or colour differential cells (Kien & Menzel, 1977). On the other hand many cells have fields without colour-sensitive features, and only a weak bias towards long or short wavelengths, and this often reflects a large but rather unselective input from rods and cones. They often possess a low-intensity threshold, and may be identified as luminosity, rather than chromaticity cells. All these types can be encountered in the perch tectum, and will be briefly described below.

Luminosity cells. Forty-four per cent of the units examined for spectral preferences showed only a weak differential response as regards either the whole RF or its constituent features. In most of these cells there was a preference for longer wavelengths, especially standard red (619 nm), and a poor or nil response to the standard blue (463 nm). Some of these units responded well to rather low light intensities ($3 \times 10^{-4}\,\mathrm{cd\,m^{-2}}$). The RFs of these cells might belong to the single-patch variety of type 1A, to type 1D or types 2 to 5, but they were less often of the well-defined concentric kind of 1A cell. However it should be pointed out that centre field cells might show very well defined colour-opponent properties or extremely weak responses.

Chromaticity cells. Eighteen per cent of cells were classified as belonging to the polychromatic or differential type; that is to say they contained features which were wavelength-dependent, but not obviously opponent on an areal basis.

Perhaps the most striking of the polychromatic cells were those with an enhanced sensitivity to deep reds (658 and 696 nm), ultraviolets (369 nm) or white light for certain features. At the start of our research it was assumed from the microspectrophotometric data of Lythgoe (see above) that the area of search within the spectrum could be safely limited to the range 460–656 nm. Subsequently we have encountered a small number of cells with stronger responses to 369 nm or 658 nm than to intermediate wavelengths. Examples illustrated here are an 'ultraviolet only' feature (Fig. 9*b*) and a field with preferences for both ultraviolet and deep red (Fig. 9*a*). The latter might be accounted simply an example of broad-band white-light inhibition, were the positive response so obviously not at the spontaneous level (which is virtually nil). Infrared preferences (700–800 nm) were looked for, but not found. Features were found in some RFs which seemed only to have their characteristic form when the stimulus was a white spot of the same intensity as that of selected wavelengths (Fig. 9*c*). The effect was not matched by responses to light peaking at 535 nm (close to the rod maximum), but at the same time the absorbance curve of the filter was a good deal narrower than that of the rod nomogram.

Of the remainder of the cells investigated 14% showed colour opponence only, and a further 24% possessed a mixture of opponent and differential features. In Fig. 10 a cell with a strong blue-sensitive 'on' centre (463 nm) is illustrated which corresponds to a red-sensitive 'off' centre (619 nm). At 535 nm (green) however, a different and more isolated pattern emerges which can be described as differential. A more regular colour-opponent RF is shown in Fig. 11. This has a red-sensitive 'on' centre corresponding to a gap between green-sensitive 'on' wings. The gap can be interpreted as a green 'off' zone.

Blue-sensitive areas (463 nm) are generally similar to green-sensitive ones, as described by Beauchamp & Lovasik (1973) in goldfish. Within the centre of the RF of this cell a red patch against a green background, such as

Fig. 9. Chromatic responses. (*a*) Wide-band bichromatic response of a single-patch field, much stronger at 369 nm and 658 nm that at intermediate wavelengths. Discrete patch field with some zonal differentiation between short- and long-wavelength response, (*b*) Specific zonal response to near-ultraviolet component (369 nm) clearly differentiated from standard blue (463 nm). (*c*) Some cells show a response to white light which differs from any restricted waveband response. A diffuse regional inhibition to white is shown here in comparison with the response to standard red (619 nm).

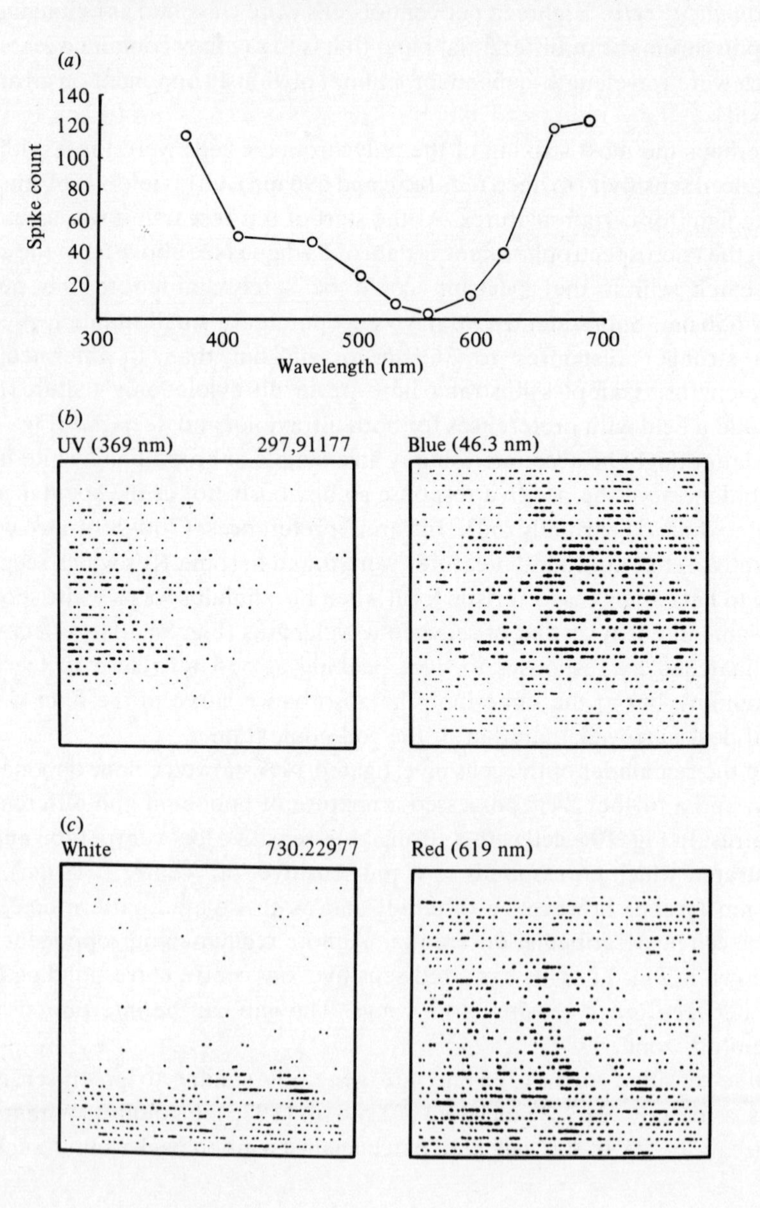

might be produced by red fins, would evoke an unequivocal excitatory effect. Red 'on' centres are most common, as in the goldfish (Spekreijse, Wagner & Wolbarsht, 1972), but blue 'on' centres were also found (striking blue features were uncommon). Spekreijse did not observe blue 'on' centres in the goldfish, but they were demonstrated by Beauchamp & Lovasik (1973), who found that they were additive rather than opponent to the green cone responses; this finding is in agreement with that of Cameron working with perch.

Although Cameron (1977) employed a more limited spectral range in his behavioural work than we have done, our findings are generally consistent with his, in that a red–green inhibitory process is seen to be present while at the same time there is high sensitivity to very short wavelengths. This short-wave sensitivity seems a little surprising in view of the absorption

Fig. 10. Opponent receptive field with polychromatic or colour differential features. The strong blue-sensitive centre visible at 463 nm corresponds to an area of very low spike density at 619 nm. At 535 nm pattern elements appear which are poorly developed or absent at longer or shorter wavelengths, that is to say they are differential features.

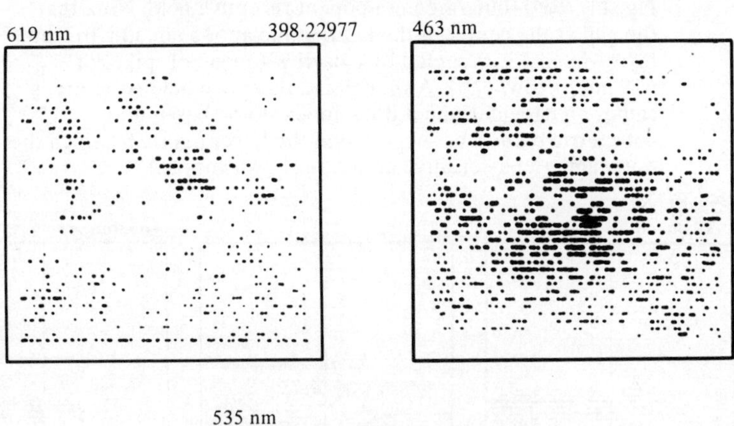

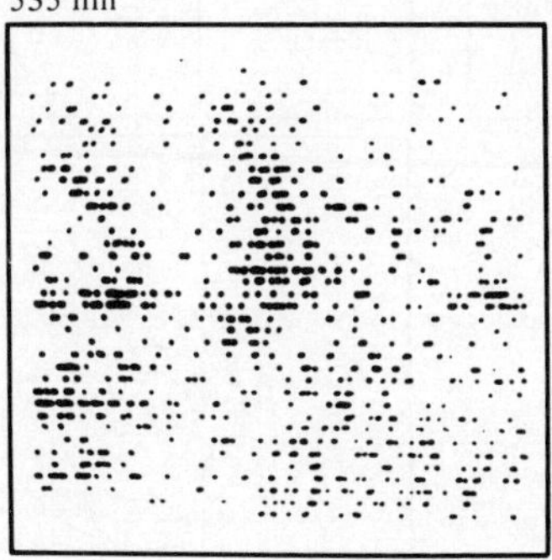

properties of the cornea mentioned earlier, but the effect of the yellow corneal filter may be to enhance the long-wavelength component of broad-band colours, thus providing a preliminary separation of colours in the short-wave region. In the perch itself the vertical bars, opercular scales and orbital skin all have a bluish colour to the human eye, and may be capable of selective stimulation of the short-wave process in the perch. In general it seems clear that the perch possesses colour discrimination of a comprehensive kind.

Computer simulation of multiannulate receptive field properties

While the discrimination of shape and contour may depend on the spatial distribution of small excited patch fields, large complex fields may also contribute to differential responses to contours. One of the properties of large multiannulate fields which might be predicted is that fields of contrasting dots would produce little effect compared with large contours. A brief computer simulation exercise has been carried out for us by Dr Derek Robinson with a number of standard shapes (circle, triangle, square, diamond) of similar area moved in single-step increments across an 'on'

Fig. 11. Red–blue/green opponent receptive field. Note that the gap at the centre of the field visible at 535 nm and 463 nm (arrowheads) is occupied by a narrow tongue of spikes at 619 nm (arrowhead). Although the receptive field structure is rather an irregular one it does appear to be based on a concentric pattern as suggested in the lower figure, in which the blue- and green-sensitive areas are shown stippled.

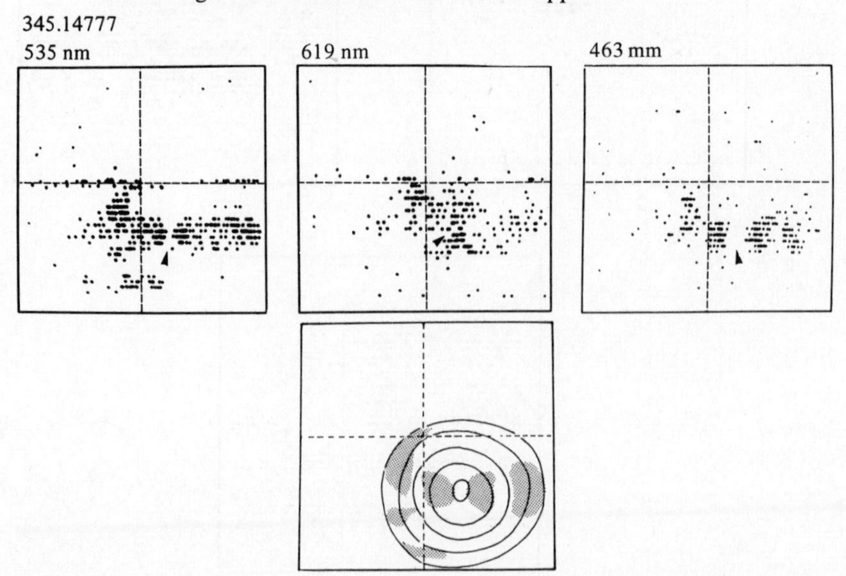

centre RF with three annuli, within a 100 × 100 grid. Each grid square in the simulated RF had a value of + 2 ('on'), 0 ('off') or + 1 (spontaneous level of RF surround). The rate and extent of increment accrual can be seen to vary according to stimulus shape, as shown in the examples illustrated in Fig. 12. As predicted, small shapes generate less obvious differences than the same shapes scaled up (Fig. 12*a* and *b*). However, where vertical contours of differing angularity are concerned – corresponding, say, to the anterior profiles of pike and perch – the response differences are especially striking. Simulated responses to 45° and 90° angular profiles are shown in Fig. 12(*c*). It must be pointed out that these response increments are not equivalent to spikes, but could be supposed to contribute to them in a proportional manner. Obviously thresholds or facilitation rates of the cell could be set so that the cell would only produce a spike output to certain amplitudes or rates of increment accrual. As far as discrimination of shapes based on rates of stimulation is concerned, cells which could give precise collateral information about object velocity would enhance the accuracy of a contour discrimination system based on multiannulate fields.

The functional properties of the nucleus reticularis mesencephali and oculomotor complex (Fig. 4)

The anatomy of the nucleus reticularis mesencephali (n.r.m.) has been briefly described earlier. The following account is drawn from unpublished observations made by Banks & Noble in our laboratory. Using standard electrophysiological methods, recording contact with individual neurons was made by means of 5–20 MΩ micropipettes, and a variety of visual stimuli, electrical stimulation of the trunk, and light mechanical stimulation of the head were used to evoke responses in the reticular cells. Cells could be fairly easily classified as belonging to one of four groups: (i) tactile cells, (ii) tactile and visual cells, (iii) visual cells, and (iv) cells which did not respond to any of the stimuli used. Approximately 200 cells were studied altogether.

Tactile cells. Neurons with exclusively tactile inputs were rare. Those that were found exhibited only weak responses which were difficult to characterize on a regional basis.

Cells responding to tactile and visual stimuli. Most of these were deep-lying units with binocular inputs. The responses to tactile stimulation were the strongest ones and derived from the skin of the head. The most sensitive areas were around the mouth, the eyes including the cornea, the cheek region and the skin over the preopercular bone.

Fig. 12. Computer simulations of the response of a tectal cell with a multiannulate receptive field to the displacement of a light figure across the receptive field, as indicated above. Matrix dimensions 100 × 100 subdivisions. Each subdivision had a value of +2 in an 'on' area, 0 in 'off' areas, or +1 in the surround when occluded by the figure. (*a*) Responses to a small square and a triangle: the differences are relatively small. (*b*) A large square and a triangle generate responses which are more clearly differentiated. (*c*) Acute and obtuse profiles produce clearly separable responses.

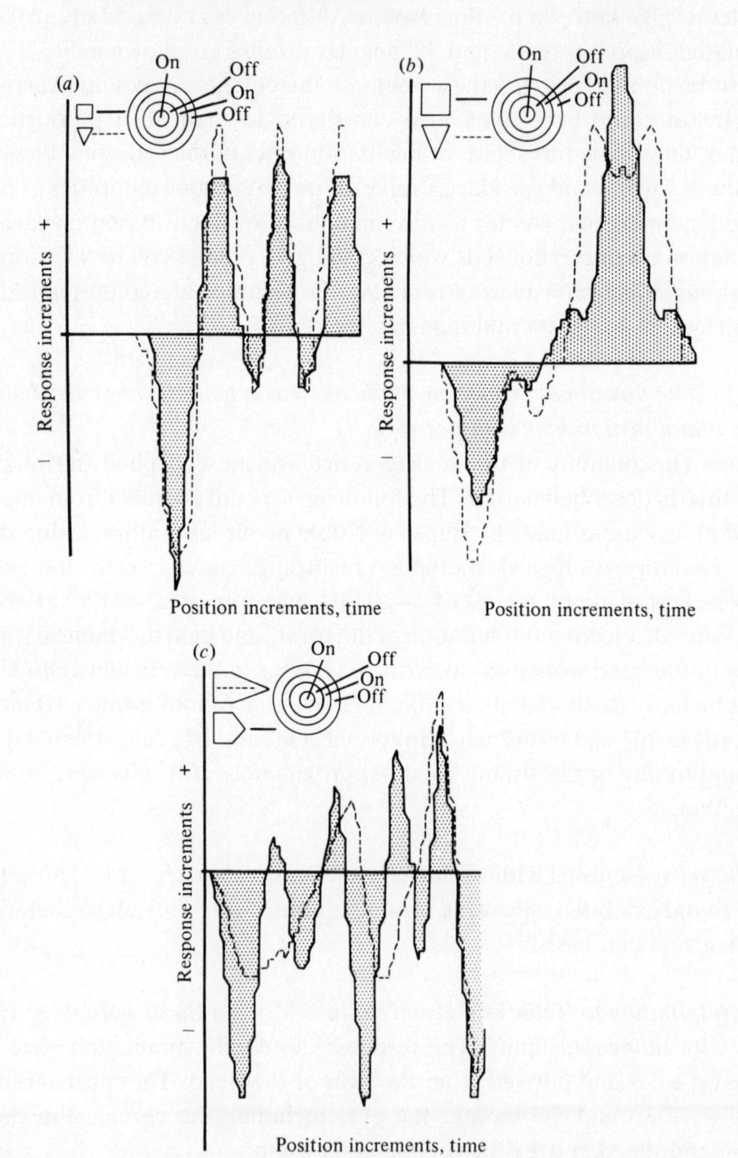

Cells with a predominantly visual input. These comprised the major group of neurons, and seemed to be medium to large cells on the basis of their extracellular potentials. About two-thirds of these cells were binocular and the remainder monocular.

Monocular cells were mostly found in the dorsal part of the nucleus 1500–2500 μm below the surface of the brain. Three kinds of input pathway could be detected: contralateral excitatory, ipsilateral excitatory and contralateral inhibitory. While most of these cells responded with regular spike trains when excited and were fairly resistant to habituation, a few responded with a characteristic bursting pattern of discharge which habituated fairly rapidly if the stimulus was repeated.

Binocular cells were usually found in the deep lateral and posterior regions of the nucleus between 2500 and 3500 μm below the surface of the brain. Of the four possible combinations of excitatory and inhibitory inputs that might be expected, only ipsilateral excitation and contralateral inhibition were absent. This is a little surprising in view of the range of monocular properties. Ocular dominance, which is a feature of cells in the mammalian visual cortex (Hubel & Wiesel, 1962), can be contralateral or ipsilateral for inputs of the same type (i.e. where both are excitatory or inhibitory). On the other hand those cells excited through the contralateral eye and inhibited through the ipsilateral one were always dominated by the latter input. Some idea of the response levels involved in binocular cells is given by the histograms in Fig. 4. Note that the fairly high spontaneous level of activity allows inhibition to show up quite clearly.

Two additional types of stimulation were occasionally found to be effective. Some cells responded to electrical stimulation of the flank, usually with an excitatory burst. Most of these were binocular cells, and had a predominantly excitatory visual input. Units also occur which respond to rotation of the visual field or to vibration. These are cells receiving binocular inputs, and the former may be oculomotor cells (see below).

Cells not responding to any of the stimuli used. These comprised about 25% of the sample, and appeared to be rather small cells lying either near the surface (1500 μm) or near the ventral margin of the nucleus (3500–4000 μm). Hermann (1971) identified three types of unit from the same general area in the goldfish brain, in relation to eye position and eye movements. The commonest type (P units) appear to have been ones in which the discharge frequency was related to the position of one of the eyes. Others (SB and SP units) were excited or inhibited at the start or the termination of a saccade. However, preferred stimuli were not demonstrated, nor was positive identification obtained of an oculomotor nucleus by retrograde staining or antidromic stimulation of the oculomotor nerve. It should be noted that most of the goldfish cells appear to be associated with monocular motor effects, while many of the perch units are essentially binocular in function. The responses of the perch cells to mechanical stimuli are also important since other spinal reticulomotor cells arc characterized by similar preferred stimuli (Martin, 1977, in lamprey; Resteaux & Satchell, 1958, in dogfish; Diamond, 1972, in teleost Mauthner cells). It seems quite likely that n.r.m. cells and oculomotor cells lie close together and cooperate in the coordination of the eye and trunk movements needed to track a moving object. The perch n.r.m. cells may also be specifically involved in generalized startle and escape responses mediated by visual and tactile stimuli.

Visual behaviour

From the preceding sections it is possible to gain some insight into the properties of the visual pathway of the perch, and we can now ask whether the characteristics of visual behaviour in the field or in training experiments complement these properties, that is to say whether similar stimulus preferences can be observed.

There is a rather scattered literature on perch behaviour, in particular work by Hoogland, Morris & Tinbergen (1956), Fabricius (1956), Thorpe (1977) and Popova & Sytina (1977), and this has been supplemented by a short-term field study made from this laboratory (Barrow & Guthrie, unpublished observations). The aspects of behaviour which are most likely to depend on vision are (i) interactions with conspecifics, especially in relation to shoaling, (ii) route recognition, and the ability to recognize features of the underwater terrain, (iii) predatory behaviour and (iv) responses to predators.

Our own studies derive from regular observations made on a single shoal of 26 fish in a shallow pool in northwest England, 1457 m² in area with a maximum depth of 1.7 m. Much of the pool is less than 1 m deep, there is one main weed-bed only, and the water remains clear for most of the year. The pool was divided by sight lines into a grid for the purposes of recording the route taken by the shoal, which could be followed with binoculars. By means of individual variations in their bar pattern about half the members of the shoal could be individually identified at a distance.

Interrelations with conspecifics

Shoal cohesion appears to depend on the ambient light levels, and is also related to water temperature. In the summer shoals fragment at night (Alabaster & Robertson, 1961) and re-form about an hour before dawn. In our experience these dawn shoals were not very dense (3–6 fish-length intervals between fish) and feeding rates were high, fish making foraging sweeps at the edge of the shoal. Shoal density was highest at noon in bright sunlight, when the shoal was more or less at a standstill (1–1½ fish-length intervals between fish). This trend was reversed as light levels dropped and there was a period of intense feeding at dusk. In the winter, feeding activity was usually centred around noon, when light intensities were between 500 and 1000 cd m^{-2} as compared with 200–1000 cd m^{-2} at dawn and dusk in summer, and shoal cohesion was never as high as at noon in summer (light levels about 2000 cd m^{-2}). At low light levels (< 500 cd m^{-2}) shoal cohesion seemed to be weakened by feeding motivation, but shoaling cohesion dominated at higher light levels and disappeared entirely at night.

In the laboratory we found that even colour discrimination persisted down to light levels equivalent to 10^{-2} cd m^{-2}, and it seems reasonable to envisage two very broad light intensity ranges: one from less than 1 cd m^{-2} (= 1 lux) to 500 cd m^{-2}, associated with feeding and exploratory behaviour, and another from 500 cd m^{-2} upwards that is dominated increasingly by close shoal cohesion. The lower of these ranges is probably mesopic rather than scotopic, and since both ranges involve several log units no very precise retinal sensitivity is involved.

Although each fish had a pattern of dark bars which differed from that of its shoal-mates there was no evidence for individual recognition, or for a dominance hierarchy. When the shoal was on the move no single fish or small group of fish could be observed to act as leader or to exert a dominant role. As Breder (1951) points out shoals of this type are leaderless. Perch shoals, like those of some other native species, are usually made up of members of the same year-class, and the absence of size differences may be associated with lack of dominance.

When the dorsal fin is down the perch is distinguishable from other freshwater species mainly on account of the regularly spaced dark vertical bars (Fig. 1). For the main range of shoaling densities (½–6 fish-length intervals between fish) the spatial frequency of the bars was equivalent to 0.15°–0.83 cycles per degree (1.2°–6.6°) for the study shoal, a fairly small range which could be the basis of a fixed-feature type of response. These values correspond to the smallest or narrowest features found in the RFs of tectal cells. It is tempting to see the vertical bars as a part of an optomotor mechanism for synchronizing shoaling movements, yet fish species with a great variety of pigment patterns, or virtually no pattern, show the same kind of following response. Harden-Jones (1963) found that perch were poorer at following rotated stripes than were sticklebacks, pike, trout and roach, although it should be noted that the test spatial frequency was very low (0.06

cycles per degree), and well below the level mentioned above. However, this low level of reflex following may be significant, and could indicate suppression of the normal response to stripes correlated with their value as a recognition feature. Clearly too, a very powerful optomotor response could reduce the independent movement of shoal members. Endler (1977) points out that a pattern like this can function as camouflage at long range and as a recognition feature at short range. The range at which this change of pattern significance might take place is difficult to judge since it depends on water turbidity, light intensity levels and the contrast of the bars. The pigment of the melanophores of the bars tends to aggregate in stress conditions or at low ambient light levels, but disperses to produce very dark bars in bright sunlight, where background contrast is also high. Vertical bar patterns like this have been regarded (Endler, 1977) as a special camouflage for life in or near beds of rushes, yet in many waters the home range of a perch shoal lies over recumbent weed cover or bare sediments (Guthrie, unpublished observations; Craig, personal communication).

The five to seven bars taper ventrally and broaden dorsally, so that the largest form Y-shaped areas (Fig. 1). The dorsal sub-bars are often fragmented or detached from the main bars. The pectoral fins unlike the other fins are quite transparent, and allow bars 1 and 2 which lie underneath them to remain perfectly visible. The bars are slightly curved (convex edge posterior), and the mean curvature is roughly equivalent to a radius of 50% body length, so that this feature appears to be polarized.

The tectal system of structured RFs includes, as we have seen earlier, concentric annuli as one of its major types of component, and due to directional preferences (especially clear in centre eye field cells) the response regions appear as slightly curved bars when scanned by horizontally moving stimuli (see Fig. 6). The bars appear bluish-grey, against a background skin colour that is pinkish or greenish according to the state of pigment adaptation, i.e. they show some degree of colour contrast. While blue versus red–green colour opponence was rather seldom seen in visual cells, it was not unknown in tectal RFs, and an example is illustrated in Fig. 10.

Where they intercept the lateral line the dark bars are roughly half the width of the intervals on either side of them; according to Paling (personal communication) stripe reversal occurs in the breeding female, so that the bar intervals become darker than the bars. Even normal blanching does not achieve this, and it may be a specific sexual signal depending on the apparent change in relative stripe width. In the percoid *Lepomis* Stacey & Chiszar (1978) also found a typical bar pattern in the breeding females.

As in other species curvature of the belly in perch is marked in breeding females, and there are sexual differences in head morphology. The extent of contour discrimination by single cells remains obscure, although some evidence for this is discussed earlier in relation to cells with large multiannulate fields.

The anterior dorsal fin seems to perform a signalling function. The 13–15 spines form an apparently continuous support for the membrane when the fin is erect, but the first three spines can be moved independently of the rest, and are probably under separate motor control. The 'anterior unit' has a distribution of pigment which emphasizes the outline of the first spine, and thus the anterior fin margin and the angle this makes with the body (see Fig. 1). The 'posterior unit' bears a spot over the membrane joining the last three spines. The spot (Fig. 1) consists of black-pigmented membrane between spines 12 and 13 and a relatively small area between spine 13 and the skin of the back. Spines 12 and 13 contain a highly reflective white pigment and their areas are increased by swellings. Measured from photographs the area of black pigment is within 5% of the apparent area of the pupil in the same fish. The swellings introduce a contour with some similarity to the one forming the anterior boundary of the pupil. Moderate levels of excitation in the fish result in the raising of the posterior spine row and the exposure of the spot. At the normal shoal densities of our study shoal the eyespot subtended angles of 0.8°–9.2°. Contrast values obtained from projected colour transparencies using a light meter were estimated on the basis of the Michaelson ratio,

$$\frac{\text{(background luminance} - \text{feature luminance)}}{\text{(background luminance} + \text{feature luminance)}}$$

to be of the order of 0.64. This is rather less than the pupil – sclerotic contrast value of 0.93, but still indicative of high contrast. The idea that the eyespot is an anti-predator device diverting attention from the head (Cott, 1940; Nursall, 1973) seems plausible but, as the latter author remarks in relation to the spot-tail shiner (*Notropis hudsonius*), it may also furnish a means of signalling to conspecifics. Experiments with *Pristella riddlei*, which has a similar black dorsal fin patch, show that the patch is important in species recognition (Keenleyside, 1955), and similar features are common amongst shoaling species. Both eyes and eye-spots appear to be nearly optimal stimuli for tectal cells with weakly directional properties and strong concentric RFs with 'off' centres. They are less so for 'off' patch fields, but such cells were not as common as 'on' centre units.

At higher states of excitability the anterior part of the fin is pulled forward exposing the first spine. In its extreme position this makes an angle of approximately 80° with the shoulder seen in lateral view, compared with 170° when only the posterior part of the fin is erect. Intermediate angles occur. The extreme position is often coupled with erection of the gill-cover spines when the fish is held, so that the perch becomes more awkward for a predator to swallow, even when held by the head. At the same time this posture can appear in any potentially hazardous or demanding situation. Aggressive displays in blennies (Eibl-Eibesfeldt, 1975) also involve the pulling forward of the anterior fin spines, as well as a 'head up' stance on the part of the dominant fish. Specific orientation detectors similar to those of mammals (Hubel & Wiesel, 1962) were not found in this study, but as pointed out earlier, leading-edge contours may be discriminated by cells with large multiannulate fields.

In the study shoal very few examples of antagonistic behaviour were seen, but these may be difficult to observe at the range involved. When one fish was foraging at the edge of the shoal it was often displaced by another shoal member from a favoured area, without any apparent confrontation or specific signalling.

The pelvic and the anal fins of the perch, and part of the caudal fin, contain like those of many other freshwater species reddish pigmentation, and a small amount of preliminary work has been done on the red pigments of the ventral fins in perch, roach, rudd and dace. Our tentative findings are that within the visible spectrum there appears to be in perch, in addition to broad-band absorption below 500 nm, a characteristic absorption peak at about 460 nm. This is absent in the three cyprinoid species, which all have a peak of similar shape near 405 nm. There are also characteristic peaks at shorter wavelengths, but problems of contamination are encountered here. It does seem, however, that there may be a characteristic pattern of absorbence or even emittance in the region between 265 and 350 nm. The evidence so far favours the idea that the red pigments in the ventral fins could be utilized in species recognition. No doubt they also have a more general function in signalling position and/or activity. The reddish hue is an effective offset or contrast colour for the green reflectance of most of our local waters, in which green algae predominate. As pointed out by Lythgoe (1974) the degree of contrast will diminish as the range increases. There seem to be two points worth noting here. The red 'on'/green 'off' opponent type of visual cell described earlier appears well suited to discriminate red fins against a green background. The ability to distinguish the subtle differences between the hue of perch and roach fins probably depends additionally on the discrimination of shorter wavelengths. That perch can do this is suggested by Cameron's discrimination tests, and our own studies demonstrate specific tectal channels within the range 369 to 481 nm. At the same time we do not know without the results of behavioural experiments, how significant these fin colours are to the perch.

A final point should be made concerning species recognition. The dorsal contour of the perch clearly differs from that of all other native freshwater species. In the only other indigenous species with two large dorsal fins – the ruffe (*Gymnocephalus cernua*) – the fins are confluent and differently pigmented.

Studies made some years ago by Sutherland (1968) on the goldfish demonstrated that the upper borders of figures were more easily discriminated than lower ones, and in one of our earlier studies (Guthrie & Banks, 1976) we demonstrated the existence of a type of tectal cell with an enhanced 'off' response to the upper border of a static figure. but we have no evidence for contour discrimination in this instance. For a mid-water fish, objects above the normal visual axis are going to be seen in silhouette against direct illumination, objects below this level against the dimly reflected background of the bottom. Colour contrast is likely to be particularly effective under the latter conditions, and achromatic contour discrimination in the former case.

In order to test shoal-mate recognition, 'foreign' perch of the same size as those in the study shoal were released from a cage staked out on the pond bottom close to the shoal's regular route, when the shoal arrived near the cage. This was only tried on two occasions, and alarm on the part of the shoal and hesitancy by the 'foreign' fish prevented a smooth introduction; however no additional shoal members were ever seen subsequently.

Route recognition

In our observation pool the shoal pursued a similar pathway round the pool on successive days: anti-clockwise in relation to the north–south axis. Sometimes several circuits were made in a single day. The route did not follow a depth contour, but was affected by local features. It avoided a fallen tree, and passed through the weed bed (in the summer and autumn when this was present). The outflow corner (east) was avoided at all times of year, perhaps because smaller prey organisms tended to be less abundant there, yet the fish normally penetrated the shallows beyond the 0.33 m contour in the north and west corners. Sometimes a particular area was subject to foraging scrutiny; for example on 19 December 1977 the shoal left the deeper water (> 1.35 m) to concentrate on an area (depth 0.33–0.66 m) some metres from the normal route. As the season advanced the route shifted to deeper water – only 35% of the pathway was below the 1.3 m mark in September as compared with 80% in December – but its form remained essentially the same.

If, as during our release experiments, the fish become slightly alarmed, they tended to return along the route they had just traversed for a short distance rather than to scatter at random. The release cage was left staked out on the route and the shoal made a wide detour round it for several weeks subsequently.

These observations suggest that the route is probably some kind of optimum pathway related to the likelihood of feeding success which is remembered from day to day. This pond it should be noted is a very small stable environment, with virtually no predators or competitors. Unfamiliar objects, like the release cage, produced strong effects on behaviour, suggesting that the perch shoal does recognize its visual environment in detail.

Predatory behaviour

There is convincing evidence (see Thorpe, 1977; Colette *et al.*, 1977; Popova & Sytina, 1977) that preferred sizes of prey increase as the perch increase in size, beginning with small planktonic animals like rotifers and proceeding eventually to quite large fish. For perch between 10 cm and 50 cm long mean prey lengths were close to 0.2 perch body length, although the range of prey sizes was large and increased with perch body length (Popova & Sytina, 1977). The relationship between eye diameter and body length falls with age from about 0.03 at 15 cm to 0.015 at 40 cm, so there is probably a non-linear relationship between retinal area and preferred prey size. In different-sized species of frogs, the RFs of Class II retinal afferents differ correspondingly (Glickman & Pomeranz, 1977) according to a non-linear function, and this can be correlated with the size of their prey.

From the description of the RFs of tectal cells given earlier it can be seen that boundary diameters range upwards from 5° to values in excess of 160°; however only the patch fields appear to be structurally as simple as those believed to mediate the prey-size discrimination system in frogs and toads (Ewert, 1976). All the others have highly responsive outer annuli with the same sign as the centre. Using the data of Popova & Sytina (1977), a 15 cm fish (the size we often use in the laboratory) would have a prey-size optimum of about 3 cm, and this would subtend angles of 23° to 2° at ranges of 0.5–6.0 times the body length (normal shoaling densities taken as a measure of visual responsiveness). This coincides fairly well with the patch field diameters of 5° to 30° mentioned earlier. Although the lower limit of field centres was around 5°, individual RF features were found down to about 1°. It should also be noted that 2° corresponds to a range of about a metre, a distance at which small objects will be poorly visible in turbid water. From these considerations we can conclude that the patch field system may well be the prime element in the visual pathway involved in the discrimination of prey-size optima, although other properties will depend on other units.

The increase in preferred prey dimensions means that when the fish reaches a certain size the type of prey alters. At a body length of approximately 10 cm (Popova & Sytina, 1977) perch become piscivorous. The strategy required to capture other fish is rather a different one to that needed to catch invertebrates. With the latter, the emphasis is on visual searching, while catching fish needs tactical and ballistic movements of a much higher order. There is some evidence of shoal involvement in cooperative predation on smaller fish (Nursall, 1973) – what Nursall terms pack hunting. The final lunge must involve the effective estimation of the range and velocity of the prey. Our information on the accommodatory mechanism of the eye, and the binocular functions of reticular neurons, reveals parts of the system likely to be involved in prey tracking. Although many visual cells will respond with progressively higher impulse frequencies to stimulus movements between 20° and 100° per second, velocity fractionation of the tectal cell system has yet to be demonstrated convincingly.

One of the problems that has to be solved in relation to predatory behaviour is the apparent suppression of normal conspecific shoaling behaviour when cannibalism occurs. Presumably the size factor dominates others in controlling responses.

Responses to predators

The major fish predator of perch in many waters is the pike (*Esox lucius*). Perch predation by pike is especially well documented by researchers studying them in Lake Windermere (Kipling & Frost, 1970), but is known to be of general occurrence (Scott & Crossman, 1973). Pike range upwards in weight to 20 kg or more and are markedly streamlined. Perch seldom attain weights above 2 kg and show a much lesser degree of streamlining. Since size and body profile are directly related to swimming speed (Bainbridge, 1958) it seems likely that pike are far swifter than perch, and the small amount of data on swimming speeds supports this contention (Bainbridge, 1958; Beamish, 1970). From this it follows that an ability on the part of the perch to detect at a distance during the initial phase of the latter's approach – what is termed 'slow motion advance' by Hoogland *et al.* (1956) – must be of considerable survival value. During this phase the pike moves itself gradually towards its prey by small fin movements alone, the body being kept straight. Perch tectal cells responding to small movements anywhere within a large part of the eye field were often encountered by us (type 3). Even though they adapt rather rapidly they may perform a vital role in attracting the attention of the perch to the pike (as indeed they would to any novel moving object). The contour of the pike seen in lateral view is very characteristic, but much less so seen approaching in frontal view, while the body has a pattern of small pale spots on a grey or green ground which appears cryptic in underwater photographs. The fins appear to the human observer to have a dull brownish hue. It seems likely that pike would be difficult to identify on the basis of colour contrast. On the other hand cells with multiannulate RFs in the perch tectum may contribute to the discrimination of the leading edge contour of a pike as demonstrated above by computer simulation. Frogs and toads (Grüsser & Grüsser-Cornehls, 1973; Ewert, 1976) show powerful predator avoidance responses to objects subtending angles over about 12° and 20° respectively. In frogs at least, a single type of retinal fibres (class 4) with 'on' centres of the critical diameter appears to mediate the response. Perch may also depend on a similar size threshold mechanism, yet we have found few cells with sharply defined 'on' centres above 10°. In toads the discrimination of large objects which trigger avoidance depends on the activation of the groups of retinal fibres, since in these animals Ewert found that the strongest avoidance responses were evoked by discs 47° in diameter, and the same could be true of perch.

Conclusion

The major discriminatory functions of vision in perch must be related to: (i) conspecific interactions, (ii) route identification, (iii) predatory behaviour and (iv) responses to predators. Conspecific interactions involve both recognition features and the signalling of activity state and intention. Recognition may entail discrimination of contour, spectral properties and relative areal dimensions. Route identification could rely largely on the relative position of objects, although clearly more precise, discriminatory functions may be necessary. Visual memory of the route produces strong reactions to new elements introduced into it. Preferred prey sizes have been demonstrated in perch, and estimation of prey size, velocity and position may be basic tectal functions.

We are some way from being able to produce a comprehensive theory of tectal cell action to explain these aspects of visual discrimination and it is possible only to make a number of tentative suggestions, as follows:

1. Discrimination of the finest detail (and therefore of smallest angular differences) is likely to depend on the small-field 'through' system of one to two million periventricular cells. These we believe have small-patch-type fields. It is attractive to see these as fine-contour discriminators, but this is unlikely from what is known about the level of shape discrimination in fish (Northmore, Volkmann & Yagar, 1978). More probably this system is used primarily in accurate estimation of prey size, position and velocity.

2. Centre eye field cells certainly respond very strongly to vertical bars, like those on the perch moving horizontally, but they may have many other functions besides that of 'perch detectors'. Their strong directional polariza-

tion may be involved in, for instance, following or eye-centring responses.

3. The possibility that cells with large multiannulate fields may have a shape discrimination function has been explored. They also make quite good candidates for 'background constancy' cells which stabilize eye movements against the effects of the displacement of backgrounds of small contrasting patches.

4. Multicentre fields could be involved in route recognition by estimation of the relative position of objects, but otherwise their probable function remains rather obscure.

5. As far as chromatic responses are concerned, the visible spectrum for perch tectal cells appears to be far wider than would be supposed from a study of cone pigments, and extends into the long ultraviolet. Strongly opponent RFs appear to be somewhat rarer than in the goldfish – less than 40% as compared with 68% (goldfish: Daw, 1968) – but this may be an artifact of our scanning method. Our studies suggest that differential spectral responses persist at relatively low light levels.

6. Study of visual processes nearer the motor control side of the system (nucleus reticularis mesencephali) indicate strongly binocular function, with a combination of visual and non-visual inputs mediating motor commands.

I would like to thank various members of Manchester University for their contributions to the research described here, most especially John Banks for his overall collaboration, in particular in the neurophysiological studies, and also Dr Derek Robinson for the computer simulations, Richard Noble for the study on the tegmentum and Steven Barrow for observations on perch behaviour. The North West Water Authority were kind enough to help us to obtain material, Dr David Moore gave invaluable aid with the pigment studies and I am grateful to Ali Al-Akell for an opportunity to quote from his studies on the anatomy of the brain.

References

Akert, K. (1949). Der visueller Greifreflex. *Helvetica Physiologica et Pharmacologica Acta,* **7**, 112–34.

Alabaster, J. S. & Robertson, K. G. (1961). The effect of diurnal changes in temperature, dissolved oxygen and illumination on the behaviour of roach, bream and perch. *Animal Behaviour,* **9**, 3–4.

Bainbridge, R. (1958). The speed of swimming of fish as related to size. *Journal of Experimental Biology,* **35**, 109–33.

Beamish, F. W. H. (1970). Swimming capacity. In *Fish Physiology*, ed. W. S. Hoar & D. J. Randall, pp. 101–72. Academic Press, New York & London.

Beauchamp, R. D. & Lovasik, J. V. (1973). Blue mechanism response of single goldfish optic fibres. *Journal of Neurophysiology,* **36**, 925–39.

Breder, C. M. Jr (1951). Studies on the structure of fish schools. *Bulletin of the American Museum of Natural History*, **98**, 1–27.

Cameron, N. D. (1974). 'Chromatic vision in a teleost fish: *Perca fluriatilis.*' PhD thesis, Sussex University.

Colette, B., Ali, M. A., Hokanson, K. E. F., Nagiee, M., Smirnov, S. A., Thorpe, J. E., Weatherly, A. H. & Willensen, J. (1977). Biology of the Percids. *Journal of the Fisheries Research Board of Canada,* **34**, 1890–9.

Cott, H. B. (1940). *Adaptive Colouration in Animals.* Methuen, London.

Daw, N. W. (1968). Colour coded ganglion cells in the goldfish retina. *Journal of Physiology,* **197**, 567–92.

Daw, N. W. (1973). Neurophysiology of colour vision. *Physiological Reviews* **53**, 571–611.

Diamond, J. (1972). The Mauthner cell. In *The Physiology of Fishes,* ed. W. S. Hoar & D. J. Randall, pp. 265–347. Academic Press, New York & London.

Eibl-Eibesfeldt, I. (1975). *Ethology: The Study of Behaviour,* 2nd edn. Holt, Rinehart & Winston, New York.

Endler, J. A. (1977). A predator's view of animal colour patterns. *Evolutionary Biology,* **11**, 319–64.

Engström, K. (1963). Cone types and cone arrangements in the retina of some cyprinids. *Acta Zoologica,* **41**, 277–95.

Ewert, J. P. (1976). The visual system of the toad. In *The Amphibian Visual System,* ed. K. Fite, pp. 141–202. Academic Press, New York & London.

Fabricius, E. (1956). Hur aborren leker. *Zoologisk Revy* **18**, 48–55.

Glickman, R. D. & Pomeranz, B. (1977). Frog retinal ganglion cells show species differences in their optimal stimulus sizes. *Nature, London,* **265**, 51–3.

Grüsser, O. J. & Grüsser-Cornehls, U. (1973). Neuronal mechanisms of visual movement perception. In *The Handbook of Sensory Physiology,* vol. 7, part 3, pp. 333–430. Springer-Verlag, Berlin.

Guthrie, D. M. & Banks, J. R. (1974). Input characteristics of the intrinsic cells of the optic tectum of teleost fish. *Comparative Biochemistry and Physiology,* **41**, 83–92.

Guthrie, D. M. & Banks, J. R. (1976). Patterned responses from widefield T2 neurones in the fish tectum. *Brain Research,* **104**, 321–4.

Guthrie, D. M. & Banks, J. R. (1978). The receptive field structure of visual cells from the optic tectum of the freshwater perch. *Brain Research,* **141**, 211–25.

Harden-Jones, F. R. (1963). The reaction of fish to moving backgrounds. *Journal of Experimental Biology,* **40**, 437–46.

Hermann, H. T. (1971). Eye movement correlated units in the mesencephalic motor complex of goldfish. *Brain Research,* **35**, 240–4.

Hester, F. J. (1968). Visual contrast thresholds of the goldfish. *Vision Research,* **8**, 1315–36.

Hoogland, R., Morris, D. & Tinbergen, N. (1956). The spines of sticklebacks as a defence against perch and pike. *Behaviour,* **10**, 205–36.

Hubel, D. E. & Wiesel, T. N. (1962). Receptive fields, binocular interaction and functional architecture in the cat's visual cortex. *Journal of Physiology,* **160**, 136–54.

Ito, H. & Kishida, R. (1978). Afferent and efferent fibre connections of

the carp torus longitudinalis. *Journal of Comparative Neurology,* **181**, 465–76.
Keenleyside, M. H. A. (1955). Some aspects of schooling in fish. *Behaviour,* **8**, 183–248.
Kien, J. & Menzel, R. (1977). Chromatic properties of interneurons in the optic lobes of the bee. *Journal of Comparative Physiology,* **113**, 17–34.
Kipling, C. & Frost, W. E. (1970). A study of the mortality, population numbers, year class strengths and food consumption of pike in Windermere, from 1944 to 1962. *Journal of Animal Ecology,* **39**, 115–55.
Leghissa, S. (1955). La struttura microscopia e la citoarchitettonica del tetto ottico dei pesci teleostei. *Zeitschrift für Anatomie und Entwicklungsgeschichte,* **118**, 427–63.
Loew, E. R. & Lythgoe, J. N. (1978). The ecology of cone pigments in teleost fishes. *Vision Research,* **18**, 715–22.
Lythgoe, J. (1974). Problems of seeing colours underwater. In *Vision in Fishes*, ed. M. Ali, pp. 619–34. Nato Advanced Study Institute Series. Plenum, New York.
Maitland, P. S. (1977). *The Freshwater Fishes of Britain and Europe.* Hamlyn, London.
Martin, J. (1977). 'The structure and function of lamprey reticular neurones'. PhD thesis, University of Liverpool.
Meek, J. & Schellart, N. A. M. (1978). A Golgi study of goldfish optic tectum. *Journal of Comparative Neurology,* **182**, 89–122.
Munk, O. (1971). On the occurrence of two lens muscles within each eye of some teleosts. *Videnskabelige Meddelelser fra Dansk naturhistorik Forening i københavn,* **134**, 7–19.
Muntz, W. R. A. (1974). The visual consequences of yellow filtering pigments in the eyes of fishes occupying different habitats. In *Light as an Ecological Factor: II,* ed. S. C. Evans, R. Bainbridge & O. Rackham, pp. 271–88, Blackwell, Oxford.
Northmore, D., Volkmann, F. C. & Yagar, D. (1978). Vision in fishes: colour and pattern. In *The Behaviour of Fish and other Aquatic Animals,* ed. D. I. Nostofsky, pp. 79–136. Academic Press, New York & London.
Nursall, J. R. (1973). Some behavioural interactions of spot-tail shiners, yellow perch, and northern pike. *Journal of the Fisheries Research Board of Canada,* **30**, 1161–77.
Popova, O. A. & Sytina, L. A. (1977). Food and feeding relations of Eurasian perch. *Journal of the Fisheries Research Board of Canada,* **34**, 1559–70.
Restieaux, N. J. & Satchell, G. H. (1958). A unitary study of the reticulomotor system of the dogfish *Squalus. Journal of Comparative Neurology,* **109**, 391–413.
Romeskie, M. & Sharma, S. C. (1979). The goldfish optic tectum: a Golgi study. *Neuroscience,* **4**, 625–42.
Schellart, N. A. M. Riemslag, F. C. C. & Spekreijse, H. (1979). Centre–surround organization and interactions in receptive fields of goldfish tectal units. *Vision Research,* **19**, 459–67.
Schellart, N. A. M. & Spekreijse, H. (1976). Shapes of receptive field centres in the optic tectum of the goldfish. *Vision Research,* **16**, 1018–20.

Scholes, J. H. (1979). Nerve fibre topography in the retinal projection to the tectum. *Nature, London,* 620–4.
Schwassman, H. O. (1974). Refractive state, accommodation and resolving power of the fish eye. In *Vision in Fishes,* ed. M. A. Ali, pp. 279–88, Plenum, New York.
Schwassman, H. & Meyer, D. L. (1971). Refractive state and accommodation in the eye of three species of *Paralabrax. Videnskabelige Meddelelser fra Dansk naturhistorik Forening i københavn,* **134**, 103–8.
Scott, W. B. & Crossman, E. J. (1973). The freshwater fishes of Canada. *Bulletin of the Fisheries Research Board of Canada,* **184**, 1–966.
Sivak, J. G. (1973). Accommodation in some species of North American fishes. *Journal of the Fisheries Research Board of Canada,* **30**, 1141–6.
Spekreijse, J., Wagner, H. G. & Wolbarsht, M. L. (1972). Spectral and spatial coding of ganglion cell responses in goldfish retina. *Journal of Neurophysiology,* **35**, 73–86.
Stacey, P. B. & Chiszar, D. (1978). Body colour pattern and aggressive behaviour of male pumpkinseed sunfish *Lepomis. Behaviour,* **64**, 271–304.
Sutherland, N. S. (1968). Shape discrimination in the goldfish. In *The Central Nervous System and Fish Behaviour,* ed. D. Ingle, 35–50. University of Chicago Press, Chicago.
Tamura, T. (1957). A study of visual perception in fish, especially on resolving power and accommodation. *Bulletin of the Japanese Society of Scientific Fisheries,* **22**, 536–57.
Tamura, T. & Niwa, H. (1967). Spectral sensitivity and colour vision of fish as indicated by S-potential. *Comparative Biochemistry and Physiology,* **22**, 745–54.
Thorpe, J. E. (1977). Morphology, physiology, behaviour and ecology of *Perca fluviatilis and P. flavescens. Journal of the Fisheries Research Board of Canada,* **34**, 1504–14.
Vanegas, H. & Ebbesson, S. O. E. (1973). Retinal projections in the perch-like teleost *Eugerres plumieri. Journal of Comparative Neurology,* **151**, 331–58.
Vanegas, H., Laufer, M. & Amat, J. (1974). The optic tectum of a perciform teleost. I. *Journal of Comparative Neurology,* **154**, 45–60.
Vanegas, H., Williams, B. & Freeman, J. A. (1979). Responses to stimulation of marginal fibres in the teleostean optic tectum. *Experimental Brain Research,* **34**, 335–49.

HORACIO VANEGAS

The teleostean optic tectum: neuronal substrates for behavioural mechanisms

The patterning of behavioural acts obviously depends on the manner in which neurons in specific regions of the brain are interconnected, on the manner in which they respond to the input from other regions (e.g. sensory organs), and on the manner in which the output of such neurons ultimately influences other neuronal pools. This chapter summarizes some aspects of the local circuitry, the input and the output of the optic tectum in bony fishes.

General morphology

The optic tectum constitutes a paired dome in the dorsal aspect of the mesencephalon, and is organized in concentric layers, namely (from surface to depth, Fig. 1*A*): stratum marginale (SM), stratum opticum (SO), stratum fibrosum et griseum superficiale (SFGS), stratum griseum centrale (SGC), stratum album centrale (SAC) and stratum periventriculare (SPV). Numerous neurons in the tectum are elongated vertically (radially) within a given layer, or, more commonly, across several layers, while other cells stretch their processes mainly horizontally. Other cells yet are multipolar in type (Vanegas, Laufer & Amat, 1974; Schroeder & Vanegas, 1977; Schroeder, Vanegas & Ebbesson, 1979).

The so-called pyramidal neuron (Fig. 1*Ad*) is typical of the teleostean tectum. From its soma, located in the SFGS, stems an apical dendritic shaft which gives rise to an ample, spiny dendritic tree across the SM. The spines make Gray type I synapses with the en-passant boutons of the marginal fibres (Fig. 1*A*; Laufer & Vanegas, 1974*a*). These are unmyelinated axons, 0.1 μm in diameter, which originate at the torus longitudinalis (Fig. 3, TL; Ito & Kishida, 1978) and run mediolaterally along the SM roughly parallel to one another (Figs. 1*A* and 2). The SM also contains the terminal arborizations of the so-called small pyriform neuron of SGC (Fig. 1*Ac*). It is not clear whether these arborizations are axon terminal branches or, rather, presynaptic dendrites; light-microscopically they are seen making contact with the pyramidal cell dendrites (Schroeder *et al*., 1979). The pyramidal cell also

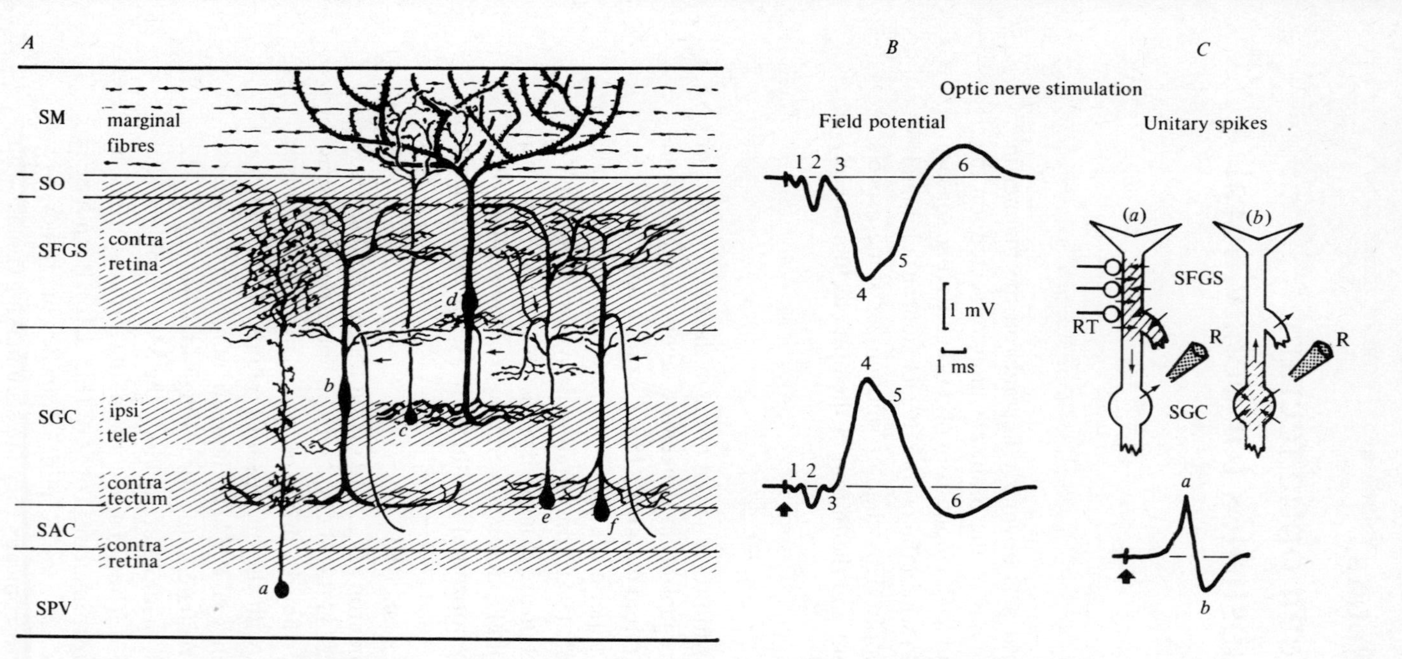

Fig. 1. *A*, Types of vertically (radially) oriented neurons in the optic tectum, and distribution of afferents from torus longitudinalis (marginal fibres) as well as ipsilateral (ipsi) or contralateral (contra) retina, telencephalon (tele) and tectum. Arrows indicate axons. The cells depicted here represent only *types*, and their exact shape varies among different teleost groups. SM, stratum marginale; SO, stratum opticum; SFGS, stratum fibrosum et griseum superficiale; SGC, stratum griseum centrale; SAC, stratum album centrale; SPV, stratum periventriculare. *B*, Field potential (drawn by hand) evoked in optic tectum by single shocks to the optic nerve (arrow). Numbers identity the various waves. Positive is upwards. Upper tracing corresponds to SFGS or more superficial layers; lower tracing corresponds to SGC or deeper layers. *C*, Diagram of a fusiform or a large pyriform neuron, Unitary extracellular recording (below, drawn by hand) is being made by a micropipette (R). In (*a*), retinotectal terminals (RT) depolarize the apical dendrite and fire the axon's initial segment (hatching). In (*b*), there is repolarization of the previously activated membrane segments, and (perhaps) depolarization of the soma (hatching). The positive (upwards) and negative phases of the recorded spike correspond to diagrams (*a*) and (*b*) respectively. Arrow indicates stimulus artifact. Calibration as in *B*.

has a basal dendritic shaft and a descending axon, both of which ramify profusely at mid-SGC.

The fusiform neuron (Fig. 1*Ab*) stretches across the SFGS and SGC. Its axon stems from the apical dendritic shaft, takes a descending course, becomes myelinated, and leaves the tectum via the SAC. The so-called large pyriform neuron (Fig. 1*Ae* and *f*) has a soma located at deep SGC, SAC or SPV, and an ascending process which gives rise to several horizontal or oblique branches as well as to a shepherd's crook axon. This axon may become myelinated and leave the tectum ('efferent' pyriforms) via the SAC, or may divide into an ascending and a descending branch, which ramify at SFGS and SGC respectively ('local' pyriforms).

Typical periventricular neurons (Fig. 1*Aa*) lack axons and have round, small somata. A slender process ascends towards the SO and gives, particularly (but not exclusively) at the SFGS, a wealth of tortuous branches laden with boutons (Schroeder *et al*., 1979). These are probably presynaptic dendrites.

The afferents from the contralateral retina (Fig. 1*A*) terminate profusely at the SO and the SFGS, and quite sparsely at the SAC (Vanegas & Ebbesson, 1973; Laufer & Vanegas, 1974*a, b*). Retinotectal terminals are large and irregular in shape, possess round vesicles and light, swollen mitochondria and make asymmetric synapses with numerous elements, including the pyramidal neuron and the fusiform neuron at the SFGS.

The afferents from the ipsilateral telencephalon (Fig. 1*A*) terminate at the SO and, particularly, as a band which overlaps at mid-SGC with the arborizations of the pyramidal neuron's axon and basal dendritic shaft (Vanegas & Ebbesson, 1976).

The tectal input

Responses to optic nerve stimulation

When the optic nerve is stimulated with single electrical shocks, three negative deflections (waves 1, 2 and 3), corresponding to fibre groups with conduction velocities of 20, 10, and 5 m s^{-1} are recorded at the optic tract and optic tectum (Fig. 1*B*; Vanegas, Essayag-Millán & Laufer, 1971). The optic nerve in teleosts is made of 100 000–200 000 axons, 96% of which are myelinated (Tapp, 1974). The 20 m s^{-1} fibre group is made of axons which originate in the tectum and normally conduct towards the eye (see below). The 10 and 5 m s^{-1} groups must thus be the retinotectal afferents.

When retinotectal afferents are stimulated by a single shock (Vanegas *et al*., 1971; Vanegas, Amat & Essayag-Millán, 1974), two successive volleys arrive at the tectum and give rise to two successive postsynaptic field potential deflections (waves 4 and 5, Fig. 1*B*). These waves are negative at the

SFGS and represent the depolarization (active inward transmembrane current) of vertically elongated neurons. Deeper than SFGS, waves 4 and 5 are positive, and represent the passive outward current from inactive segments of such neurons. The depolarization of vertically elongated neurons at the SFGS triggers in these neurons an action potential which also arises at the SFGS. This means either that the vertical dendritic elements are capable of generating action potentials, or that the vertical neurons which generate waves 4 and 5 possess an axon whose initial segment is located within SFGS. Such is indeed the case for the pyramidal neuron, the fusiform neuron and the large pyriform neuron (Fig. 1*A*). The last two have their somata at the SGC or deeper strata. Recordings from neuronal segments (probably somata) located at such strata, and obtained by means of micropipettes which record extracellularly but very close to the membrane, show positive–negative unitary spikes monosynaptically evoked by optic nerve stimulation (Fig. 1*C*). The shape of such spikes can be interpreted in the following manner (Fig. 1*C*). Depolarization of the apical dendrites by optic nerve terminals fires the axon's initial segment; all of this occurs at the SFGS. The inward current flow at this initial segment causes a passive outward flow in front of the micropipette's tip, located near the soma, thus giving rise to the initial, fast, positive phase of the unitary spike (Fig. 1*Ca*). Subsequent depolarization of the axon reverses the current flow, causing the negative, slower phase of the spike, conceivably (but by no means certainly) aided by propagation of the action potential to the soma or region located near the micropipette's tip (Fig. 1*Cb*).

In summary, the arrival of two successive afferent volleys (waves 2 and 3) causes two successive depolarizations (waves 4 and 5) of apical segments of vertically elongated neurons, with subsequent firing of their axon's initial segment, all of which takes place at the SFGS. The direct (monosynaptic) recipients of retinofugal impulses are the pyramidal neuron, the fusiform neuron and most likely also the large pyriform neuron. Other neurons might of course be monosynaptically excited by retinotectal afferents but, due to horizontal or multipolar distribution of their processes, do not contribute to the generation of waves 4 and 5.

Wave 6 of the field potential (Fig. 1*B*) could represent an outward transmembrane current caused either by the repolarization, at the SFGS, of the neuronal segments which generated waves 4 and 5, or by disynaptic inhibitory phenomena at such segments. Wave 6 might also result from excitatory phenomena elicited by slow, deep retinotectal axons synapsing at SGC or deeper strata upon vertically elongated neurons (Schmidt, 1977). All these mechanisms could of course contribute to wave 6.

Inhibition of neuronal firing after optic nerve stimulation (Vanegas, Amat

& Essayag-Millán, 1974) coincides with wave 6 and thus has a latency longer than that of the excitatory phenomena. This suggests that retinotectal synapses are purely excitatory.

Responses to stimulation of marginal fibres

Marginal fibres can be excited by electrical stimulation at the tectal surface or at the torus longitudinalis (Fig. 2; Vanegas, Williams & Freeman, 1979). Their compound action potential (spike M) can be extracellularly recorded 'on beam' at the SM (Fig. 2*a*) with a velocity of about 0.18 m s^{-1}. This is followed by field potential wave S, which is negative at the SM and positive at SFGS, and probably represents the depolarization of the pyramidal neuron's apical dendritic tree by the marginal fibre synapses (Fig. 2*b*). Wave S is followed by field potential wave N (Fig. 2*c*), which is negative at the depth (about SO) where the apical branches converge to form the spineless apical dendritic shaft of the pyramidal neuron. Wave N could represent an action potential at this convergence point. It is followed by a negative field potential wave at the SGC (wave D), sometimes of spike-like character, which probably represents the activation of the pyramidal neuron's axon (Fig. 2*d*). At the same time at the SM, a long-lasting, negative, field potential deflection (wave L) is recorded. This might be a remnant of the depolarization which gave rise to wave S (interrupted by wave N, which is positive at SM) or might be a new depolarization of the apical dendritic tree caused by, for example, the arborization of the small pyriform neuron of SGC (Fig. 1*Ac*). Only speculative proposals can presently be given regarding the nature of wave L.

The marginal fibres and apical dendritic tree of pyramidal neurons somewhat resemble the parallel fibres and their postsynaptic structures in the cerebellar cortex. Also, the M-spike and the S-wave are, respectively, reminiscent of the parallel fibre's action potential and the negative wave produced by the depolarization of the Purkinje cells' apical dendritic tree. Interestingly enough, the marginal fibres originate from the torus longitudinalis, whose main input arises from the valvula cerebelli (Ito & Kishida, 1978). Since the pyramidal neuron receives retinotectal terminals, the marginal fibre system provides a means whereby cerebellar influences meet the monosynaptic arrival of retinofugal information.

The tectal output

The ascending tectal projections (Fig. 3*a*; Ebbesson & Vanegas, 1976) terminate ipsilaterally upon the dorsolateral thalamic nucleus (Schnitzlein, 1962; dorsomedial optic nucleus of Ebbesson, 1972) and the nucleus rotundus (Schnitzlein, 1962) or prethalamicus (Meader, 1934). The

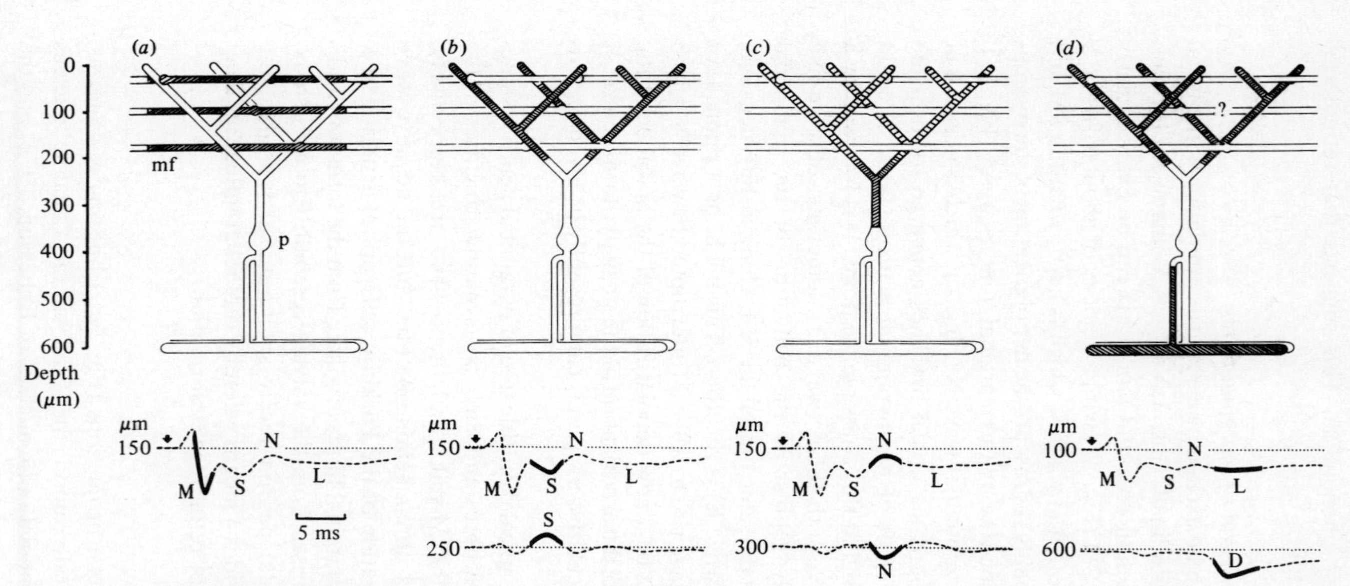

Fig. 2. Tentative diagram of phenomena which possibly underline the 'on-beam' field potential deflections elicited by single pulse stimulation of the tectal surface. Hatching represents the activation (inward transmembrane current), or sinks, at marginal fibres (mf) and pyramidal neuron (p). Current-source density profiles obtained at several depths are shown below in dash tracings (current sources are upwards, sinks downwards); in each case, thick segments correspond to the phenomenon suggested above. Arrows indicate stimulus shock. (Modified from Vanegas *et al*., 1979.)

Fig. 3. (*a*) Degenerative fibres (dashes) and terminals (dots) resulting from a lesion (L) at the optic tectum. The right side of the lower drawing shows the position of stimulating (St) and recording (R) electrodes that produced the recordings shown in (*c*). DL, nucleus dorsolateralis thalami; G, corpus glomerulosum; GLi, nucleus geniculatus lateralis ipsum; GPv, nucleus geniculatus posterior pars verntralis; H, habenula; I, nucleus isthmi; PD, predorsal bundle; PTh, nucleus prethalamicus; SFGS, stratum fibrosum et griseum superficiale; SO, stratum opticum; SPV, stratum periventriculare; TL, torus longitudinalis; TrO, tractus opticus; TS, torus semicircularis. (*b*) Hand drawings showing field potential (upper tracing) and unitary recording (lower tracing) in an experimental design diagrammed below, where arrows indicate *normal* direction of conduction. Cb, cerebellum; T, telencephalon; TeO, tectum opticum. When the optic nerve is stimulated at St, the axon of the neuron being recorded from by R conducts an antidromic action potential (at 20 m/s^{-1}) which invades the soma (AD) with a latency of 0.6 ms, i.e. similar to that of wave 1 of the field potential. The stimulus also triggers an action potential in retinotectal axons. This is conducted orthodromically (at 10 m/s^{-1}) and gives rise to a monosynaptic depolarization (Syn) of the neuron, coincident with wave 4 of the field potential. (*c*) Recordings (drawn by hand) from the nucleus isthmi as depicted in lower diagram of (*a*). Electrical stimulation of the tectum opticum (TeO) elicits a field potential (upper tracing) that resembles the similarly evoked spike bursts of isthmic neuron (middle tracing). Electrical stimulation of optic nerve (ON) elicits similar bursts in same unit, but with longer latency.

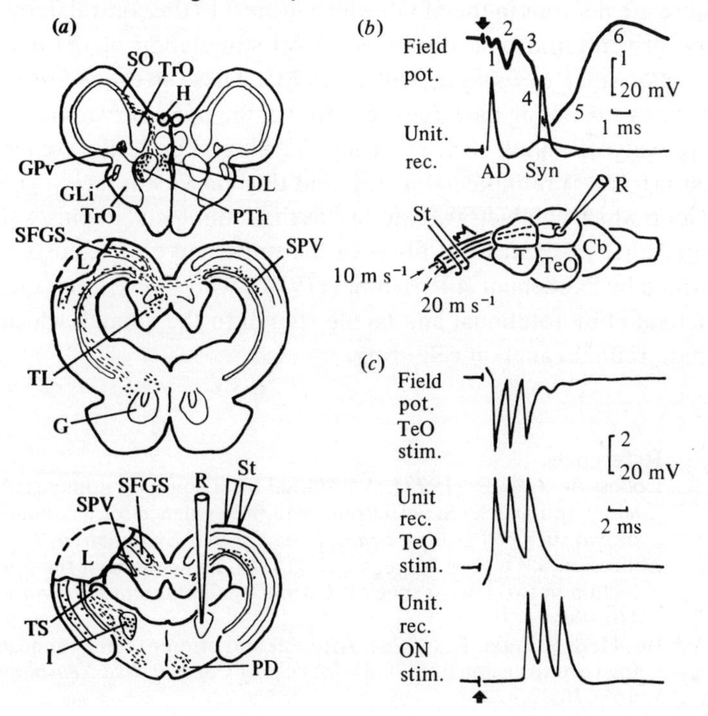

dorsolateral thalamic nucleus also receives contralateral retinofugal axons and ipsilateral telencephalofugal axons. Neurons of the nucleus rotundus or prethalamicus can be driven by electrical stimulation of the optic tectum (B. Williams & H. Vanegas, unpublished). This nucleus probably projects to the ipsilateral telencephalon.

The tectum projects medialwards to the torus longitudinalis and, homotopically, to the deep SGC of the contralateral tectum via the tectal commissure (Fig. 3*a*).

The descending tectal projection terminates ipsilaterally in dorsolateral cell groups of the mesencephalic tegmentum, in the nucleus isthmi and in the lateral reticular formation of the brainstem. The contralateral projection (predorsal bundle) terminates in the medial reticular formation of the brainstem (Fig. 3*a*). Neurons in the nucleus isthmi fire short latency bursts of spikes as a response to electrical stimulation of the ipsilateral optic tectum or – with about 3.5 ms longer latency – the contralateral optic nerve (Fig. 3*c*; B. Williams & H. Vanegas, unpublished). The field potential recorded in the nucleus isthmi shows large, negative deflections which resemble the unitary spike bursts. This suggests the existence of a mechanism which gives rise to synchronous, repetitive firing of nucleus isthmi neurons when the tecto–isthmic axons are activated.

There are neurons in the SFGS which project to the contralateral eye and can be driven antidromically by electrical stimulation of the optic nerve (Vanegas, Amat & Essayag-Millán, 1973). These neurons are large and their axons are among the largest-calibred of the optic nerve, with a conduction velocity of about $20\,\mathrm{m\,s^{-1}}$. They are monsynaptically excited by the fastest ($10\,\mathrm{m\,s^{-1}}$) retinotectal axons, and thus provide a tecto-ocular feedback loop which conducts twice as fast as the retinotectal impulses (Fig. 3*b*). The presence of centrifugal fibres in the optic nerve of teleosts has been confirmed by Sandeman & Rosenthal (1974), who showed that these fibres are activated by rotational and tactile stimuli to the animal, which in turn facilitate retinal ganglion cell firing.

References

Ebbesson, S. O. E. (1972). A proposal for a common nomenclature for some optic nuclei in vertebrates and the evidence for a common origin of two such cell groups. *Brain, Behavior and Evolution*, **6**, 75–91.

Ebbesson, S. O. E. & Vanegas, H. (1976). Projections of the optic tectum in two teleost species. *Journal of Comparative Neurology*, **165**, 161–80.

Ito, H. & Kishida, R. (1978). Afferent and efferent fiber connections of the carp torus longitudinalis. *Journal of Comparative Neurology*, **181**, 465–76.

Laufer, M. & Vanegas, H. (1974*a*). The optic tectum of a perciform teleost. II. Fine structure. *Journal of Comparative Neurology*, **154**, 61–96.
Laufer, M. & Vanegas, H. (1974*b*). The optic tectum of a perciform teleost. III. Electron-microscopy of degenerating retino-tectal terminals. *Journal of Comparative Neurology*, **154**, 97–116.
Meader, R. G. (1934). The optic system of the teleost, *Holocentrus*. I. The primary optic pathways and the corpus geniculatum complex. *Journal of Comparative Neurology*, **60**, 361–407.
Sandeman, D. C. & Rosenthal, N. P. (1974). Efferent axons in fish optic nerve and their effect on retinal ganglion cells. *Brain Research*, **68**, 41–54.
Schmidt, J. T. (1977). The synaptic organization of optic afferents to the normal goldfish tectum. *Society for Neuroscience Abstracts*, **3**, 94.
Schnitzlein, H. N. (1962). The habenula and the dorsal thalamus of some teleosts. *Journal of Comparative Neurology*, **118**, 225–67.
Schroeder, D. M. & Vanegas, H. (1977). Cytoarchitecture of the tectum mesencephali in two types of siluroid teleosts. *Journal of Comparative Neurology*, **175**, 287–300.
Schroeder, D. M., Vanegas, H. & Ebbesson, S. O. E. (1979). The cytoarchitecture of the optic tectum of the squirrel-fish, *Holocentrus*. *Journal of Comparative Neurology*, in press.
Tapp, R. (1974). Axon numbers and distribution, myelin thickness, and the reconstruction of the compound action potential in the optic nerve of the teleost: *Eugerres plumieri*. *Journal of Comparative Neurology*, **153**, 267–74.
Vanegas, H., Amat, J. & Essayag-Millán, E. (1973). Electrophysiological evidence of tectal efferents to the fish eye. *Brain Research*, **54**, 309–13.
Vanegas, H., Amat, J. & Essayag-Millán, E. (1974). Postsynaptic phenomena in optic tectum neurons following optic nerve stimulation in fish. *Brain Research*, **77**, 25–38.
Vanegas, H. & Ebbesson, S. O. E. (1973). Retinal projections in the perch-like teleost *Eugerres plumieri*. *Journal of Comparative Neurology*, **151**, 331–58.
Vanegas, H. & Ebbesson, S. O. E. (1976). Telencephalic projections in two teleost species. *Journal of Comparative Neurology*, **165**, 181–96.
Vanegas, H., Essayag-Millán, E. & Laufer, M. (1971). Responses of the optic tectum to stimulation of the optic nerve in the teleost *Eugerres plumieri*. *Brain Research*, **31**, 107–18.
Vanegas, H., Laufer, M. & Amat, J. (1974). The optic tectum of a perciform teleost. I. General configuration and cytoarchitecture. *Journal of Comparative Neurology*, **154**, 43–60.
Vanegas, H., Williams, B. & Freeman, J. A. (1979). Responses to stimulation of marginal fibers in the teleostean optic tectum. *Experimental Brain Research*, **34**, 335–49.

TOSHIAKI J. HARA

Behavioural and electrophysiological studies of chemosensory reactions in fish

Fish detect various chemical stimuli in the aquatic environment through two principal channels, olfaction and taste, and respond by changing their behavioural patterns. These may be characterized as either avoidance, preference, or changes in activity level, depending upon the stimuli perceived. However, it is not well understood at present which chemicals are detected through which chemosensory system and what types of behaviour patterns they cause. Although recent electrophysiological studies have revealed that certain chemicals are extremely effective olfactory stimuli and may play an important role in olfactory communication in fishes, there has been little investigation of their behavioural correlates (Hara, 1975). It has long been known that fishes avoid certain chemicals at one concentration and prefer the same chemicals at others (Jones, 1947; Sprague & Drury, 1969). The purpose of this paper is to present new evidence concerning the bimodal behaviour reactions of fish to a chemical, copper sulphate, and to examine electrophysiologically the role of olfaction in the behaviour and its underlying brain mechanisms.

Behavioural reactions of fish to chemicals

Avoidance and preference reactions of whitefish to copper sulphate

The behavioural reactions of lake whitefish (*Coregonus clupeaformis*), 12–17 cm in body length, to copper sulphate ($CuSO_4$) were tested in an avoidance/preference trough (plexiglass, 17 × 120 × 17 cm). This trough is similar to those originally described by Jones (1947) and modified later by Sprague (1964) and Scherer & Nowak (1973). Water flows into each end and out of the centre of the trough. A test solution can be introduced at one end or the other (Fig. 1*a*). With an appropriate flow rate, a distinct separation between the two bodies of water at the centre can be achieved. Following initial acclimatization for 5 min, fish generally swam back and forth across the boundary, with only occasional stops. The movement of fish from end to end in the trough under normal conditions is a prerequisite of this type of test. Avoidance and preference reactions of fish were best measured by comparing the amount of time that a fish spent in clean versus treated water. In the following experiments, therefore, only the accumulated times spent by fish in each end during the control and test period were measured.

The test procedure adopted was as follows: (1) with clean water in both halves of the trough, the time which a fish spent in one side was calculated as a percentage of the total time of 5 min (control recording), and (2) after 5 min with a test chemical in one or the other side, the percentage of time spent by the fish in the clean side was calculated as in (1). A time-response of 50% was considered to be a neutral response, one of over 50% to be avoidance, and one of less than 50% to be preference. Control recordings with clean water in both halves indicated no bias; the mean value of time-responses for all tests was close to 50% (50.8 ± 9.3; $n = 95$). Test solutions were introduced into either side at random. Because the first experience of the chemical by fish affected their subsequent responses, each fish was tested only once at one concentration.

The reactions of fish to chemicals in the trough were characterized by

Fig. 1. (*a*) Schematic illustration of an avoidance/preference trough. Water flows into each end and out of the centre of the trough; a test solution is introduced at one end or the other. (*b*) An example of preference reaction of whitefish to 5×10^{-4}M copper sulphate. (*c*) An example of avoidance reaction to 5×10^{-6}M copper sulphate. In both recordings, chemical solutions were introduced at the left end (arrows).

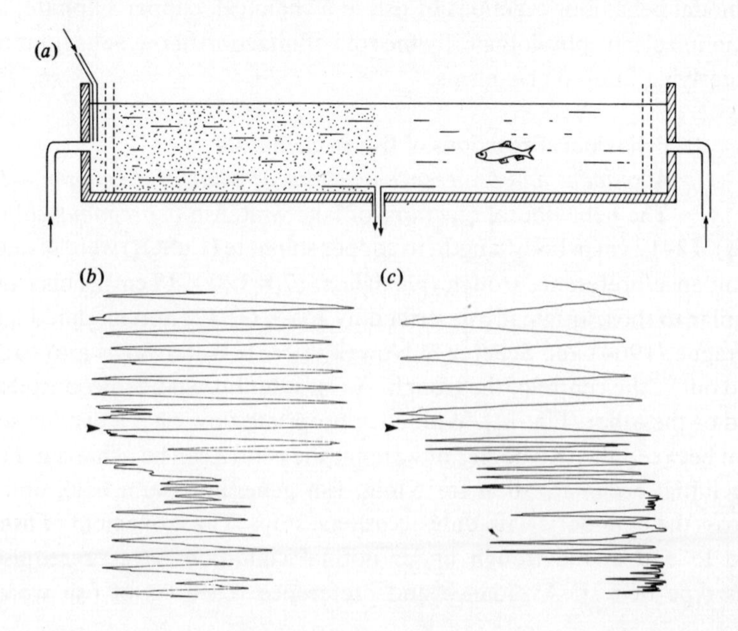

sudden stops accompanied by a sharp turnabout movement in the case of repellents, and by a nosing and exploring behaviour followed by a stationary posture for attractants. Also, an increase in the general swimming activity was often observed for repellents. Fig. 1 (*b*) and (*c*) illustrate tracings of typical preference and avoidance reactions to copper sulphate. Results of all tests with copper sulphate are shown in Fig. 2, with the mean responses joined by straight lines. At the lowest concentration tested all except three fish showed avoidance; mean time-response was 59.2 ± 10.0. The lowest concentration of copper sulphate to induce an appreciable avoidance reaction was estimated at approximately 10^{-7} M. With an increase in the concentration the degree of avoidance increased, with the maximum avoidance at 5×10^{-6} and 1×10^{-5} M (mean time-responses 78.9 ± 4.1 and 78.6 ± 6.4, respectively). When the copper sulphate concentration was increased further, the time-response decreased and fell to below the 50% level, indicating preference. This reversal in the reaction took place between

Fig. 2. Avoidance and preference reactions of whitefish to copper sulphate solutions. The lines connect mean values at each concentration. Open circles represent the control response of fish with clean water in both halves of the trough.

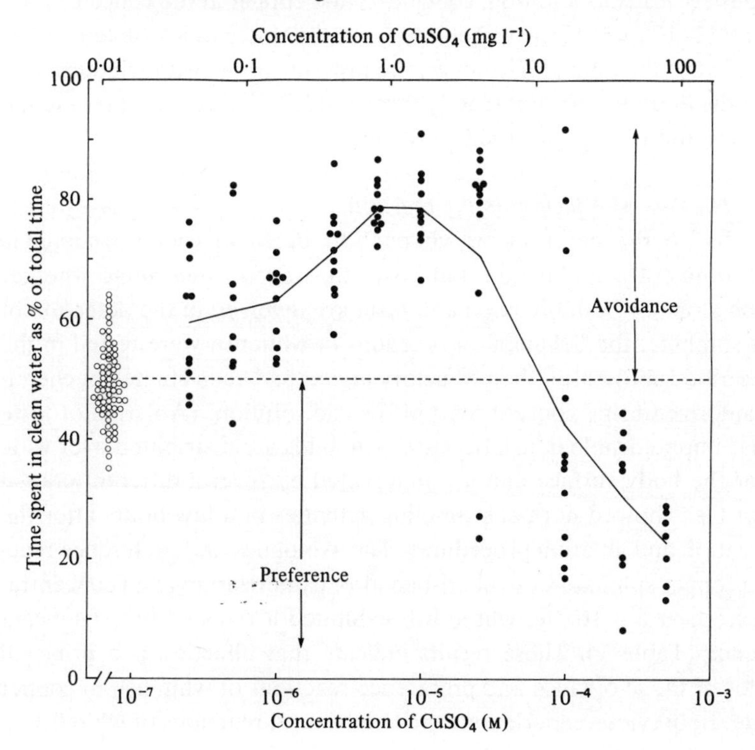

2.5×10^{-5} and 1×10^{-4} M copper sulphate. At these 'critical' concentrations some fish showed either strong preference or avoidance, and others responded by initial avoidance followed by preference. The opposing reactions thus offsetting each other resulted in a neutral response. When the copper sulphate concentration was higher than 10^{-4} M, soon after fish entered the chemical side they swam back and forth only within the one side. Time and time again fish reached, or crossed, the boundary into clean water and quickly turned back into the chemical side (see Fig. 1*b*). There were some indications of a stupefying effect by copper sulphate during the 5 min test period. A typical bimodal behavioural reaction of fish to a chemical is clearly shown in these experiments.

The above observations with whitefish seem to be at variance with other studies of behavioural responses of fish to the same chemical (Jones, 1947; Kleerekoper *et al.*, 1972). Jones (1947) reported that ten-spined sticklebacks, *Pygosteus pungitius*, were attracted to copper sulphate solutions and became stupefied and motionless at concentrations of 5×10^{-4} and 5×10^{-3} M. He concluded that the preference for copper is not due to the fish deliberately selecting the solution, but is because they cannot perceive the substance, which therefore acts as a trap. Kleerekoper *et al.* (1972), working on goldfish (*Carassius auratus*), concluded that copper at the concentration range of 11–17 $\mu g\,l^{-1}$ was either attractive to this species or led to the entrapment of the animal. However, the present results with whitefish and those in the literature are not readily comparable, because the experimental techniques and fish species used were very different.

The role of olfaction in the reactions

In fish, the major sensory channels for detecting chemical stimuli in the environment are olfaction and taste. In order to determine whether olfaction alone or both olfaction and taste are involved in the detection of copper sulphate, the behavioural reactions in whitefish were tested in the way described above after their olfactory organs had been ablated or chemically cauterized using concentrated nitric acid solution. (Ablation of taste organs is impracticable in fish because taste buds are distributed over wide areas of the body surface and are innervated by several different cranial nerves.) Fish showed normal swimming activities in a few hours after the cauterization and ablation procedures. The avoidance and preference reactions to copper sulphate were abolished after cauterization at all concentrations except at 5×10^{-4} M, where fish exhibited a reduced but significant preference (Table 1). These results indicate that olfaction is a principal mediator of the avoidance and preference reactions of whitefish to copper sulphate. In previous experiments the behavioural reactions of whitefish to an amino acid, glycine, were also found to be mediated by olfaction (Hara, 1977*b*). Preference observed at 5×10^{-4} M copper sulphate may be due either to incomplete cauterization of the olfactory organs (i.e. increased threshold) or to the functions of other chemical senses, taste and lateral-line organs. A potentiating effect of 10^{-3} M copper sulphate on the taste responses has been reported in Atlantic salmon (*Salmo salar*) parr (Sutterlin & Sutterlin, 1970). The lateral-line organs of teleosts are also known to be stimulated by various mono- and divalent metallic ions (Katsuki, Hashimoto & Kendall, 1971).

Olfactory responses to copper sulphate

Olfactory system of whitefish

It is evident from the above observations that olfaction is the major sensory channel mediating the behavioural reactions of whitefish to copper sulphate. It is therefore important to examine their olfactory system and its neural functions when the receptors are exposed to known odorous chemicals and copper sulphate.

In whitefish, like many other fish species, the paired nares are located in front of the eye close to the anterior wall of the cranial cavity (Fig. 3). Lining the nasal cavities is the olfactory epithelium, which is generally raised from the floor of the organ into a complicated series of folds, or lamellae, to form a rosette. The size and number of the olfactory lamellae vary greatly between different species, and between different-size fish of the same species. The olfactory epithelial system consists of receptor cells, supporting cells and basal cells. The receptor cell, which is a bipolar primary neuron, sends a slender cylindrical process or dendrite towards the surface of the epithelium,

Table 1. *Effects of ablation or cauterization of the olfactory organs of whitefish on their behavioural reactions to copper sulphate*

	Time-response (%)[a]		
		Copper sulphate	
	Control	5×10^{-6} M	5×10^{-4} M
Normal fish	50.8±9.3 (n=19)	78.9±4.1 (n=11)	23.2±4.8 (n=8)
Operated fish	50.4±8.3 (n=19)	51.1±9.7 (n=9)	34.9±11.4 (n=10)

[a] Time-response is expressed as time spent by fish in clean water as percentage of total test time. 50% = neutral; <50% = preference; >50% = avoidance.

where it is in direct contact with the external environment. The fine axons arise from the basal pole of the receptor cells. They pass through the basement membrane, become grouped in the submucosa, and form the olfactory nerve, which runs posteriorly to end in the olfactory bulb. The olfactory nerve fibres terminate in the olfactory bulb and make a special synaptic contact with the bulbar secondary neurons in the glomeruli. Information filtered in the first relay station, the olfactory bulb, is then transferred through the olfactory tract to higher nervous centres (for details see Hara, 1971, 1975).

Olfactory bulbar electrical responses to chemical stimuli

When recording electrodes are placed on the surface of the olfactory bulb spontaneous electrical activities can be recorded (intrinsic waves). Infusion of odorous chemicals into the nares induces large rhythmic oscillations (induced waves) which terminate on cessation of the stimulus. Although the nature of the induced potentials is not yet fully understood, they have been widely adopted as one of the most sensitive monitors of olfactory response of the brain in vertebrates (Hara, 1975). In a series of experiments in the frog (*Rana temporaria*) Ottoson (1956) analysed the induced slow potential and rhythmic waves in the olfactory bulb and presented strong evidence that the latter are the result of postsynaptic discharge of secondary olfactory neurons in response to excitation of receptors in the olfactory mucosa. Furthermore, recent biochemical studies have shown that a close correlation exists between binding of stimulus molecules

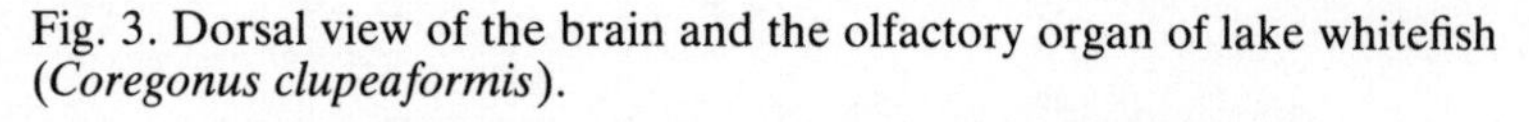

Fig. 3. Dorsal view of the brain and the olfactory organ of lake whitefish (*Coregonus clupeaformis*).

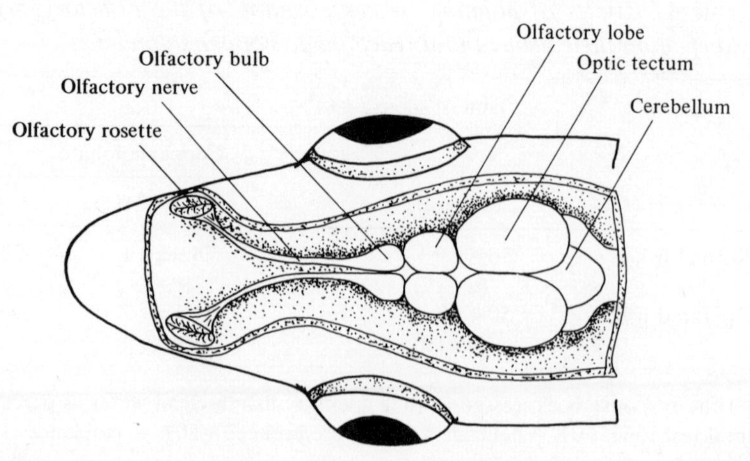

with olfactory receptor preparations and the bulbar induced responses in trout (Cagan & Zeiger, 1978; Brown & Hara, unpublished).

Fig. 4 illustrates a record of olfactory bulbar responses of whitefish induced by L-serine, a typical stimulatory amino acid, and its concentration–response relationships. Quantitative features of the bulbar response vary greatly depending upon the effectiveness of stimulants (Hara, 1973, 1976*a*, 1977*a*). Response spectra of whitefish to representative amino acids are similar to those of other salmonid species (Hara, Law & Hobden, 1973). It is not known whether amino acids causing a bulbar response are attractive or repellent to the intact fish. However, glycine acts as an attractant to whitefish over the concentration of 10^{-7}–10^{-4} M (Hara, 1977*b*). Also, preliminary behavioural study indicates that L-serine acts as a stimulant or attractant to this species (Hara, unpublished).

Olfactory bulbar response to copper sulphate

Fig. 5 illustrates a series of records of the bulbar responses when copper sulphate solutions were infused into the nares in the same manner as

Fig. 4. Concentration–magnitude relationships of the olfactory bulbar response to an amino acid, L-serine, in whitefish. The insert shows a typical recording of the response at 10^{-5}M. The top record indicates the integration of the lower one. A 1 s time scale is shown at the bottom, the heavy line indicating period of stimulation.

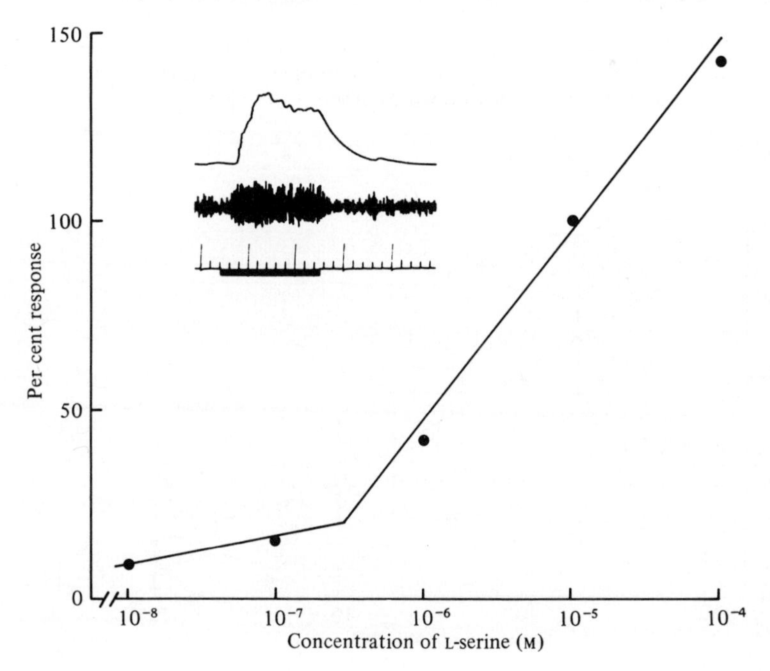

was L-serine. The response to lower concentrations of copper sulphate is essentially a depression of the spontaneous background electrical activity. The depression progressed as the concentration of copper sulphate was increased (Fig. 6). The lowest concentration to cause minimum effect was estimated at 10^{-7} M, which is in good agreement with that determined for

Fig. 5. Olfactory bulbar responses to water containing copper sulphate applied for 10 s. (*a*) Control, response to water alone; (*b*) 10^{-7}M, (*c*) 10^{-6}M, (*d*) 10^{-5}M, and (*e*) 10^{-4}M copper sulphate. In each pair the top record is the integrated response of the lower. Period of stimulation is shown by heavy lines.

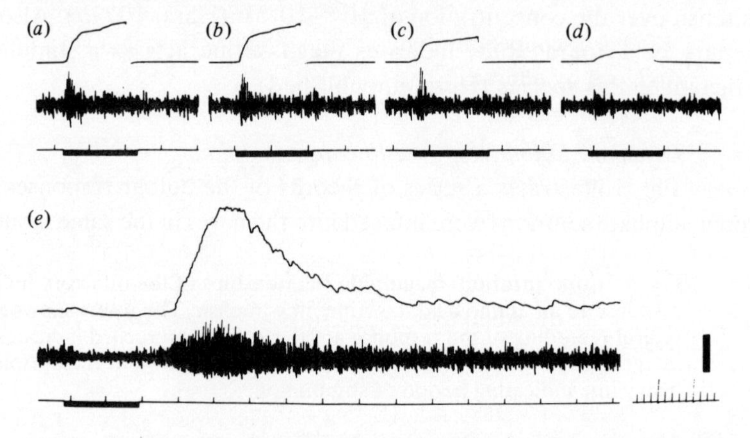

Fig. 6. Concentration–magnitude relationships of the bulbar response to copper sulphate recorded as in Fig. 5. Double circle indicates the control background activity level with water alone.

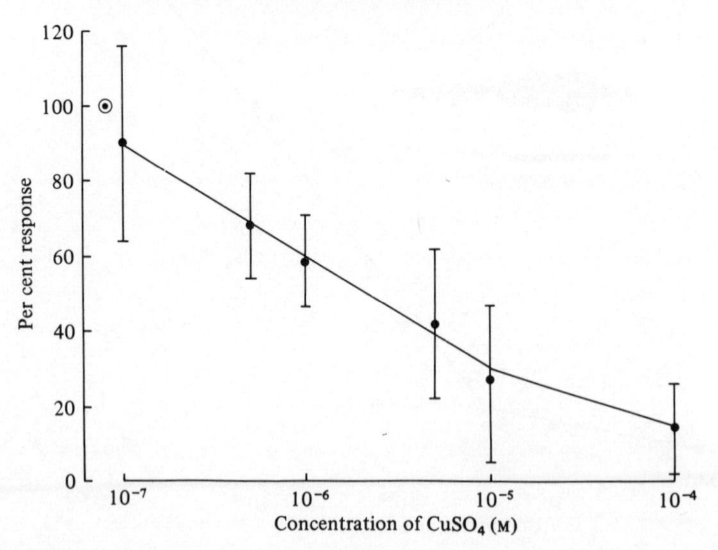

behavioural reactions (see Fig. 2). At 10^{-4} M the background activity was depressed upon stimulation, and then followed by a large oscillatory potential with continuous stimulation or upon the commencement of rinsing (Fig. 5*e*). This off-response increased in magnitude and duration with repeated stimulation at 2 min intervals. This effect caused by 10^{-4} M copper sulphate is similar to the phenomenon observed with 10^{-1} M morpholine, which has been used in the study of artificial imprinting of migratory salmonids (Hara, 1974; Hara & Macdonald, 1975; Hasler, Scholz & Horrall, 1978; Hara & Brown, 1979). This morpholine effect is assumed to be caused by a mechanism which is not directly associated with the normal olfactory function.

Effects of copper sulphate on the olfactory bulbar responses induced by odorants

Spontaneous background activities recorded from the olfactory bulb may depend partly upon the activities of receptor neurons responding to naturally occurring stimulants such as amino acids in water perfusing the nares. The depression of the spontaneous activity by brief exposure of the nares to copper sulphate could be caused by interference of stimulant molecule–receptor interaction by the heavy metal ions. The following experiments were designed to examine further this possibility; the olfactory bulbar responses to a standard stimulant, 10^{-5} M L-serine, were recorded at

Fig. 7. Olfactory bulbar responses to the standard stimulant 10^{-5}M L-serine before, during and after exposure of the nares to copper sulphate solutions. *A*, 5×10^{-6}M copper sulphate: (*a*) control, (*b*) 2 min after exposure, (*c*) 10 min exposure, and (*d*) 10 min after rinsing. *B*, 10^{-4}M copper sulphate: (*a*) control, (*b*) 8 min after exposure, (*c*) 2 min after rinsing, and (*d*) 2 min after re-exposure to copper sulphate.

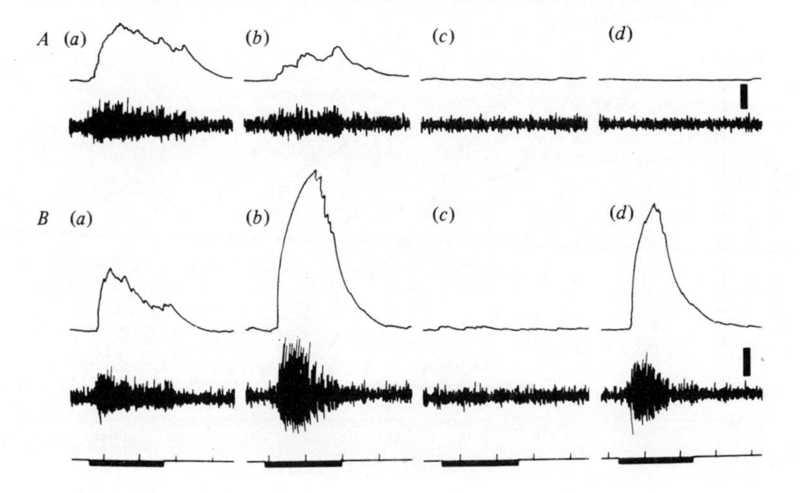

2 min intervals, while the nares were perfused with water containing copper sulphate at various concentrations. The method employed was essentially similar to that described previously (Hara, Law & Macdonald, 1976).

Fig. 7 shows examples of records of the olfactory bulbar responses to L-serine before, during and after treatment of the nares with 5×10^{-6} and 1×10^{-4} M copper sulphate. The effect of copper sulphate at lower concentrations, with the threshold of approximately 10^{-7} M, is essentially an inhibition of L-serine response. The degree of inhibition increased with an increase of the concentration up to 5×10^{-5} M (Fig. 8). With concentrations of 10^{-4} M or higher, however, the responses were augmented during perfusion, and sharply decreased upon rinsing. The responses reappeared, though decreased in magnitude, when the nares were treated repeatedly with the same copper sulphate solution (Figs. 7*Bd* and 9). Thus, the effect of lower concentrations of copper sulphate is, as in the case of depression of the spontaneous activity, primarily the inhibition of the bulbar response induced by stimuli. The presence of excess copper sulphate augments the bulbar response, probably by facilitating chemical transduction processes through irreversibly modified ionic permeability mechanisms in the receptor membranes.

Discussion and conclusion

It has been shown that fish can react to the same chemical, copper sulphate, by two opposing behavioural patterns, avoidance and preference,

Fig. 8. Bulbar responses to the standard stimulant 10^{-5}M L-serine over time when the nares were treated with 5×10^{-7}M (closed circles) and 5×10^{-6}M (open circles) copper sulphate. The bar indicates the duration of the treatment.

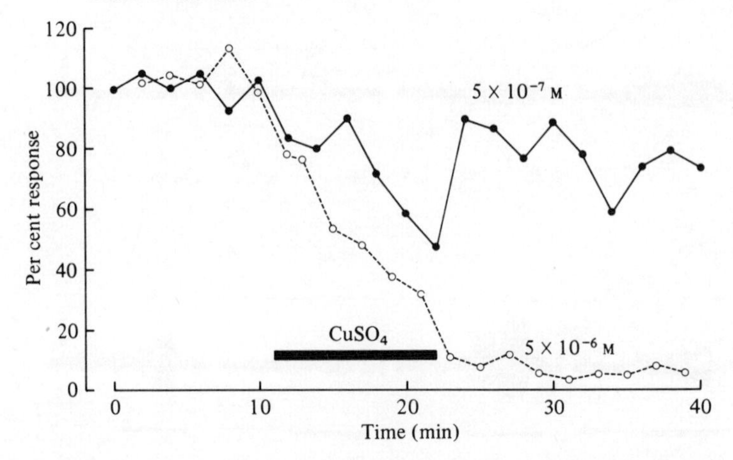

depending upon the concentration, and that both reactions are primarily mediated through olfaction. It has been also shown from electrophysiological experiments that the olfactory bulbar activities, whether spontaneous or stimulant-induced, are inhibited by copper sulphate at the same concentrations at which the fish avoid it behaviourally, and augmented when they prefer it. Thus, do fish avoid all chemicals that inhibit the olfactory activities, and prefer those which augment them? Hiatt, Naughton & Matthews (1953*a*,*b*), working on chemical repellents of schooling fish, *Kuhlia sandvicensis*, suggested that the most effective repellents for this species are inhibitors of sulphydryl groups in enzyme systems of sensory receptors: (1) mercaptide-forming agents such as heavy metals, (2) oxidizing agents, and (3) alkylating agents. Certainly, heavy metals such as mercury, cadmium and zinc are repellents to salmonids, and, at the same time, strong inhibitors of the olfactory functions (Hara *et al.*, 1976; Thompson & Hara, 1977; Hara & Scherer, unpublished). No systematic study has been done on chemicals belonging to the other categories of Hiatt *et al.* (1953*a*,*b*).

There seem to be at least three possible explanations for the preference

Fig. 9. Bulbar responses to the standard stimulant 10^{-5}M L-serine over time when the nares were treated with 10^{-4}M copper sulphate. The bars indicate the duration of the treatment.

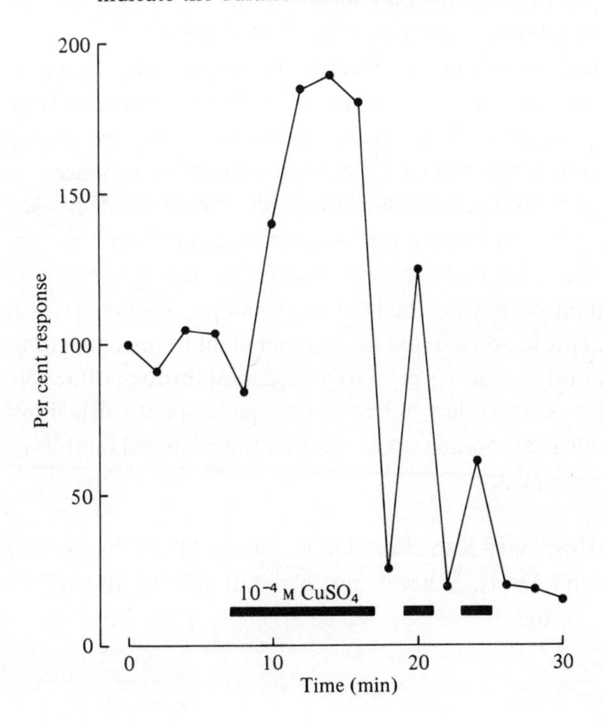

reaction to high concentrations of copper solution exhibited by whitefish: (1) the preference for altered water quality caused by addition of copper sulphate, e.g. pH, (2) changes in reaction pattern due to impaired olfactory organs, and (3) positive preference for copper ions. Addition of copper sulphate into the trough resulted in a decrease in the pH value of water (at 5×10^{-4} M, pH dropped from 7.7 to 7.2), due to the liberation of carbonic acid from bicarbonates (Doudoroff, 1965). Since both speciation of copper in natural water and the olfactory activities of fish are known to be dependent upon the pH of the environment (Sylva, 1976; Hara, 1976*b*), the pH is likely to be a factor regulating the chemosensory-mediated behavioural reactions. According to Jones (1948), however, the stickleback (*Gasterosteus aculeatus*) showed no or vaguely positive reactions to water of pH range 5.8–11.2. No comparable data are available for whitefish.

As shown in Figs. 8 and 9, brief exposure of the olfactory organs to high concentrations of copper sulphate resulted in immediate impairment of the responsiveness of the receptors. It should be considered, therefore, that under such circumstances the normal olfactory mechanisms of fish are largely dysfunctional. Nevertheless, the olfactory centres of fish would continue to receive elevated signals for as long as the animals stayed within the high copper side of the tank. Once the fish comes into contact with clean water all the signals disappear. This may motivate the fish to keep in the concentrated copper solutions, even if they are highly toxic.

In the present species, biologically significant olfactory stimuli, e.g. amino acids, cause the electrical activity of the olfactory bulb to increase in amplitude and reduce in frequency whilst the animal shows greater swimming activity. In contrast when this type of electrical activity is produced in response to high copper sulphate concentrations the fish show decreased behavioural activity such as adopting a stationary posture and stupefaction.

In summary, when whitefish had a choice between clean water and water treated with copper sulphate they avoided the copper sulphate when it was at low concentrations but preferred it to the clean water at high concentrations. This bimodal behavioural reaction is primarily mediated through olfaction. Evidence is presented to suggest that, at least for copper sulphate, fish either avoid or prefer a chemical depending upon whether the chemical inhibits or augments the olfactory inputs.

I wish to thank Miss Rosemary Kamchen, University of Manitoba, for her technical assistance, and Dr E. Scherer and his staff for the use of the avoidance/preference trough.

References

Cagan, R. H. & Zeiger, W. N. (1978). Biochemical studies of olfaction: binding specificity of radioactively labelled stimuli to an isolated olfactory preparation from rainbow trout (*Salmo gairdneri*). *Proceedings of the National Academy of Sciences, USA*, **75**, 4679–83.

Doudoroff, P. (1965). 'Formal discussions', in the paper by S. Ishio, 'Behaviour of fish exposed to toxic substances'. In *Water Pollution Research: Proceedings of the 2nd International Conference*, ed. O. Jaag, vol. 1, pp. 19–40. Pergamon Press, Oxford.

Hara, T. J. (1971). Chemoreception. In *Fish Physiology*, ed. W. S. Hoar & D. J. Randall, vol. 5, pp. 79–120. Academic Press, New York & London.

Hara, T. J. (1973). Olfactory responses to amino acids in rainbow trout, *Salmo gairdneri*. *Comparative Biochemistry and Physiology*, **44A**, 407–16.

Hara, T. J. (1974). Is morpholine an effective olfactory stimulus in fish? *Journal of the Fisheries Research Board of Canada*, **31**, 1547–50.

Hara, T. J. (1975). Olfaction in fish. In *Progress in Neurobiology*, ed. G. A. Kerkut & J. W. Phillis, vol. 5, pp. 271–335. Pergamon Press, Oxford.

Hara, T. J. (1976*a*). Structure–activity relationships of amino acids in fish olfaction. *Comparative Biochemistry and Physiology*, **54A**, 31–6.

Hara, T. J. (1976*b*). Effects of pH on the olfactory responses to amino acids in rainbow trout, *Salmo gairdneri*. *Comparative Biochemistry and Physiology*, **54A**, 37–9.

Hara, T. J. (1977*a*). Further studies on the structure–activity relationships of amino acids in fish olfaction. *Comparative Biochemistry and Physiology*, **65A**, 559–65.

Hara, T. J. (1977*b*). Olfactory discrimination between glycine and deuterated glycine by fish. *Experientia*, **33**, 618–19.

Hara, T. J. & Brown, S. B. (1979). Olfactory bulbar electrical responses of rainbow trout (*Salmo gairdneri*) exposed to morpholine during smoltification. *Journal of the Fisheries Research Board of Canada*, **36**, 1186–90.

Hara, T. J., Law, Y. M. C. & Hobden, B. R. (1973). Comparison of the olfactory response to amino acids in rainbow trout, brook trout, and whitefish. *Comparative Biochemistry and Physiology*, **45A**, 969–77.

Hara, T. J., Law, Y. M. C. & Macdonald, S. (1976). Effects of mercury and copper on the olfactory response in rainbow trout, *Salmo gairdneri*. *Journal of the Fisheries Research Board of Canada*, **33**, 1568–73.

Hara, T. J. & Macdonald, S. (1975). Morpholine as olfactory stimulus in fish. *Science, Washington*, **187**, 81–2.

Hasler, A. D., Scholz, A. T. & Horrall, R. M. (1978). Olfactory imprinting and homing in salmon. *American Scientist*, **66**, 347–55.

Hiatt, R. W., Naughton, J. J. & Matthews, D. C. (1953*a*). Effects of chemicals on a schooling fish, *Kuhlia sandvicensis*. *Biological Bulletin*, **104**, 28–44.

Hiatt, R. W., Naughton, J. J. & Matthews, D. C. (1953*b*). Relation of chemical structure to irritant responses in marine fish. *Nature, London*, **172**, 904–5.

Jones, J. R. E. (1947). The reactions of *Pygosteus pungitius* L. to toxic solutions. *Journal of Experimental Biology*, **24**, 110–22.

Jones, J. R. E. (1948). A further study of the reactions of fish to toxic solutions. *Journal of Experimental Biology*, **25**, 22–34.
Katsuki, Y., Hashimoto, T. & Kendall, J. E. (1971). The chemoreception in the lateral-line organs of teleosts. *Japanese Journal of Physiology*, **21**, 99–118.
Kleerekoper, H., Westlake, G. F., Matis, J. H. & Gensler, P. J. (1972). Orientation of goldfish (*Carassius auratus*) in response to a shallow gradient of sublethal concentration of copper in an open field. *Journal of the Fisheries Research Board of Canada*, **29**, 45–54.
Ottoson, D. (1956). Analysis of the electrical activity of the olfactory epithelium. *Acta Physiologica Scandinavica*, **35**, *Suppl. 122*, 1–83.
Scherer, E. & Nowak, S. (1973). Apparatus for recording avoidance movements of fish. *Journal of the Fisheries Research Board of Canada*, **30**, 1594–6.
Sprague, J. B. (1964). Avoidance of copper–zinc solutions by young salmon in the laboratory. *Journal of the Water Pollution Control Federation*, **36**, 990–1004.
Sprague, J. B. & Drury, D. E. (1969). Avoidance reactions of salmonid fish to representative pollutants. In *Water Pollution Research: Proceedings of the 4th International Conference*, ed. S. H. Jenkins, pp. 169–79. Pergamon Press, Oxford.
Sutterlin, A. M. & Sutterlin, N. (1970). Taste responses in Atlantic salmon (*Salmo salar*) parr. *Journal of the Fisheries Research Board of Canada*, **27**, 1927–42.
Sylva, R. N. (1976). The environmental chemistry of copper (II) in aquatic systems. *Water Research*, **10**, 789–92.
Thompson, B. E. & Hara, T. J. (1977). Chemosensory bioassay of toxicity of lake waters contaminated with heavy metals from mining effluents. In *Proceedings of the 12th Canadian Symposium 1977: Water Pollution Research Canada*, ed. P. H. Jones, pp. 179–89.

JÖRG-PETER EWERT

Neural coding of 'worms' and 'antiworms' in the brain of toads: the question of hardwired and softwired systems

Introduction

In comparative ethology, those stimuli which activate fixed patterns of behavioural responses are called key stimuli, because they relate to the responses in the way a key is related to its lock. The central nervous filtering system which decides whether the key fits the lock is called the releasing mechanism (RM). Releasing mechanisms may have an innate basis. A classical example of a fixed behavioural response, from Tinbergen (1948), is that young turkeys show escape behaviour towards airborne predators. The key stimuli, filtered by an innate releasing mechanism (IRM), are at first relatively unspecific: any large shadow (e.g. of a bird) moving over them releases escape responses (Fig. 1*A*). The shape plays no specific role. In the course of time, following some kind of learning, the IRM becomes selective for a certain configuration of stimulus. The turkeys become habituated to goose-like birds (long neck, short tail) flying overhead repeatedly in their environment, whereas the less frequently seen birds of prey (short neck, long tail) continue to be avoided. This can be tested most convincingly by experiments using dummies (Fig. 1*B*). When the bird dummy is moved with the short end leading, it symbolizes a hawk and releases avoidance behaviour; when the same dummy is moved in the opposite direction it symbolizes, with its long neck, an inoffensive goose and hence elicits no response. In this instance the stimulus *configuration* plays an important role in decision-making by the modified IRM (Schleidt, 1962).

To study quantitatively the neural bases of gestalt perception I selected a simpler example (Ewert, 1968). The common toad *Bufo bufo* responds to small moving objects with prey-catching behaviour. Experiments with a dummy consisting of a small narrow stripe that contrasted with the background showed that the key stimulus 'prey' consists mainly of two characteristics: (1) the movement and (2) the shape relative to the direction of movement. For example, when a stripe is moved parallel to its longer axis in a *worm-like* fashion (Fig. 1*C*) it signals 'prey', but when the longer axis of the same stripe is oriented perpendicular to the direction of movement – so that

Fig. 1. Configurational key stimuli of predator and prey in turkeys and toads. *A*, In young turkeys any large shadow moving in the dorsal visual field elicits escape behaviour. *B*, Effect of a bird dummy on the escape response of adult turkeys. The releasing character depends on the movement direction (arrows) of the dummy and, thus, its configuration. (Modified from Tinbergen, 1948.) *C*, A small, e.g. 2 mm × 30 mm, stripe moving in the direction of its long axis signals prey for the toad; if the same stripe is moved perpendicular to its long axis it signals non-prey or even threat. (Modified from Ewert, 1968.)

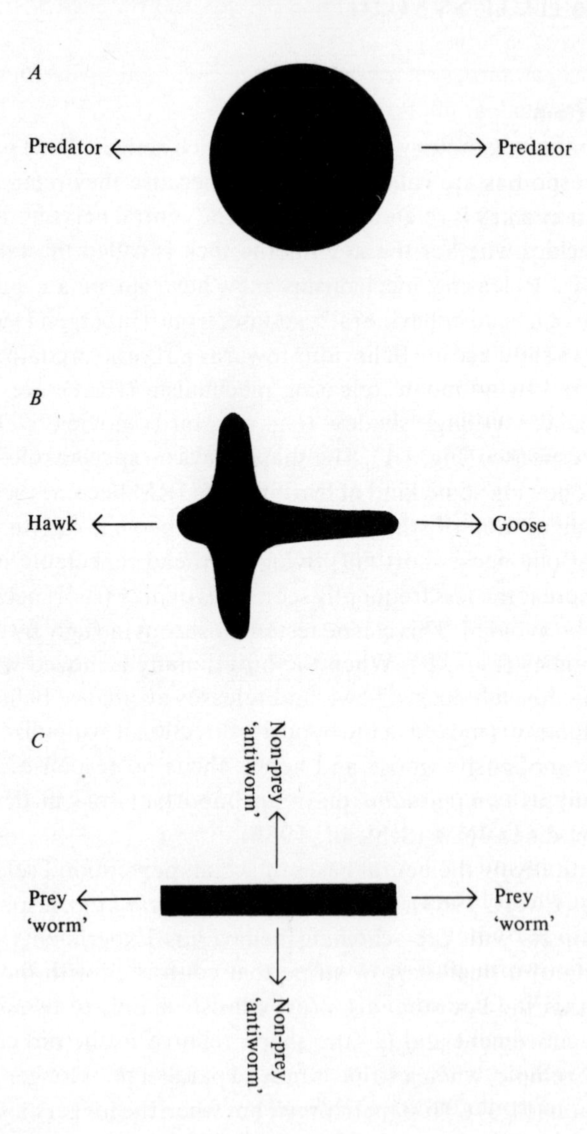

we might speak of an 'antiworm' configuration – the stimulus loses its key feature as 'prey' and may even signal 'threat'.

In this paper the following questions will be investigated:

1. Is the worm/antiworm discrimination in toads already established by the beginning of the animal's terrestrial life and thus based on a neuronal '*hardwired* system'?
2. Can the configurational selectivity with respect to worm-like and antiworm-like stimuli be modified by experience in the framework of '*softwired* systems'?
3. Is the worm/antiworm discrimination invariant with changes in other stimulus parameters?
4. What is the neural basis of coding worms and antiworms?
5. What is the basis of the decision 'prey or non-prey' in terms of neuronal events?
6. Can the worm/antiworm recognition be restored after brain lesions?

Form discrimination during early life

The prey-catching behaviour of various species of anuran amphibians, including its ontogenetic aspects, was studied with respect to differences in the shapes of moving stimulus patterns (Ewert, 1968; Ewert & Burghagen 1979*a*,*b*). Prey-catching in common toads (*Bufo bufo*) during the first hours and days of their terrestrial life was tested in terms of their response to small stripes moving in either a worm-like or an antiworm-like manner (Traud & Ewert, unpublished). Almost all of the investigated prey-naive toads preferred the worm configuration of a $2 \times 16°$ (visual angle) stripe as prey. During the course of ontogeny the configurational selectivity showed a remarkable increase. There was no statistically significant effect on prey-shape selection in postmetamorphic toads if the tadpoles were raised in differently structured environments consisting either of black vertically or of black horizontally striped patterns that were stationary or moving against a white background (Traud & Ewert, unpublished). The toads preferred worms to antiworms independently of the structure of the artificial visual environments in which they were raised.

We may conclude from these experiments that basic gestalt recognition – e.g. the worm/antiworm phenomenon – in common toads is innate and mediated by an IRM. Of course, most of the preys – such as worms, bugs, slugs – have shapes which are elongated in the direction of their movement. But there is no specific innate 'prey image' related to particular prey objects.

The increase in configurational selectivity during ontogeny may not necessarily be due to processes based on prey experience. During experiments with midwife toads *Alytes obstetricans*, which also display this phenomenon of maturation of prey selection, animals were fed exclusively with the stumpy-winged *Drosophila melanogaster*, that hardly display a worm shape. Himstedt, Freidank & Singer (1976) discovered a comparable phenomenon in salamanders, namely that the particular filter properties of the IRM are temporarily preprogrammed for the different environmental situations that the animal will encounter in the water and on land.

Pomeranz (1972) has reported significant changes in ranid frog ganglion cell physiology and anatomy before and after metamorphosis, which shows one aspect of programming in neuronal circuits. The ontogenetically late morphological differentiation of the dorsolateral visual area of the anuran thalamus – which outlasts the end of metamorphosis – is another aspect (Clairambault, 1976). It might be correlated with maturation of (i) configurational prey selection (Ewert & Burghagen, 1979*a*; Traud & Ewert, unpublished) and (ii) estimation of absolute prey size (Ewert & Burghagen, 1979*b*).

Finally we should also consider the possibility that experience in early life might influence, to some extent, general developments of neuronal mechanisms (Wiens, 1970, 1972). Indeed, in the next section we shall see that in adult toads the selectivity of the IRM for prey-catching can be modified by experience (Ewert, 1968; Brzoska & Schneider, 1978).

Modification of prey selection by experience

Habituation

The variety of the shapes and patterns of natural prey objects should allow toads to link individual experience to particular visual cues and to store this information in order to recall it when faced with the appropriate stimulus (Freisling, 1948; Eibl-Eibesfeldt, 1951). This phenomenon can be studied in habituation experiments (Birukow & Meng, 1955). With the repeated release of the orienting reaction by the same prey dummy form during a long-term series of presentations, the toad habituates to the stimulus. If immediately afterwards a prey dummy of another shape appears and the toad again reacts, it must have been able to differentiate the latter stimulus from the preceding one.

The stimulus habituation paradigm is suitable for studying in detail the ability of the toad to discriminate patterns by experience (Ewert & Kehl, 1978). Fig. 2*A* shows a compilation of different shapes in a hierarchy illustrating their stimulus effectiveness as prey. The main features governing prey selection by habituation are, as far as has been investigated: (i) area components, (ii) tips leading the stimulus in the direction of movement, (iii) isolated dots, (iv) striped patterns. Indeed patterns such as these are components of cues provided by natural prey objects. These experiments clearly demonstrate that toads are able to discriminate, as a result of experience, detailed structures within the 'innate' prey image – which is in general a worm-like shape.

Conditioning

Can toads be trained to prefer an antiworm as prey rather than the worm configuration?

In a first experimental group adult toads were exposed to the familiar odour of mealworm excrements (Ewert, 1968). During a process of self-

Fig. 2. Possibilities for modifying, or rather extending, the innate releasing mechanism of prey-catching in the toad *B. bufo* by experience. *A*, Hierarchical ordering of features governing prey selection on the basis of individual experience by habituation. (After Ewert & Kehl, 1978.) *B*, In the presence of known prey odour the efficacy of visual stimuli may be greatly enhanced (see arrows). Curves of means from experiments with 15 toads.

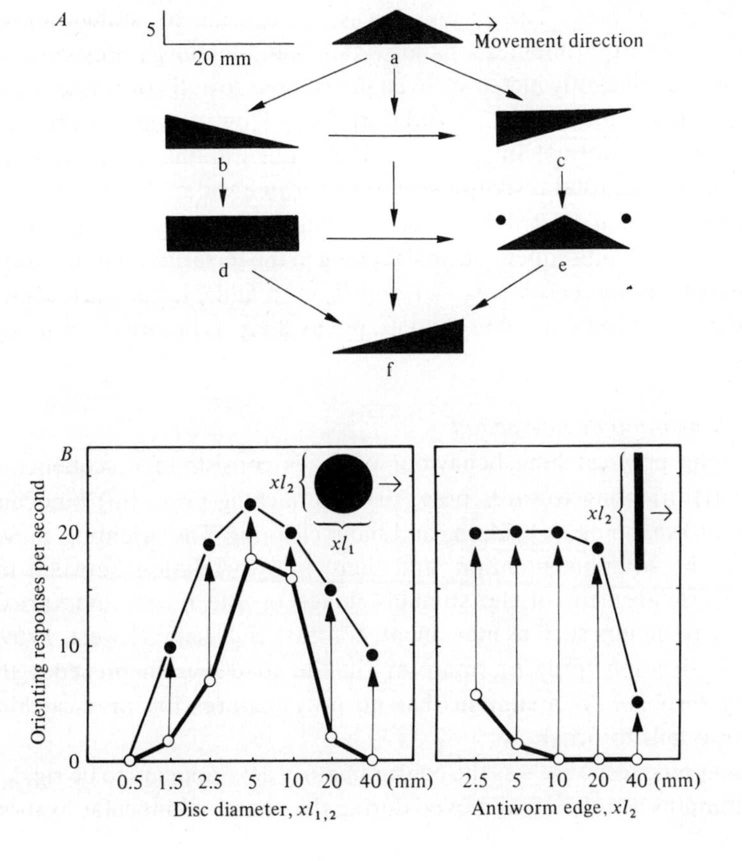

training toads associated this odour with the visual prey. If this known odour was then offered in the experimental situation along with a visual dummy, the stimulus efficacy was thereby increased, even for those stimulus configurations which previously had poorly or never resembled prey, such as long antiworms (Fig. 2*B*).

In a second experimental group adult animals were fed with mealworms from a particular holder (Burghagen & Ewert, unpublished) that had the shape of an antiworm-like stripe moving to and fro. Thus, during feeding, toads were presented with two kinds of stimuli: the natural prey stimulus and the antiworm which should serve as a conditioned stimulus. It was not possible by means of this procedure to alter the worm preference in favour of antiworm-like objects presented alone.

In a third experimental group adult toads were allowed to eat prey objects out of the experimenter's hand (Brzoska & Schneider, 1978; Ewert & Burghagen, unpublished). In the course of time the animals came to associate the presence of the hand of the experimenter with food and they finally responded to the moving hand alone. This was generalized to include other large objects as prey. The response was not specific to configurational features of the experimenter's hand! After 'hand-feeding', prey-catching activity was significantly increased even in response to antiworm-like stripes and large square objects (Fig. 3*A* and *B*, cf. 2–6). However, antiworms were not preferred to worms! In summary, the configurational selectivity was decreased and the total response spectrum became wider. The accuracy of differentiation of the IRM had been modified in relation to individual experience. After subsequent normal feeding in the terrarium the configurational sensitivity increased (Fig. 3*A* and *B*, cf. 6 and 7); but even after 4 months selectivity was, in some animals, not as sharp as before the training.

Sensorimotor interaction

The prey-catching behaviour of toads consists of a sequence of actions: (i) orienting towards prey, (ii) approaching prey, (iii) binocular fixation, (iv) snapping, (v) gulping and snout cleaning. The orienting movement fits the stimulus situation, and there is a correlation between the effective displacement of the stimulus image on the retina and various stimulus parameters such as movement, contrast and shape (Ewert, 1969). Thus, the decision 'prey or non-prey' in the toad's brain precedes the orienting response. If a stimulus has no prey features the prey-catching orientation fails to occur.

The snapping response – as the consummatory act – appears to be rigid. If a prey stimulus is suddenly removed during the phase of binocular fixation, the toad snaps into the vacuum (Hinsche, 1935). Snapping mainly depends upon adequate stimulation of the retinas of both eyes during the period of fixation. Toads snap at small objects with a tongue flip, but they grasp large ones with the jaws. However, these motor patterns do not fit precisely the stimulus situation (Eikmanns, 1955): toads previously fed with small objects will subsequently snap at large objects with a tongue flip, while the same animals when used to being fed on large objects will subsequently also grasp small objects with the jaws. Thus, it appears that toads must recall the appropriate *pattern* of snapping, e.g. by a previous mistake.

The question of invariants in gestalt perception

The human visual system shows functional invariants and enables recognition of the 'gestalt' (configuration) even when other visual parameters are varied within limits. The construction of these invariants appears to be a basic precondition for 'gestalt' perception. The ability of common toads to distinguish between the same stripe moving either as worm or as antiworm – and their preference for the worm configuration – has been proved to be invariant with respect to changes in various stimulus parameters, such as:

1. Movement direction in the *x–y–z* coordinates (Fig. 4*B*; Beck & Ewert, 1979; Ewert *et al.*, 1979*a*).

Fig. 3. Modification of prey selection in *B. bufo* after hand-feeding with mealworms. Two representative examples, *A* and *B*, are shown. The prey-catching activity was measured in response to 2.5 mm stripes of different lengths (mm) oriented either in the direction of movement (*a*: worm configuration) or perpendicular to the movement direction (*b*: antiworm configuration); in another test series squares (*c*) of different edge length were presented. All black stimuli were moved at constant angular velocity ($v = 15° s^{-1}$) at constant distance ($d = 7$ cm) around the toad in a standard procedure (Ewert, 1968). 1, Response before hand-feeding; 2, response after 4 months of hand-feeding, 3–6, repeat of tests each being separated by 1 month of hand-feeding; 7, test after 4 months of normal feeding (presenting only mealworms). For explanation see text. (Burghagen & Ewert, unpublished.)

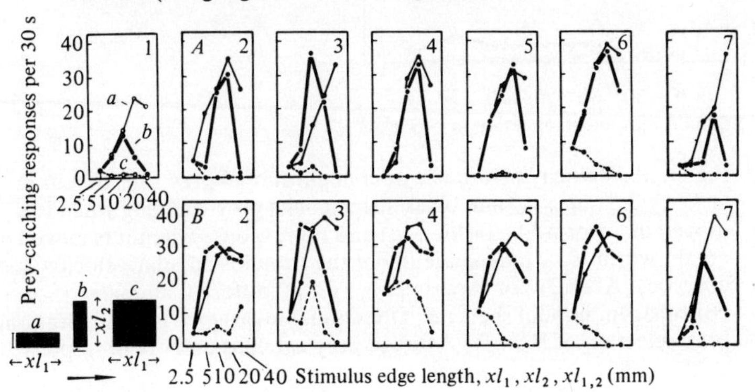

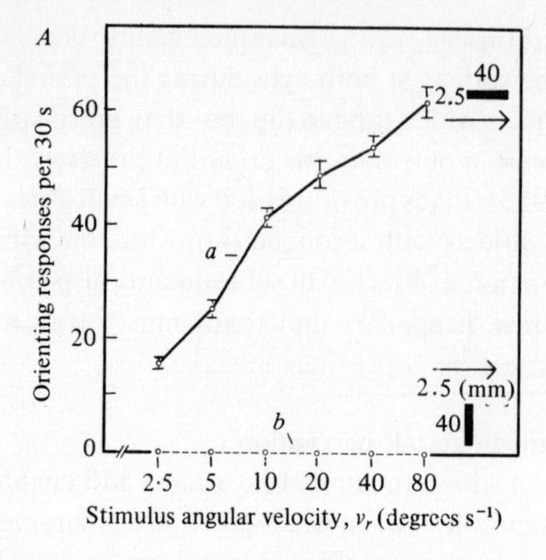

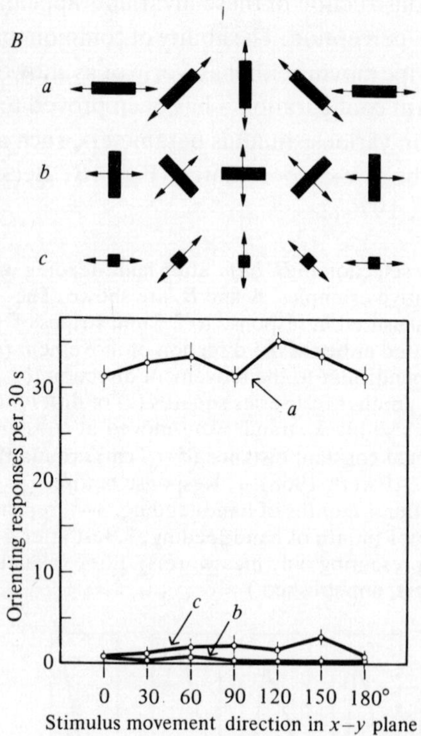

Fig. 4. *A*, Velocity invariance of configurational prey selection in *B. bufo*. A 2.5 mm × 40 mm black stripe elicits prey-catching when it is moved in a worm-like fashion (*a*) and no response when it is moved as an antiworm (*b*) – independently of the stimulus angular velocity (see abscissa). Averages of experiments with 20 different animals. (Burghagen, unpublished.) *B*, Directional invariance in configurational prey selection of *B. bufo*. Average prey-catching activity in response to

2. Angular velocity (Fig. 4*A*; Borchers, Burghagen & Ewert, 1978).
3. Movement pattern (continuous or stepwise) (Borchers *et al.*, 1978).
4. Presence of black stimuli against white background or vice versa (Ewert, 1968).
5. Background structure (Ewert, Albrecht, Burghagen & Kepper, unpublished).
6. Perceptual depth of objects (Gantner & Ewert, unpublished).

However, all of these parameters have an influence on the general strength of a visual stimulus as prey.

With respect to 1 it must be emphasized that object movements in $y(+)$ and $z(-)$ directions of the visual field are better for eliciting prey-catching than movements in the reverse directions. Concerning 2, in a certain range prey-catching activity increases with increasing movement velocity of a prey stimulus. Regarding 3 it was found that steps at 1–2 Hz are the best frequencies for the activation of prey-catching. With regard to 4 it is most interesting to note that black configurational objects moving against white are more selectively discriminated than white objects moving against a black background. Regarding 5, the configurational selectivity of small black moving stimuli is increased to some extent if the stimuli are moved on a stationary structured background. Small white moving objects, however, are partly masked by a structural background (Ewert, Albrecht, Burghagen & Kepper, unpublished).

Central mechanisms of gestalt perception

Methodology

The following section describes investigations into whether there are neurons in the visual system that show sensitivity to or selectivity for worm-like and antiworm-like objects. More specifically, in these experiments the *configuration* of a rectangular stimulus is studied in two ranges of possible transformations: (1) area extension *in* the direction of stimulus movement, and (2) extension *perpendicular* to the direction of movement. At constant stimulus velocity, transformation 1 changes stimulus features in

---

different configurational stimuli (*a*–*c*) moving in different directions of the visual field *x*–y coördinates in front of the animal, to and fro (see double arrows). *a*, A 2.5 mm × 30 mm stripe oriented with its main axis in the direction of movement (worm-like); *b*, a 2.5 mm × 30 mm stripe oriented with its main axis perpendicular to the direction of movement (antiworm-like); *c*, a 2.5 mm × 2.5 mm square. The black stimuli were moved against a white background at 25 mm s^{-1} at a distance of 70 mm from the toad's eyes in a standard 'belt procedure'. (After Ewert *et al.*, 1979*a*.)

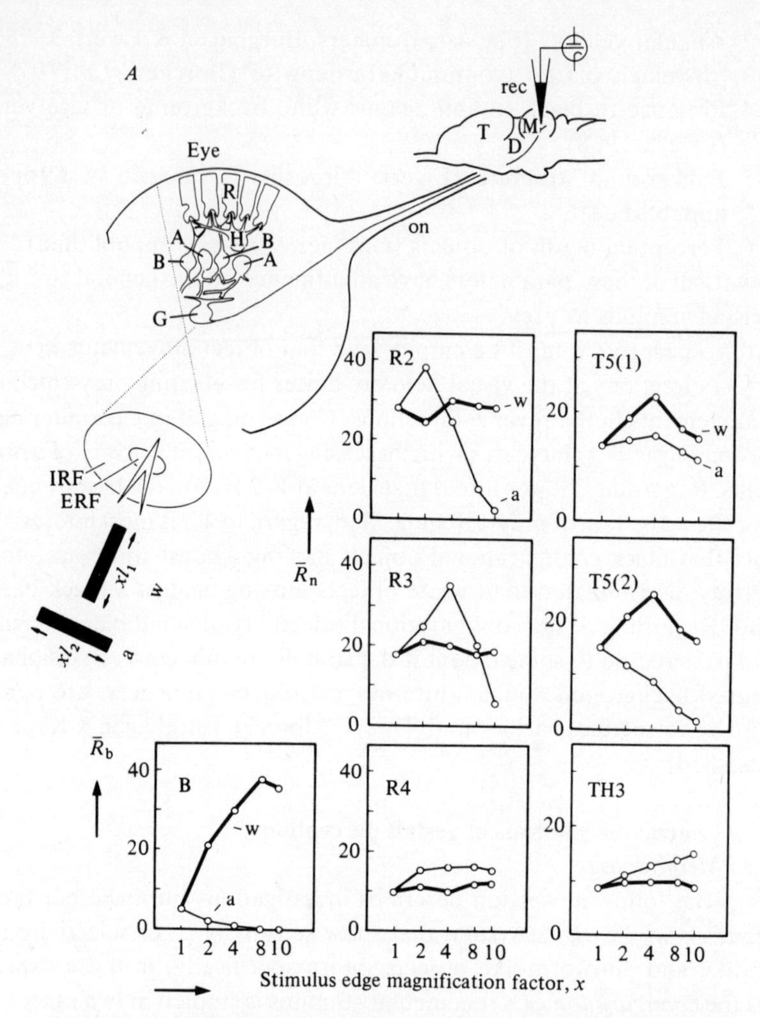

Fig. 5. *A*, Average prey-catching activity, $\bar{R}_b$ (number of orienting responses per minute), and average neuronal discharge frequency, $\bar{R}_n$ (impulses per second), of toads *B. bufo* in response to worm-like (w) and antiworm-like (a) moving stimuli; of different length xl_1 (l_2 = constant) and xl_2 (l_1 = constant). *B*, Discriminate values $D_{w,a} = (\bar{R}_w - \bar{R}_a)(\bar{R}_w + \bar{R}_a)^{-1}$ for selection between worm-like (w) and antiworm-like (a) stripes of different length. $\bar{R}$ is the average prey-catching activity or the neuronal activity, respectively. The values were calculated for: prey-catching behaviour (B), retinal ganglion cells (classes R2, R3, R4; $n = 10$ each), thalamic pretectal neurons (class TH3; $n = 21$), and tectal neurons (classes T5(1) and T5(2); $n = 20$ and 18). The values were measured in the range of $1 \leqslant x \leqslant 10$ for $l_1 = l_2 = 2.5$ mm (B) or 2° visual angle (R2 to T5(2). All stimuli were black and were moved in a horizontal direction at 20° s^{-1} (B) or 7.6° s^{-1} (R2 to T5(2)) against a white background by means of a standard 'rotating procedure' or a

the *space* domain, and 2 changes features in *space and time* domains (Ewert, 1968). The geometry of these stimuli is linked to the movement direction. The entire population of all possible rectangles can be represented in a two-dimensional array, which is called in gestalt psychology a 'pattern system' (Gibson, 1950). The aims of the following experimental strategy are (i) to evaluate in toads the behavioural significance of elements belonging to this *pattern system* (i.e. degree of resemblance with prey), and (ii) to analyse their transformation by neurons at different points of the visual system. Together these constitute the precondition for investigating a *pattern recognition system* (Ewert, 1974).

Definition of input signals. The visual stimuli tested were rectangular black stripes, of length xl_1 in the direction of movement and of length xl_2 perpendicular to the direction of movement. In one stimulus series the edge xl_1 of a

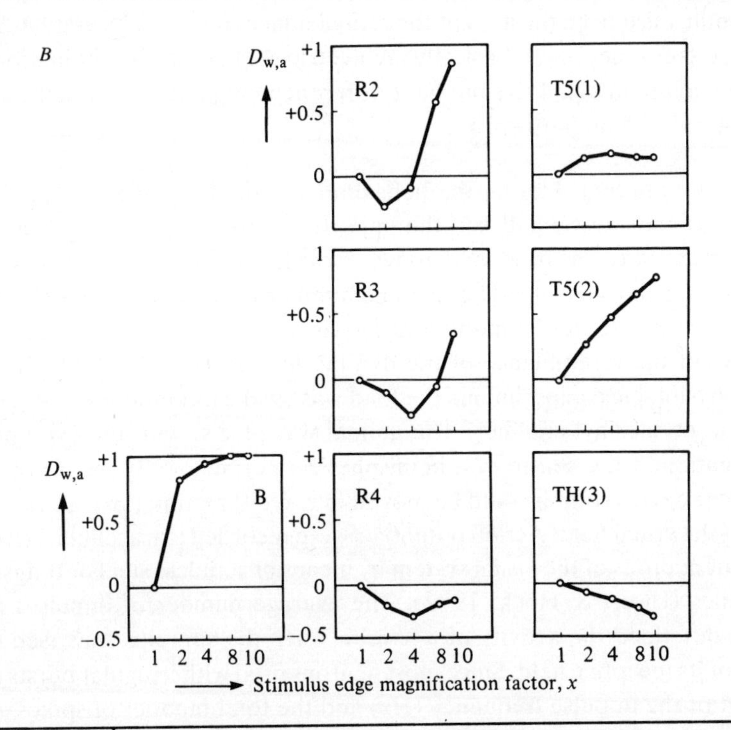

special perimeter for neurophysiological experiments. The top of *A* shows a schematic presentation of a receptive field of a retinal ganglion cell. ERF, excitatory receptive field; IRF, inhibitory receptive field; R, receptor cell; H, horizontal cell; A, amacrine cell; B, bipolar cell; G, ganglion cell; ON, optic nerve; T, telencephalon; D, diencephalon; M, mesencephalon; rec, recording electrode. (Modified from Ewert, Borchers & von Wietersheim, 1978.)

small $l_1 \times l_2$ square was elongated in the direction of movement (worm-like objects; l_2= constant), and in another series the edge xl_2 was elongated perpendicular to the direction of movement (antiworm-like objects; l_1= constant). Magnification steps were $x = 1, 2, 4, 8, 10$ (Fig. 5). In behavioural experiments the values l_1 and l_2 corresponded to 2.5 mm, equivalent to a visual angle of 2° used in the neurophysiological studies. The stimulus movement velocity was held constant at $v = 25\ \mathrm{mm\,s^{-1}}$ in the behavioural experiments and at 7.6 (or 18) degrees per second in the neurophysiological recording studies.

In behavioural experiments frogs and toads display 'size-constancy', which means that they are judging the absolute size of a visual object by estimating its distance (Ingle, 1968; Ewert & Gebauer, 1973). However, this phenomenon has not yet been seen in the response of neurons. Visually sensitive neurons investigated so far in immobilized amphibians determine the stimulus size from the area of the retinal image. It should be emphasized that 'size-constancy' is not basically involved in the general discrimination of stimuli that are of equal size but have different configurations, as in the case of a 'worm' and an 'antiworm'.

Definition of output. During the behavioural experiment the toad sat in a cylindrical glass vessel and had the opportunity to respond towards a rectangular stimulus – as described above – which circled around the vessel at constant angular velocity. The average number of prey-catching orienting responses towards the stimulus during an interval of 1 min served as a measure of the resemblance of the dummy to prey (Ewert, 1969). In the neurophysiological experiments the toad was awake but immobilized (after injection of succinyl-choline). The animal was placed with the eye under investigation in the centre of a hemispherical perimeter. By means of the perimeter device stimuli could be moved in *x, y, z* direction through various parts of the visual field. Action potentials were recorded extracellularly from different neurons of the visual system by means of stainless steel or tungsten electrodes (Ewert & Hock, 1972). The average number of impulses per second described the activity of a neuron when the stimulus traversed the centre of its receptive field. Since most neurons fired with irregular bursts the product of the impulse frequency (Hz) and the total number of spikes was also considered; this corresponds to the power of the output.

Discriminate value $D_{\mathrm{w,a}}$. A quantitative measure for selection between worm-like (w) and antiworm-like (a) objects of equivalent sizes is given by the form-contrast $D_{\mathrm{w,a}} = (\bar{R}_{\mathrm{w}} - \bar{R}_{\mathrm{a}})\ (\bar{R}_{\mathrm{w}} + \bar{R}_{\mathrm{a}})^{-1}$, where $\bar{R}_{\mathrm{w}}$ is the average response to a worm-like stripe and $\bar{R}_{\mathrm{a}}$ the average response to the same stripe moving in antiworm-like fashion (Ewert *et al.*, 1978). $D_{w,a}$ values of behavioural and neuronal responses are plotted against the stimulus parameters xl_1 (with l_2 constant) and xl_2 (with l_1 constant). The values of $D_{w,a}$ can be expected to be between +1 and −1. For cases where worm-like objects are preferable to the same stimuli presented as antiworms, $D_{w,a}$ is positive. When $D_{w,a}$ is +1 or −1 the two stimuli are distinguished by a clear 'yes/no' decision. In Fig. 5*B* the $D_{w,a}$ values of behavioural and neuronal responses are plotted against the parameters xl_1 or xl_2.

Discrimination analysis. The neuronal transformation of worm-like stimuli (xl_1) can be described by the response function $R = f_1(xl_1)$. Similarly the antiworm-transformation can be described by the function $R = f_2(xl_2)$. The correlation coefficient $r_{w,a}$, derived from a Pearson wave-form analysis, thus describes the similarity between the transformations of the neuronal systems. A non-discrimination between worm and antiworm ($f_1 = f_2$) would give a $r_{w,a} = +1$ and perfect discrimination $r_{w,a} = -1$ (Borchers & Ewert, 1979).

Double correlation. The stimulus–response relationships measured for prey-catching behaviour, $R_b = f_b\ (xl_1)$ and $R_b = f_b^*\ (xl_2)$, can be compared statistically with the corresponding neurophysiological stimulus–response relationships, $R_n = f_n\ (xl_1)$ and $R_n = f_n^*\ (xl_2)$. The question of correlation between behavioural and neurophysiological response activities has to be analysed for both parameters xl_1 (with l_2 constant) and xl_2 (with l_1 constant). Therefore in the case of perfect correlation a pair of $\{r_w; r_a\} = \{1;1\}$ should be expected (Borchers & Ewert, 1979).

Prey-catching behaviour

The $D_{w,a}$ values show a strong increase with increasing stripe length xl_1 and xl_2 (Fig. 5*B*), which means that the worm-like stripe is progressively preferred and the antiworm-like one neglected as the stripe is made longer. A relatively high negative value of the correlation coefficient, $r_{w,a} = -0.9$, indicates that the different configurational stimuli are properly distinguished.

Retinal ganglion cells

Histology and receptive field organization. The amphibian retina consists of four sequentially connected cell types: receptor cells, bipolar cells, amacrine cells and ganglion cells (Fig. 5*A*). Horizontal connections are made by horizontal cells and amacrine cells. The output of retinal information processing is fed into the visual centres of the toad's brain by the axons of at least three types of ganglion cells – R2, R3 and R4 neurons – which are distinguished by the diameter of their excitatory receptive fields (ERF): $\approx 4°$ (class R2), $\approx 8°$ (class R3) and 12–16° (class R4) (Fig. 5). (For results in frogs see Lettvin *et al.*, 1959.) The responses of these ganglion cells can be recorded extracellularly from their axon terminals in the optic tectum. The ERF of a ganglion cell is defined as that area of the visual field in which a moving stimulus elicits a neuronal response. The ERF is surrounded by an inhibitory receptive field (IRF), defined as that region in which a moving stimulus inhibits the response elicited by simultaneous stimulation of the ERF. The strength of the IRF decreases from class R2 to R4. Class R3 and R4 neurons are movement-sensitive, whereas the response of class 2 neurons is movement-specific. (For summary see Grüsser & Grüsser-Cornehls, 1976.)

Sensitivity to moving configurational stimuli. As Fig. 5 demonstrates, the first steps of information processing with regard to the stimulus configuration have already taken place at the retinal level. Among the different ganglion cells investigated the class R2 neurons show the best selective response to worm-like and antiworm-like objects. Here a correlation coefficient of $r_{w,a} = -0.9$ is found, in contrast to those for the other ganglion cell classes which even show positive values ($r_{w,a} = 0.6$ (R3) and $r_{w,a} = 0.5$ (R4)). The correlation analysis gives a pair of $\{r_w; r_a\} = \{0.5; 0.6\}$ for R2 neurons, $\{0.2; -0.7\}$ for R3 neurons and $\{0.6; -0.9\}$ for R4 neurons. In R3 and R4 neurons the correlation coefficients are negative if responses to antiworm-like stimuli are compared. From an analysis of the $D_{w,a}$ values it becomes evident that the antiworm response is predominant if the stimulus does not extend beyond the ERF. This is valid in the ranges $1 \leqslant x \leqslant 4$ (for class R2), $1 \leqslant x \leqslant 8$ (class R3) and $1 \leqslant x \leqslant 10$ (class R4). If the stimulus approaches the IRF the worm configuration is preferred: $x > 4$ (class R2) and $x > 8$ (class R3). Thus worm/antiworm discrimination at retinal level is mainly based on properties of the IRF.

The stimulus response relationships were found to be relatively constant over time and they were generally not altered by either the movement direction or the angular velocity of the stimulus (Ewert, Krug & Schönitz, 1979*b*). In class R2 neurons a seasonal dependence in the response to the direction of the stimulus background contrast was observed, quite similar to that obtained in the prey-catching behaviour (Ewert & Siefert, 1974).

If the behavioural and neurophysiological results for small worm-like and antiworm-like stimuli ($x < 4$) are compared, configurational prey selection cannot be explained solely in terms of the properties of one of the retinal ganglion cell classes, so far as known neurons from the retinotectal projection are concerned (Fig. 5).

Retino-thalamic-pretectal nerve nets

Histology. It is known that lesions in the caudal thalamus and the pretectum (TP region) have a remarkable effect on prey selection in frogs and toads. Animals lose all caution and they snap at a variety of moving visual objects, even at predators. The configurational prey recognition system is not working in TP-lesioned toads and frogs (Ewert, 1967, 1968). Therefore, it might be reasonable to look for neurons in this region which are sensitive to configurational cues of moving stimuli. The posterior thalamic nucleus is situated in the pretectal region and has the appearance of a vertically oriented cylindrical body (Lázár, 1971; Scalia & Fite, 1974). It consists of two parts, the posterocentral (pc) and posterolateral (pl) complexes (Fig. 6*A* and *B*). The retina projects to the 'pretectal' neuropil of the contralateral diencephalon topographically (Fig. 7*B*). All of the contralateral retinal projections develop in early larval stages. It is assumed that neurons of pl receive inputs both from retinal R3 and from R4 neurons (Ewert, 1971; Grüsser & Grüsser-Cornehls, 1976).

Whereas the differentiation of *ventral* thalamus takes place during larval life and is finished long before the end of metamorphosis, differentiation of *dorsal* thalamus starts before metamorphosis and is completed 6 months to 1 year after metamorphosis (Clairambault, 1976). Formation of dorsal thalamus proceeds in two steps: first there is differentiation of the area dorsomedialis; then by cell migration shortly before the middle of metamorphosis, this area gives rise to an area dorsolateralis (Clairambault, 1976). The differentiation of the dorsomedial area is obviously related to the establishment of tectothalamic optic connections. It develops only in the presence of the rostral optic tectum (Straznicky & Gaze, 1972).

Sensitivity to moving configurational stimuli. Various classes of neurons could be distinguished in the caudal thalamic-pretectal region (Ewert, 1971). Among them the class TH3 neurons (type 3 of Ewert, 1971) appear to be first- or second-order postsynaptic neurons of contralateral optic fibres (see nomenclature of Grüsser & Grüsser-Cornehls, 1976). These neurons are monocularly driven and they have receptive fields of about 47° diameter. The selectivity of these neurons in the differentiation of worm-like and antiworm-like stripes is relatively weak ($r_{w,a} = 0.8$). If neurophysiological and behavioural responses are compared, TH3 neurons show, in a similar manner to retinal R4 neurons, a negative correlation coefficient for antiworm-like stripes $\{r_w; r_a\} = \{0.9; -0.9\}$. The phenomenon of negative correlation is also seen if the $D_{w,a}$ values are compared (cf. Fig. 5*B*, B and TH3). Taking the results together it can be concluded that TH3 neurons are

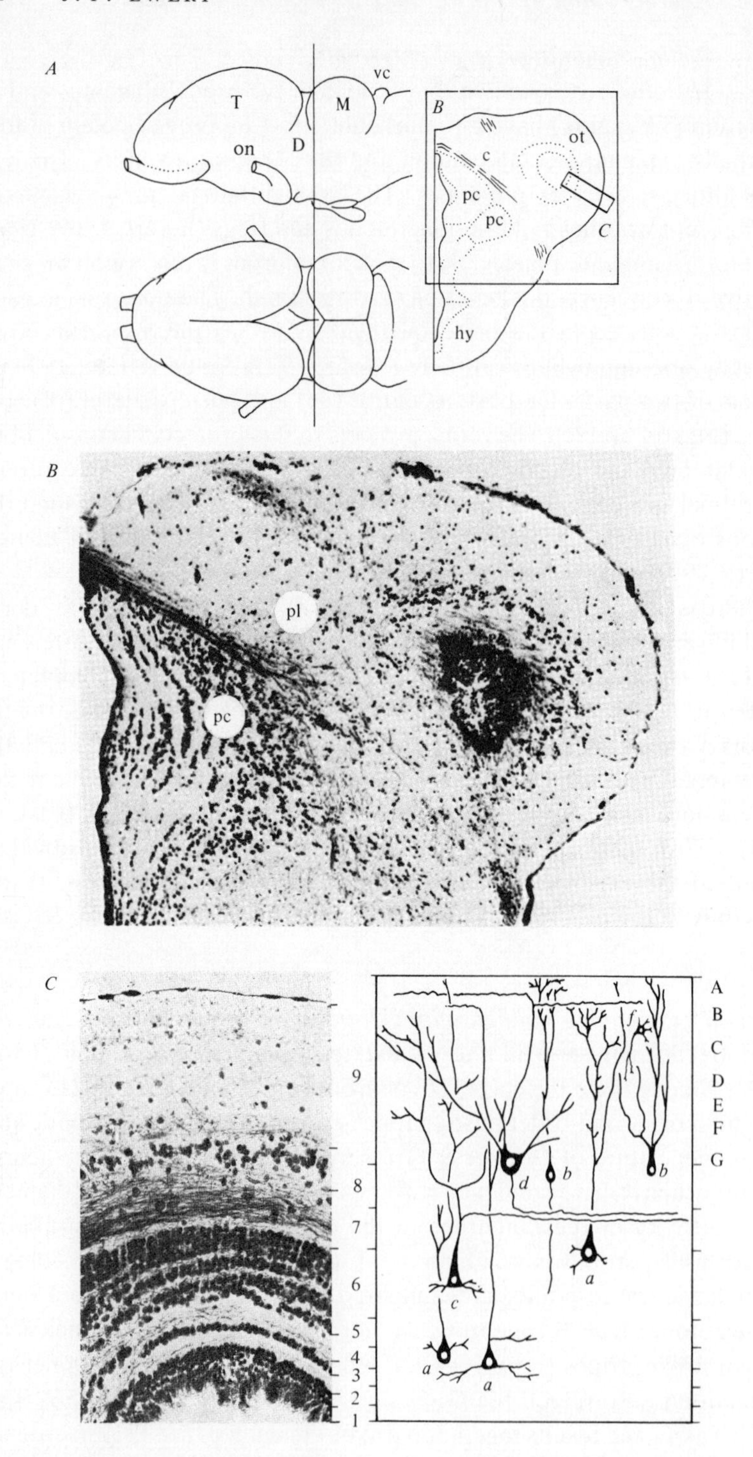
A
T
M
vc
D
on
B
ot
c
pc
pc
C
hy
B
pl
pc
C
9
8
7
6
5
4
3
2
1
A
B
C
D
E
F
G
d
b
b
f
a
c
a
a

sensitive to the extension of an object perpendicular to the direction of movement.

Since toads are able to recognize prey and predators with one eye, neurons with monocular inputs were studied in the investigation of basic gestalt perception principles. In general we know that *movement perception* (gestalt evaluation) is linked predominantly to monocular vision whereas *position detection* depends to a large extent upon binocular vision. Thus, there are particular processes in the vision and behaviour of anurans which are mediated by neurons with monocular inputs and others which are guided by cells with binocular inputs.

Retino-tectal nerve nets

Histology. The optic tectum – which is the main projection field of crossed optic nerve terminals – can be divided into nine alternating cellular and plexiform layers (Fig. 6*C*) (Gaupp, 1899). Six different types of neurons are morphologically distinguished (Fig. 6*C*): (*a*) large pear-shaped cells, (*b*) small pear-shaped cells, (*c*) large pyramidal cells, (*d*) large ganglionic cells, (*e*) bipolar cells, (*f*) small stellate and amacrine cells (Székely, Setalo & Lázár, 1973; Székely & Lázár, 1976). The retinal input to the optic tectum is fed into three (*Bufo bufo*) or four (*Rana pipiens*) different laminae (B, D, F, G) of layer 9 (Knapp, Scalia & Riss, 1965). Afferent fibres of tegmental, contralateral tectal and ipsilateral thalamic origin terminate in layers 3, 5 and 6 of the optic tectum (Lázár, 1969; Trachtenberg & Ingle, 1974). Diencephalic terminals were also identified in layers 3 and 9 including the most superficial lamina A. Both pyramidal cells and ganglionic cells have efferent axons. Fibres of layer 7 mediate the main output from the optic tectum (Fig. 6*C*).

It is assumed that tectal neurons are organized in vertically oriented functional units, called 'sensory columns' (Székely & Lázár, 1976). The axons of retinal ganglion cells terminate in the contralateral optic tectum in a retinotopic manner. In relation to the sagittal axis the retinal projection

Fig. 6. *A* and *B*, Histology of the thalamic-pretectal region and the optic tectum of *B. bufo*. Brain in lateral view: D, diencephalon; M, mesencephalon; ON, optic nerve; T, telencephalon; vc, valvula cerebelli. Section through the posterior thalamus (Klüver-Barrera stain): c, posterior commissure; hy, hypothalamus; ot, optic tectum; pc, posterocentral nucleus; pl, posterolateral nucleus. (Modified from Ewert, 1971.) *C*. *Left*, Section of the optic tectum of the common toad (Klüver-Barrera stain): *1–9*, alternating cellular and plexiform layers by Gaupp (1899); *A–G*, laminae of layer 9. *Right*, Main cell types of the frog optic tectum: *a*, large pear-shaped cell; *b*, small pear-shaped cell; *c*, large pyramidal cell; *d*, large ganglionic cell; *f*, small stellate cell. (Modified from Székely & Lázár, 1976.)

diagram in the TP region is like a mirror image of the projection diagram in the optic tectum (Figs. 7*A* and *B* and 13*C right*) (Ewert, 1971; Scalia & Fite, 1974; Ewert, Hock & von Wietersheim, 1974).

Sensitivity to moving configurational stimuli. In toads the three retinal ganglion cell classes send their axons to three laminae of layer 9. Various classes of different tectal neurons have been recorded from the optic tectum (for

Fig. 7. Topographic relations between ERF positions of tectal T5 neurons (open circles of visual field *A*) and thalamic TH3 neurons (solid circles of visual field *B*). The recording positions of the stainless steel microelectrodes in the optic tectum (*A*, open circles) or posterior thalamus (*B*, solid circles) were identified – after passing anodal DC current – in histological brain sections (*a–f*) by staining the iron deposit with the Prussian blue reaction. p, pretectum; te, optic tectum; th, posterior thalamus. (After Ewert, Hock & Wietersheim, 1974.)

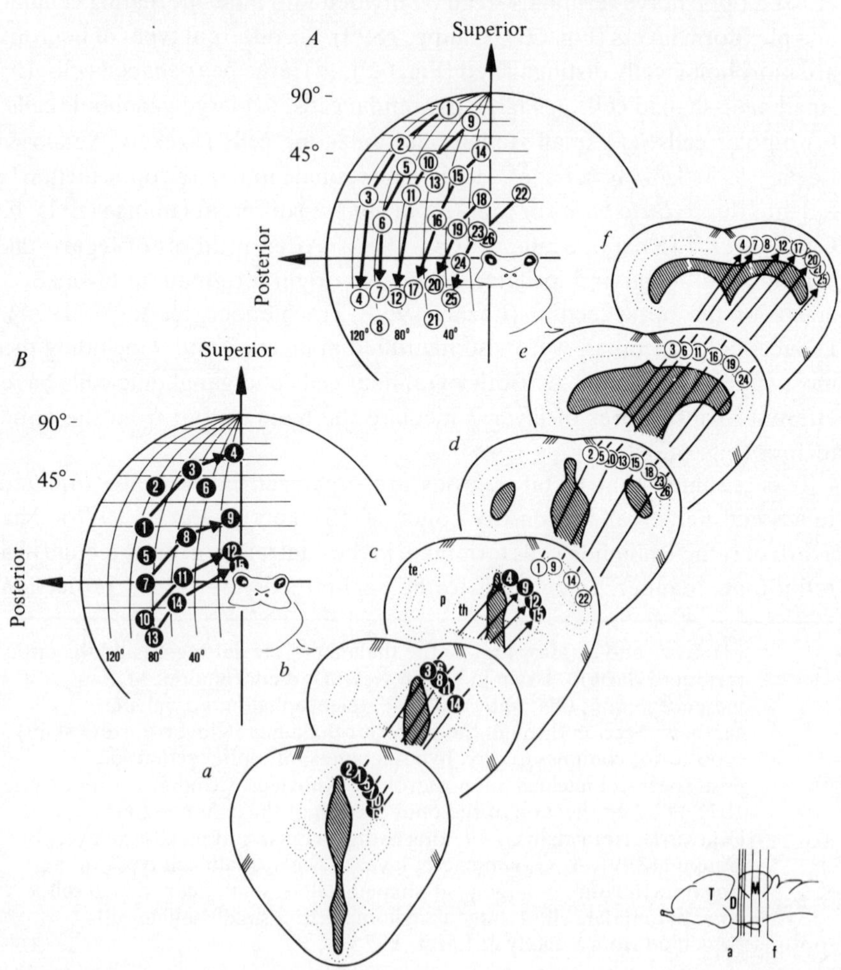

review see Grüsser & Grüsser-Cornehls, 1976). The quantitatively investigated T5 neurons (for nomenclature see Grüsser & Grüsser-Cornehls, 1976) have monocular inputs and – according to their localization in layers 7 and 8 – they presumably correspond to the small pear-shaped cells and the pyramidal cells. The horizontal diameter of the ERFs of T5 neurons is about 26°. At least two main types of neurons could be distinguished by means of physiological criteria. T5(1) neurons (type 1 of Ewert, 1974) showed relatively weak differentiation between worm-like and antiworm-like stimuli ($r_{w,a} = 0.6$). No correlation was found with the prey-catching activity: $\{r_{w}; R_{a}\} = \{0.8; 0.0\}$. If the $D_{w,a}$ values are considered (Fig. 5*B*) it appears that T5(1) neurons are mainly sensitive to extension of an object in the direction of movement.

There is at least one other type of T5 neurons, called T5(2) (type 2 of Ewert, 1974), which exhibits selective response to the configurational stimuli tested ($r_{w,a} = -0.6$). Relatively high values for positive correlation with the prey-catching activity were found, $\{r_w; r_a\} = \{0.7; 0.9\}$. This is also confirmed if the $D_{w,a}$ values are considered (Fig. 5). However, the configurational selection based upon the properties of a single neuron is not as sharp as found in the prey-catching behaviour of the toad (cf. Fig. 5*B*, B and T5(2)).

A classification of T5 neurons into at least two main types is also obtained when the $D_{w,a}$ values are calculated for stimuli traversing the ERF in different directions of the x–y coordinates (Ewert, Borchers & von Wietersheim, 1979*c*). T5(2) neurons (Fig. 8*A*) show a relatively high degree of selectivity in response to the worm-like and antiworm-like stripes tested. This relationship is found to be invariant for the direction of movement and for the stimulus angular velocity. Furthermore T5(2) neurons appear to be relatively constant according to these selective characteristics. T5(1) neurons, on the other hand, may show various degrees of configurational sensitivities for different movement directions (Fig. 8*B*–*E*). From the tectal class T5 layer responses were also recorded exhibiting properties of class TH3 neurons (Fig. 8*F*). At least in some of the T5(1) neurons the directional dependence of configurational sensitivity appears to change with time (cf. Fig. 9*A*–*C*).

Retino-tectal-thalamic integration

Configurational selectivity in the visual pathway. It is reasonable to suggest that some first steps of information processing with regard to evaluation of the configuration of a moving stimulus are performed at the retinal level (Lettvin *et al.*, 1959). The information is further processed in neuronal populations beyond the retina. Class TH3 and T5(1) neurons show different

Fig. 8. *A–F*, Discriminate values, $D_{w,a}$, for selection between worm-like (w) and antiworm-like (a) stripes of 2° × 8° size moving in different directions of the x–y coordinates through the receptive field centres of a tectal T5(2) neuron (*A*) and of T5(1) neurons (*B–F*) in the toad *B. bufo*. The values are plotted in polar coordinates. Values inside of the zero circle are negative (antiworm preference) and values outside are positive (worm preference). Thin lines correspond to the average neuronal impulse frequency and thick lines to the product of the frequency and the total number of spikes recorded during a stimulus traverse of the excitatory receptive field. (From Ewert *et al.*, 1979*c*.)

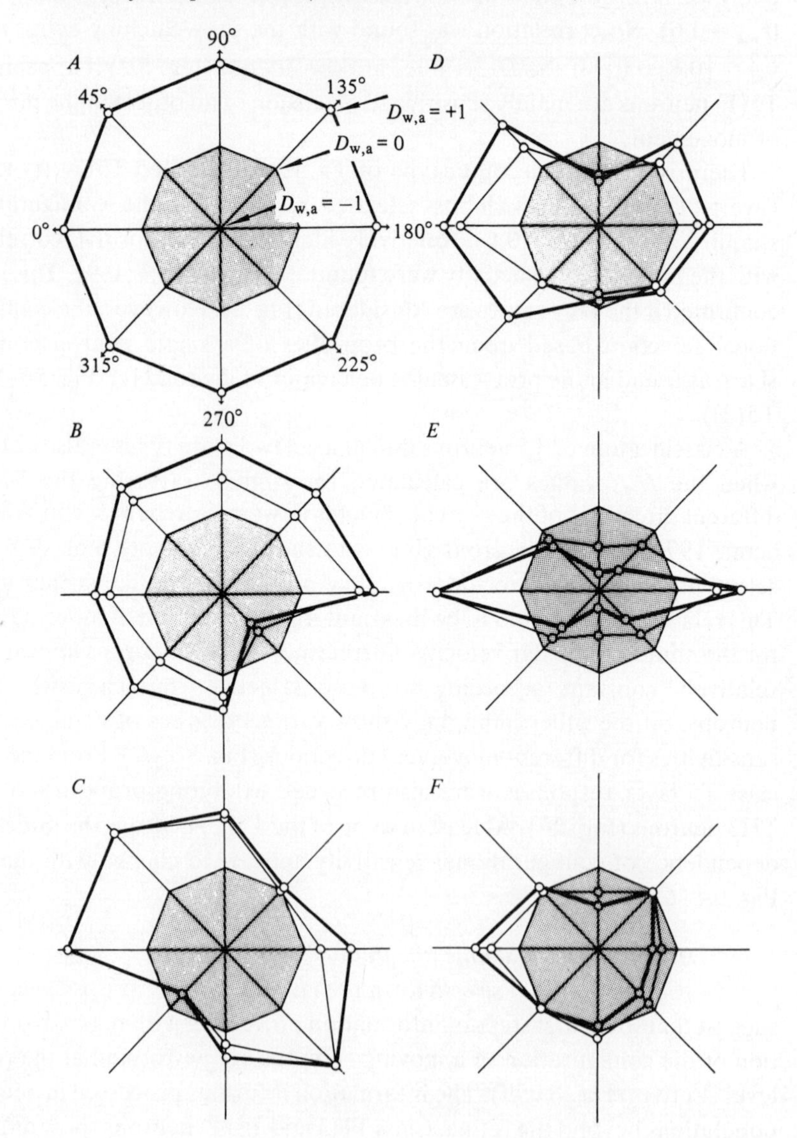

kinds of *sensitivity* to moving configurational stimuli. T5(2) neurons exhibit *selective* responses to worm-like and antiworm-like stimuli, showing worm preference. However, these neurons exhibit no worm *specificity* and they may have additional discrimination and storage properties which have not yet been quantitatively investigated.

The activity of T5(2) neurons in response to moving configurational stimuli reflects approximately the probability that a stimulus fits the prey

Fig. 9. *A–C*, Variability of $D_{w,a}$ values in the *x–y* coordinates of the receptive field from a tectal T5(1) neuron in the toad *B. bufo* (for explanation see also Fig. 8). The stimulation programme was repeated here five times, each programme lasting for 50 min. The figure shows $D_{w,a}$ plots of the first (*A*), third (*B*) and fifth (*C*) trials. (From Ewert *et al*., 1979*c*.)

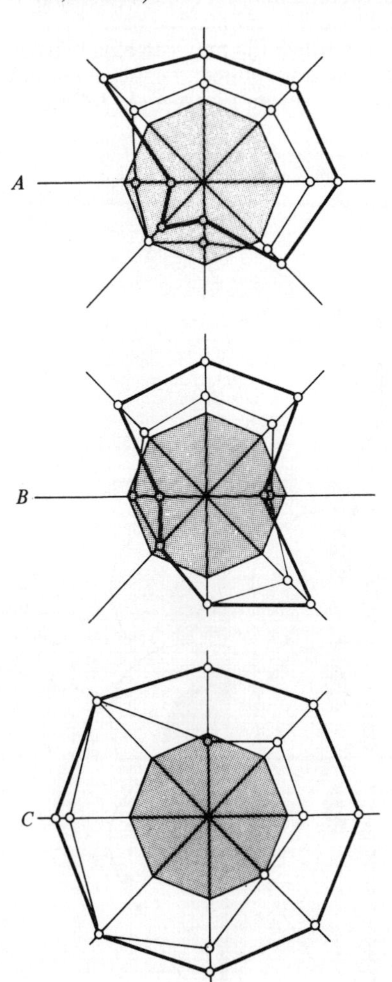

category (Ewert *et al*., 1978). However, these neurons do not appear to fulfil all of the conditions of a command neuron (Kupferman & Weiss, 1978). The estimation of the absolute stimulus size, for example, is performed in association with further neuronal populations. Furthermore the toad's configurational selectivity measured in terms of behaviour is sharper than in terms of activity of a single T5(2) neuron. It seems likely, therefore, that T5(2) *populations* are associated with a *system* which recognizes prey and commands the orienting turn of the prey-catching sequence (Fig. 13*A*).

Thalamo-tectal inhibitory actions. There is evidence that the configurational sensitivity or selectivity of T5 neurons is determined by inhibitory inputs

Fig. 10. *A–D*, Relationships between unilateral lesions (black) in the thalamic-pretectal region of the toad *B. bufo* and the part of the contralateral visual field (white) in which the prey-catching behaviour in response to moving visual stimuli is 'disinhibited', i.e. behaviourally relevant prey selection fails to occur. (From Ewert *et al*., 1974.)

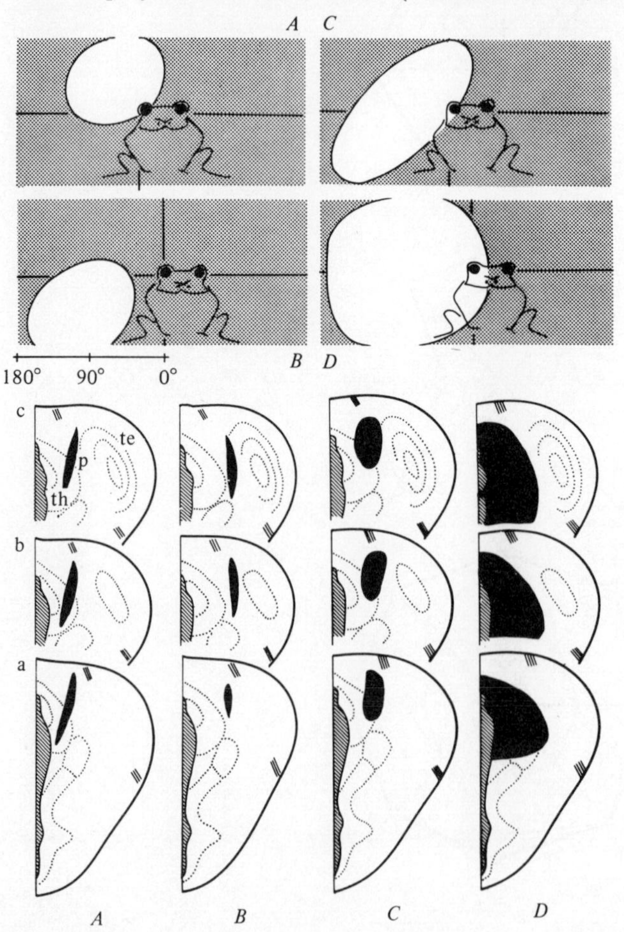

from the TP region (Fig. 12*C*) – presumably from TH3 neurons which are more strongly activated by antiworm-like than by worm-like objects. Neuroanatomical projections from pretectum to optic tectum in frogs have been demonstrated by Trachtenberg & Ingle (1974) and Wilczynski & Northcutt (1977). A schematic diagram is shown in Fig. 13*C*. Following *unilateral* TP lesions (Fig. 10) the configurational selectivity in the prey-catching behaviour is abolished in response to objects moving in the contralateral visual field (Fig. 11, cf. *A* and *B*) (Ewert, 1968). Here any moving object, irrespective of configuration, elicits prey-catching. The prey-catching response is 'disinhibited'. Also the selectivity of T5 neurons (Fig. 12*A*) located in the tectum ipsilaterally to the TP lesion, is abolished (Fig. 12*B*) (Ewert & von Wietersheim, 1974*a*).

Functional plasticity. After small TP lesions the prey-catching behaviour of the toad is disinhibited as regards objects moving in a small portion of the visual field and may become 'normal' after some hours (Fig. 10*A*, *B*). Following large TP lesions prey-catching is disinhibited for large parts of the visual field and becomes 'normal' after days or weeks (Fig. 10*D*).

Fig. 11*A*–*O* shows the prey-catching activity of a toad in response to different configurational moving stimuli prior to bilateral TP lesioning and on successive days after the lesion (*B*–*O*). In the first postoperative days

Fig. 11. *A*–*O*, Functional recovery of prey selection in the toad *B. bufo* following thalamic-pretectal lesions (*B*–*O*; *A*, prelesion). The prey-catching activity was measured in response to worm-like stripes (*a*) of different length xl_1 ($l_2 = 2.5$ mm), to antiworm-like stripes (*b*) of different length xl_2 ($l_1 = 2.5$ mm) and to squares (*c*) of varying edge length $xl_{1,2}$. All stimuli were black and were moved around the toad on a white background in a horizontal direction at an angular velocity of $15° \mathrm{s}^{-1}$ in a standard procedure. (Burghagen & Ewert, unpublished.)

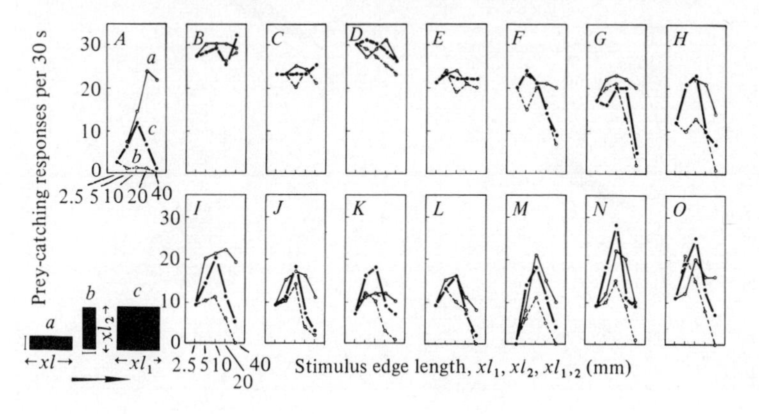

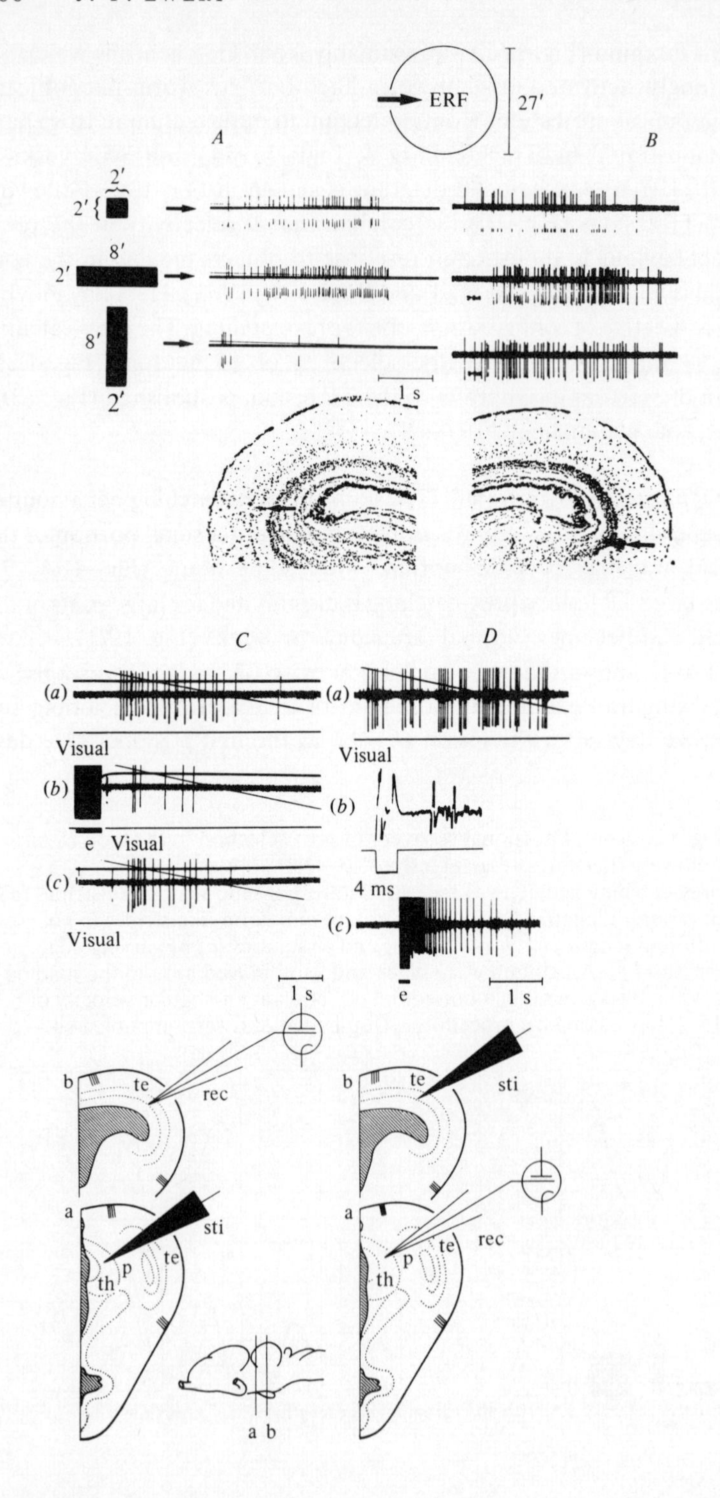
ERF
27′
A
B
2′
2′
8′
2′
8′
2′
1 s
C
D
(a)
Visual
(b)
e
Visual
(c)
Visual
(a)
Visual
(b)
4 ms
(c)
e
1 s
1 s
b
te
rec
a
sti
p
te
th
b
te
sti
a
rec
te
p
th
a b

(*B–E*) square stimuli, worm-like and antiworm-like stripes of any size elicit equally high prey-catching activity. Later (*F–O*) the activity decreases in response to very small and very large stimuli. The stimulus responses become 'normal'. However, the precise configurational selectivity of the animal prior to the lesion will not be regained! We assume that the phenomena of fast recovery after relatively small TP lesions are based on functional plasticity in the thalamo-tectal wiring system (Fig. 13*B*) and are not a result of neuroplasticity due to sprouting of axons.

There is some evidence that thalamo-tectal inhibition is mediated by a cholinergic system: after local application of curare to the optic tectum the activities of T5 neurons and of prey-catching behaviour are 'disinhibited' in response to moving objects in a way quite similar to that seen after TP lesions (for results in frogs see Stevens, 1973; in toads see Ewert & von Wietersheim, 1974*b*). Further experiments must be carried out to investigate possible intrinsic inhibitory mechanisms in the optic tectum itself. A facilitation of those mechanisms after TP lesions could explain the phenomena of functional recovery. An alternative explanation is shown in Fig. 13*B*.

Neuronal response variability. We assume that not only T5(2) but also T5(1) neurons are influenced by inhibitory inputs from the TP region (Ewert, 1980). These inputs appear to be stronger for T5(2) than for T5(1) neurons and they may be modulated, a phenomenon occurring predominantly in some of the latter (Fig. 9). What might cause variability of gestalt selectivity in T5(1) neurons?

Fig. 12. Thalamo-tectal interconnections. *A*, Records of a T5(2) neuron (see arrow) in response to three different configurational moving stimuli: *a*, 2° square; *b*, 2° × 8° worm-like stripe; c, 2° × 8° antiworm-like stripe. *B*, Records of a T5 neuron (see arrow) in response to the same stimuli after ipsilateral lesion of the thalamic-pretectal region. *C* and *D*, Physiological evidence of connections between optic tectum and thalamus-pretectum (TP) in *B. bufo*. *C(a)* Response of a tectal T5(2) neuron to a black 4° square moving at 7.6° s^{-1} against white background. *C(b)* Three minutes later; weak response to the same visual stimulus after previous electrical (e) point stimulation of the TP region with a train of negative square-wave pulses of 50 c.p.s., 5 ms pulse duration and intensity of 20 μA. *C(c)* Recovery of the visual response 30 s later. *D(a)* Response of a thalamic-pretectal large-field TH4 neuron to a 8° moving visual stimulus. *D(b)* Response to a single electrical square-wave pulse. *D(c)* Activation of the same neuron by electrical point stimulation of the optic tectum with a pulse train. p, pretectum; rec, recording electrode; sti, stimulus; te, tectum; th, posterior thalamus. (From Ewert *et al*., 1974, and Ewert, 1980.)

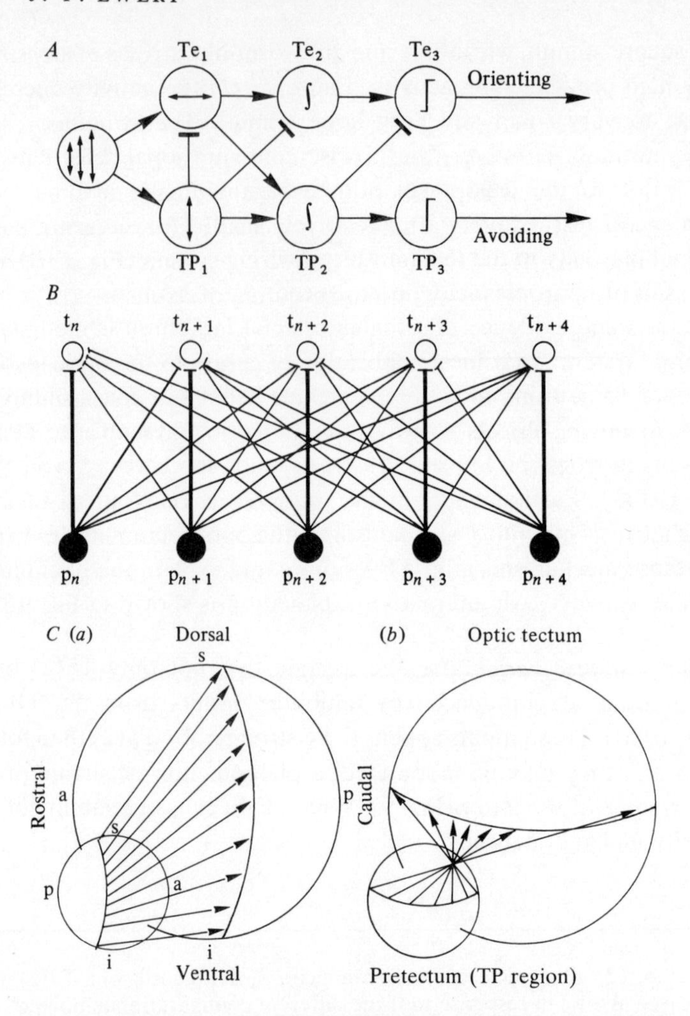

Fig. 13. Schematic representations of tectal-thalamic integration. *A*, Simple model explaining thalamo-tectal interactions in configurational prey selection of toads *B. bufo*. The first steps of worm/antiworm discrimination are interpreted as the result of inhibitory (barred lines) and excitatory (arrows) interactions between neurons in the TP region (TH3 neurons, which are sensitive to expansion of the stimulus object perpendicular to the direction of movement) and those in the optic tectum, Te_1 (T5(1) neurons, which are sensitive to object expansion in the direction of movement). The response characteristic of T5(2) neurons (Te_2) may result from excitatory inputs from T5(1) neurons (Te_1) and inhibitory inputs from TH3 neurons (TP_1). For explanations see text. (From Ewert, 1968, and Ewert & von Seelen, 1974). *B*, Schematic illustration of assumed pretectal–tectal (p–t) projections in a framework of lateral inhibition explaining functional recovery after small pretectal lesions. It is supposed that inhibition of a p-cell (p_{n+2}) acts not only on its projection cell in the optic tectum (t_{n+2}) but also spreads via

(i) The TP region itself could be a source of modulatory influence. Many neurons with changing response characteristics were found in this region, including so-called memory units (Ewert, 1971). However, nothing is known at present as to whether their (spontaneous) effects on the tectum change over time, or whether their elicited activity changes.

(ii) Variability in pattern discrimination can also be due to stimulus-specific habituation. The TP region appears to be a source involved in these processes: following TP lesions stimulus-specific habituation is remarkably decreased in prey-catching (Ewert, 1967) as well as in T5 neurons (Ingle, 1973).

(iii) Thalamic inputs to the optic tectum could be modulated by telencephalic structures. There are also projections from the telencephalon to the central layers of the optic tectum (Wilczynski, 1978).

(iv) From training experiments it is known that olfactory cues associated with feeding can modify configurational prey selection (Fig. 2*B*). Furthermore, current observations suggest that the general response level of T5(2) neurons is state-dependent (e.g. hunger state, time of day, season). This might imply an interaction of tectal neurons with telencephalic-hypothalamic systems. Projections from telencephalon via hypothalamus to the optic tectum were recently described by Wilczynski & Northcutt (1977). It is interesting to note that visually guided prey-catching behaviour fails to occur after total forebrain ablations, whereas escape avoidance behaviour, even towards small moving visual stimuli, is hyperexcited (Ewert, 1967). However, after additional TP lesions visual avoidance behaviour is abolished and prey-catching in response to any moving visual objects is hyperexcited.

Decision making in terms of neuronal events. What is the simplest model explaining worm/antiworm (prey/non-prey) discrimination (Fig. 13*A*)? After comparing the neurophysiological and behavioural results for each

---

axon collaterals over neighbouring cells of the optic tectum (t_{n+1}; t_n; t_{n+3}, t_{n+4} . . .). Then p_{n+2} strongly inhibits t_{n+2}, but t_{n+2} may also receive 'silent' inhibitory inputs from p_{n+3}, p_{n+4}; p_{n+1}, p_n . . . . After lesion of p_{n+2} the tectal area t_{n+2} is disinhibited. It is assumed that functional recovery occurs to some extent by facilitation of the 'silent' pathways from p-cells (p_{n+3}, p_{n+4}; p_{n+1}, p_n . . .) in the neighbourhood of the lesion. *C*, Schematic representation of the visual field quadrants (s, superior; i, inferior; a, anterior; p, posterior) in the optic tectum (large circle) and in the pretectal region (small circle). Part (*a*) illustrates the pretecto-tectal projection (arrows) concerning superior–inferior visual field positions; (*b*) shows the relationships for anterior–posterior visual field positions. (From Ewert *et al*., 1974.)

visual pattern tested (Fig. 5), it is evident that prey recognition cannot be the result of either retinal, or retinal-tectal, or retinal-pretectal information processing alone. Prey recognition obviously proceeds in steps: (i) a preliminary transformation of different stimulus parameters, such as size, velocity, and contrast at the retinal level; (ii) a particular amount of gestalt analysis by means of different neuronal populations (TP_1: pretectal TH3 neurons; Te_1: tectal T5(1) neurons; cf. Fig. 13*A*); (iii) the gestalt recognition process based on inhibitory and excitatory interactions between TP and Te neurons (Fig. 13*A*). For example, if a prey stimulus traverses the visual field of the toad the tectal Te_1 system is well activated whereas the pretectal TP_1 system shows only weak responses. Te_1 in turn activates another tectal system (Te_2: T5(2) neurons), which is part of a command system for the orienting movement towards prey. In the case of a moving antiworm stimulus the TP_1 system is more strongly activated than Te_1. As a result of inhibitory connections from TP_1 to Te_2, the threshold excitation for the Te_2 system will not be reached. In the case of a large moving stimulus (predator) the TP_1 and Te_1 systems are both well activated, TP_1 more strongly than Te_1. Because $TP_1 > Te_1$, the Te_2 system is inhibited by TP_1, whereas both TP_1 and Te_1 activate another pretectal TP_2 system which, functioning as an 'AND gate', may be part of a command system for avoidance or escape behaviour. There is anatomical as well as physiological evidence for inhibitory and excitatory connections between TP and Te (e.g. Fig. 12*A–D*).

Conclusions

We shall keep in mind that the worm/antiworm phenomenon reported here is a useful tool for the experimenter to test pattern recognition with simple behaviourally meaningful stimuli ('key stimuli'). However, this should not imply the assumption that the central nervous system of the toad is as simple as illustrated in Fig. 13*A*, though such wiring obviously resembles an important feature of a configurational recognition system. On the other hand it would be trivial to overemphasize that the prey recognition system is more complex than has been shown here. Indeed the results on stimulus-specific habituation and learning cannot be simply explained by our present knowledge of the response properties of retinal, tectal and thalamic neurons and thus by the two-stage filter model (Fig. 13*A*).

More generally, however, we assume that configurational prey recognition in the common toad is partly based on preprogrammed 'hardwired systems' forming the fundamentals of the innate releasing mechanism (Lorenz, 1954). Other parts may be based on 'softwired systems', opening the possibility of modification (Schleidt, 1962) that enables classification of stimulus distributions in the environment into learned classes of functional significance. Whether those systems exist as parallel structures or whether there is one fundamental structure, which is modified according to individual experience, is a question which is being investigated.

It has been shown how the toad recognizes certain features of its prey both from a behavioural and a neurophysiological point of view. Further research may help to classify the relationship between the neuronal analysis of sensory information and the ultimate response of the animal in terms of behaviour.

This work was supported by the Deutsche Forschungsgemeinschaft EW 7/1–6 and the Foundations Fund for Research in Psychiatry No. 669–461.

References

Beck, A. & Ewert, J.-P. (1979). Prey selection by toads (*Bufo bufo* L.) in response to configurational stimuli moved in the visual field *z*, *y*-coordinates. *Journal of Comparative Physiology*, **129**, 207–9.

Birukow, G. & Meng, M. (1955). Eine neue Methode zur Prüfung des Gesichtssinnes bei Amphibien. *Naturwissenschaften*, **42**, 652–3.

Borchers, H.-W. & Ewert, J.-P. (1979). Correlation between behavioral and neuronal activities of toads *Bufo bufo* (L.) in response to moving configurational prey stimuli. *Behavioral Processes*, **4**, 99–106.

Borchers, H.-W., Burghagen, H. & Ewert, J.-P. (1978). Key stimuli of prey for toads: configuration and movement pattern. *Journal of Comparative Physiology*, **128**, 189–92.

Brzoska, J. & Schneider, H. (1978). Modification of prey-catching behavior by learning in the common toad (*Bufo b. bufo* (L.), Anura, Amphibia): changes in responses to visual objects and effects of auditory stimuli. *Behavioral Processes*, **3**, 125–36.

Clairambault, P. (1976). Development of the prosencephalon. In *Frog neurobiology*, ed. R. Llinàs & W. Precht, pp. 926–45. Springer, Berlin, Heidelberg & New York.

Eibl-Eibesfeldt, I. (1951). Nahrungserwerb und Beuteschema der Erdkröte (*Bufo bufo* L.). *Behaviour*, **4**, 1–35.

Eikmanns, K.-H. (1955). Verhaltensphysiologische Untersuchungen über den Beutefang und das Bewegungssehen der Erdkröte (*Bufo bufo* L.). *Zeitschrift für Tierpsychologie*, **12**, 229–53.

Ewert, J.-P. (1967). Untersuchungen über die Anteile zentralnervöser Aktionen an der taxisspezifischen Ermüdung beim Beutefang der Erdkröte (*Bufo bufo* L.). *Zeitschrift für vergleichende Physiologie*, **57**, 263–98.

Ewert, J.-P. (1968). Der Einfluss von Zwischenhirndefekten auf die Visuomotorik im Beutefang- und Fluchtverhalten der Erdkröte (*Bufo bufo* L.). *Zeitschrift für vergleichende Physiologie*, **61**, 41–70.

Ewert, J.-P. (1969). Quantitative Analyse der Reiz-Reaktionsbeziehungen bei visuellem Auslösen der Beutefang-Wendereaktion der Erdkröte (*Bufo bufo* L.). *Pflügers Archiv*, **308**, 225–43.

Ewert, J.-P. (1971). Single unit response of the toad's (*Bufo americanus*) caudal thalamus to visual objects. *Zeitschrift für vergleichende Physiologie*, **74**, 81–102.

Ewert, J.-P. (1974). The neural basis of visually guided behavior. *Scientific American*, **230**, 34–42.

Ewert, J.-P. (1980). *Neuroethology. An Introduction to the Fundamentals of Behavior*. Springer, Berlin, Heidelberg & New York.

Ewert, J.-P. & Burghagen, H. (1979*a*). Configurational prey selection by *Bufo, Alytes, Bombina* and *Hyla. Brain, Behavior and Evolution*, **16**, 157–75.

Ewert, J.-P. & Burghagen, H. (1979*b*). Ontogenetic aspects on visual 'size-constancy' phenomena in the midwife toad *Alytes obstetricans* (Laur.). *Brain, Behavior and Evolution*, **16**, 99–112.

Ewert, J.-P. & Gebauer, L. (1973). Grössenkonstanzphänomene im Beutefangverhalten der Erdkröte (*Bufo bufo* L.). *Zeitschrift für vergleichende Physiologie*, **85**, 303–15.

Ewert, J.-P. & Hock, F. J. (1972). Movement sensitive neurons in the toad's retina. *Experimental Brain Research*, **16**, 41–59.

Ewert, J.-P. & Kehl, W. (1978). Configurational prey-selection by individual experience in the toad *Bufo bufo. Journal of Comparative Physiology*, **126**, 105–14.

Ewert, J.-P. & Seelen, W. von (1974). Neurobiologie und System-Theorie eines visuellen Muster-Erkennungsmechanismus bei Kröten. *Kybernetik*, **14**, 167–83.

Ewert, J.-P. & Siefert, G. (1974). Seasonal change of contrast-detection in the toad's (*Bufo bufo* L.) visual system. *Journal of Comparative Physiology*, **94**, 177–86.

Ewert, J.-P. & Wietersheim, A. von (1974*a*). Musterauswertung durch Tectum- und Thalamus/Praetectum-Neurone im visuellen System der Kröte (*Bufo bufo* L.). *Journal of Comparative Physiology*, **92**, 131–48.

Ewert, J.-P. & Wietersheim, A. von (1974*b*). Der Einfluss von Thalamus/Praetectum-Defekten auf die Antwort von Tectum-Neuronen gegenüber visuellen Mustern bei der Kröte (*Bufo bufo* L.). *Journal of Comparative Physiology*, **92**, 149–60.

Ewert, J.-P., Arend, B., Becker, V. & Borchers, H.-W. (1979*a*). Invariants in configurational prey selection by *Bufo bufo* (L.). *Brain, Behavior and Evolution*, **16**, 38–51.

Ewert, J.-P., Borchers, H.-W. & Wietersheim, A. von (1978). Question of prey feature detectors in the toad's *Bufo bufo* (L.) visual system. *Journal of Comparative Physiology*, **126**, 43–7.

Ewert, J.-P., Borchers, H.-W. & Wietersheim, A. von (1979*c*). Directional sensitivity, invariance and variability of tectal T5 neurons in response to moving configurational stimuli in the toad *Bufo bufo* (L.). *Journal of Comparative Physiology*, **132**, 191–201.

Ewert, J.-P., Hock, F. J. & Wietersheim, A. von (1974). Thalamus/Praetectum/Tectum: Retinale Topographie und physiologische Interaktionen bei der Kröte (*Bufo bufo* L.). *Journal of Comparative Physiology*, **92**, 343–56.

Ewert, J.-P., Krug, H. & Schönitz, G. (1979*b*). Activity of retinal class R3 ganglion cells in the toad *Bufo bufo* (L.) in response to moving configurational stimuli: influence of the movement direction. *Journal of Comparative Physiology*, **129**, 211–15.

Freisling, J. (1948). Studien zur Biologie und Physiologie der Wechselkröte (*Bufo viridis* Laur.). *Österreichische Zoologische Zeitschrift* (*Vienna*), **1**, 383–440.

Gaupp, E., Ecker, A. & Wiedersheim, R. (1899). *Anatomie des Frosches*. Friedrich Vieweg and Son, Braunschweig.
Gibson, J. J. (1950). The perception of visual surfaces. *American Journal of Psychology*, **63**, 367–84.
Grüsser, O.-J. & Grüsser-Cornehls, U. (1976). Neurophysiologie of the anuran visual system. In *Frog Neurobiology*, ed. R. Llinás & W. Precht, pp. 297–385. Springer, Berlin, Heidelberg & New York.
Himstedt, W., Freidank, U. & Singer, E. (1976). Die Veränderung eines Auslösmechanismus im Beutefangverhalten während der Entwicklung von *Salamandra salamandra* (L.). *Zeitschrift für Tierpsychologie*, **41**, 235–43.
Hinsche, G. (1935). Ein Schnappreflex nach 'Nichts' bei Anuren. *Zoologischer Anzeiger*, **111**, 113–22.
Ingle, D. (1968). Visual releasers of prey-catching behavior in frogs and toads. *Brain, Behavior and Evolution*, **1**, 500–18.
Ingle, D. (1973). Disinhibition of tectal neurons by pretectal lesions in the frog. *Science*, **180**, 422–4.
Knapp, H., Scalia, F. & Riss, W. (1965). The optic tracts of *Rana pipiens. Acta Neurologica Scandinavica*, **41**, 325–55.
Kupferman, I. & Weiss, K. R. (1978). The command neuron concept. *Behavioral and Brain Science*, **1**, 3–39.
Lázár, Gy. (1969). Efferent pathways of the optic tectum in the frog. *Acta Biologica Academiae Scientarum Hungaricae*, **20**, 171–83.
Lázár, Gy. (1971). The projection of the retinal quadrants on the optic centers in the frog. *Acta Morphologica Academiae Scientarum Hungaricae*, **19**, 325–34.
Lettvin, J. Y., Maturana, H. R., McCulloch, W. S. & Pitts, W. H. (1959). What the frog's eye tells the frog's brain. *Proceedings of the IRE*, **47**, 1940–51.
Lorenz, K. (1954). Das angeborene Erkennen. *Natur und Volk*, **84**, 285–95.
Pomeranz, B. (1972). Metamorphosis of frog vision. Changes in ganglion cell physiology and anatomy. *Experimental Neurology*, **34**, 187–99.
Scalia, F. & Fite, K. V. (1974). A retinotopic analysis of the central connections of the optic nerve in the frog. *Journal of Comparative Neurology*, **158**, 455–78.
Schleidt, W. (1962). Die historische Entwicklung der Begriffe 'Angeborenes auslösendes Schema' und 'Angeborener Auslösmechanismus' in der Ethologie. *Zeitschrift für Tierpsychologie*, **19**, 697–722.
Stevens, R. J. (1973). A cholinergic inhibitory system in the frog optic tectum: its role in visual electrical responses and feeding behavior. *Brain Research*, **49**, 309–23.
Straznicky, K. & Gaze, R. M. (1972). The development of the tectum in *Xenopus laevis*: an autoradiographic study. *Journal of Embryology and Experimental Morphology*, **26**, 67–79.
Székely, G. & Lázár, Gy. (1976). Cellular and synaptic architecture of the optic tectum. In *Frog Neurobiology*, ed. R. Llinás & W. Precht, pp. 407–34. Springer, Berlin, Heidelberg & New York.
Székely, G., Setalo, G. & Lázár, Gy. (1973). Fine structure of the frog's optic tectum: optic fibre termination layers. *Journal für Hirnforschung*, **14**, 189–225.

Tinbergen, N. (1948). Social releasers and the experimental method required for their study. *Wilson Bulletin*, **60**, 6–52.
Trachtenberg, M. C. & Ingle, D. (1974). Thalamus-tectal projections in the frog. *Brain Research*, **79**, 419–30.
Wiens, J. A. (1970). Effects of early experience on substrate pattern selection in *Rana aurora* tadpoles. *Copeia*, **3**, 543–8.
Wiens, J. A. (1972). Anuran habitat selection: early experience and substrate selection in *Rana cascadae* tadpoles. *Animal Behaviour*, **20**, 218–20.
Wilczynski, W. (1978). 'Connection of the midbrain auditory center in the bullfrog *Rana catesbeiana*'. PhD dissertation, University of Michigan, Ann Arbor.
Wilczynski, W. & Northcutt, R. G. (1977). Afferents to the optic tectum of the leopard frog: an HRP study. *Journal of Comparative Neurology*, **173**, 219–29.

PART III

Sleep, wakefulness and arousal

COLIN M. SHAPIRO, CLIFFORD J. WOOLF & D. BORSOOK

Sleep ontogeny in fish

Introduction

Why man or any other animal sleeps is yet unknown. One of the potentially valuable approaches to this problem is a phylogenetic one. Although it is generally accepted that mammals and birds sleep and have both dreaming (REM) and non-REM sleep, the position *vis-à-vis* reptiles, amphibians and fish is not clear. A few studies on reptiles have shown sleep to occur (Flanigan, Wilcox & Rechtschaffen, 1973), but contradictory claims have been made (Walker & Berger, 1973). Different species of amphibians have also yielded contradictory results (Hobson, Goin & Goin, 1968; Lazarev, 1978*a*, *b*). There are perhaps less than half a dozen studies on the subject of fish sleep (for references see Shapiro & Hepburn, 1976) although numerous anecdotal accounts of what is perceived as fish sleep are available from fishermen and tropical-fish keepers. Circadian rhythms have been recognized to occur in fish (Aschoff, 1960). Diurnal and nocturnal activity of different species of fish were recognized from early fish-catch (Hart, 1931) and laboratory studies (Spencer, 1939). The questions of fish sleep is of interest because it may add further understanding to various proposed general theories of sleep function. These theories include: sleep as a function of thermoregulatory control (Allison & Van Twyver, 1970), sleep as a function of development and plasticity of binocular vision and visual acuity (Berger, 1969; Tauber & Weitzman, 1969), sleep as a restorative process (Adam & Oswald, 1977), and sleep as a function of memory processing (Fishbein, 1970; Empson & Clarke, 1970). With respect to the last theory it has been shown in our laboratory (Shapiro & Girdwood, 1981), amongst others (Hartmann & Stern, 1972), that the rebound in REM sleep following deprivation is associated with increased rates of protein synthesis in various sites within the brain. It has been suggested that REM deprivation causes defective functioning of central catecholamine neuronal systems which in turn leads to poorer learning in this state (Bobillier, Sakai & Jouvet, 1974). The phylogeny of memory development indicates that only short-term memory is observed in fish (Beritashvili, 1973; Borsook *et al.*, this volume). From this one might expect (if the sleep–memory hypothesis is correct) that there would be some form of sleep in fish but that this would not be fully developed.

Lazarev has suggested on the basis of electroencephalogram (EEG) patterns in frogs that elements of paradoxical (REM) sleep evolved in poikilotherms as a precursor of REM sleep in homoiotherms. It would seem unlikely that homoiothermy is associated with REM sleep as there is a malfunctioning of temperature regulation during this type of sleep (Shapiro *et al*., 1974; Parmeggiani *et al*., 1977). The question of homoiothermy and sleep is further complicated by the observation (Fry & Hochachka, 1970) that there is partial regulation of body temperature in certain fish. On the basis of phylogenetic studies, Ruckenbusch & Toutain (1977) conclude that sleep is 'an adaptive non-responding system which appears when feed need and security are satisfied'. Although they point to studies of rest–activity cycles in invertebrates (e.g. octopus) they are only satisfied that EEG criteria of sleep are accepted in birds and mammals. This paper deals with the ontogenetic development of behavioural sleep in fish. This has not been studied previously, but in general terms Gibson, Blaxter & de Groot (1978) have concluded (in considering activity rhythms) that 'the type of rhythmic pattern exhibited by fish at each stage in their life history is appropriate to their particular environment'.

We have previously shown that sleeping behaviour occurs in the schooling fish *Tilapia mossambica* (Shapiro & Hepburn, 1976). The definition of sleep used in this study was the criteria of Flanigan *et al*. (1973), which include: behavioural quiescence, stereotypic posture, elevated behavioural thresholds, and very rapid state-reversibility following relatively intense stimulation. The reasons for using these criteria of sleep when studying fish are threefold. Firstly, they agree with a popular conception of what sleep is, irrespective of species; secondly, the use of EEG techniques (the usual mammalian standard) has certain problems in a free-ranging aquatic species (these are dealt with in the following two chapters) and studies of electrical activity in fish have yielded contradictory conclusions (Marshall, 1972; Peyrethon & Dusan-Peyrethon, 1967); and thirdly, it has been shown that the anatomical centres seemingly responsible for mammalian sleep are not present in fish (Broughton, 1972; Rukenbusch & Toutain, 1977). Recently Voronov *et al*. (1977) have shown the presence of 'cholinoreceptive brain neurons' which are involved in the functional systems which generate 'swimming-automatism – rhythmic movements of the tail'. They consider this swimming-automatism to be the precursor of paradoxical sleep.

In our earlier experiments we found that immature fish did not show the behaviour patterns of sleep that we observed in adults of the species

(Shapiro & Hepburn, 1976), and studies of the ontogenetic development of sleep have shown marked differences in the sleep patterns of individuals of a species at various ages (Roffwarg, Muzio & Dement, 1966). This variation relates to the amount of sleep, the order of occurrence of the different stages of sleep, and the type of sleep. We therefore thought it worthwhile to study the development of sleep behaviour in a species of fish previously studied, as well as the interaction of adult and juvenile individuals in this schooling species, as group behaviour had previously appeared to influence the sleep patterns of individuals. Olla & Studholme (1978) have reported differences in activity cycles in adult and young of a species.

Method

Observations in other experiments have established that the species used in this study, *T. mossambica* (Perciformes: Cichlidae), is very sensitive to movements such as are caused by persons walking near the tank (Shapiro & Hepburn, 1976). For this reason the 50-l experimental tanks were made opaque on three sides and the front fitted with a sheet of solar-shield one-way glass. In addition to preventing the fish from detecting outside movements this glass has the advantage of allowing direct observation of the fish when the room lighting is dim with respect to tank illumination. The tanks were covered with a close-fitting roof with light bulbs placed at each end. A small opening at one end of the tank provided entry for an air-supply, water-filtering and food.

There were two tanks each containing 15 fish, which were fed daily on hand-rolled bread pellets. This method of feeding had been found to be satisfactory in other experiments (Hepburn *et al.*, 1973). Feeding experiments were conducted only after the fish had been trained to accept the pellets and had been adapted to a 15-h light/9-h dark photoperiod. Except during specific study periods the fish were fed shortly after the onset of the light period, at a rate of 1 pellet per 6 s for 1 min with a 1-min gap between each series of 10 pellets. The number of pellets falling to the bottom of the tank uneaten was noted for each series, and once this reached five or more in a set of ten it was assumed that the fish were replete and feeding was stopped. The fish were disturbed only for tank cleaning, twice a week.

To observe the fish under laboratory conditions at night, the tank lights were connected to a variac so that the lights could be suitably dimmed at the end of the (artificial) light periods. The tanks were kept in a photographic darkroom to obviate undesirable light leaks. Lines were painted on the back walls of the tanks thus dividing them into quadrants. This allowed a crude quantitative measure of fish activity in terms of the displacement of any one fish from any quadrant to another per unit time.

A simple stimulator connected to two wire leads with metal plates at their ends that were placed at diagonally opposite ends of the tank was used to measure the effects of electrical stimuli during different activity periods. Previous experiments (Shapiro & Hepburn, 1976) had established that the first discernible response to electrical stimuli involved gulping, an alternate opening and closing of the mouth. This response was used to measure response threshold, which was taken as the lowest stimulus level at which any of the fish displayed gulping. In the test for latency of feeding, the first pellet eaten by any fish in the tank was taken as the latency value. It should be clearly noted that in no case were individual animals studied. Only populations of fish were considered because it was felt that this approach would be more natural in view of the fact that *T. mossambica* is day-active and forms schools in its normal environment.

To evaluate development of sleep patterns two approaches were used. Firstly, two groups of 15 juvenile fish were placed in two separate tanks and observed with respect to the development of sleep behaviour. Secondly, a further two groups of 10 juvenile fish were placed in two tanks into which five adult fish (living under a similar light/dark regime) were introduced at fortnightly intervals for periods of 4 days. The former situation has the advantage of depriving the juvenile fish of the adult behaviour pattern which may be an influencing factor in the development of sleep behaviour in the normal situation. The groups of juvenile fish with adults periodically introduced had the disadvantage of a change in the number present in the 'tank society' on the fourteenth to eighteenth day of each 18-day cycle but the advantage of the possibility of modelling their behaviour on that of the adults.

Results

Under conditions of a 15-h light/9-h dark cycle the gross behaviour of *T. mossambica* adults can be divided as follows. During the light phase two behavioural patterns were observed: a very active pattern, referred to as state A, characterized by much swimming (more than 10 fish displacements per min), and a second more passive phase (state B) of three or fewer displacements per min with less than three fish resting on the bottom of the tank at any time. Fish in state A exhibited higher respiratory activity, as judged by gill movement, than did state B fish. These two phases alternated throughout the light period with an average of a 20 min active state A behaviour followed by 10 min of the passive state B behaviour. However, both of these states were quite labile and readily altered by environmental stimuli such as feeding.

At the onset of the dark period there was a marked change in the behaviour of the fish compared with that in the light period. For an initial period of 90± 30 min there was a general reduction in swimming and a gradual downward vertical movement of the fish. This was followed by an extended period of some 6.5 to 7.5 h during which at least 12 of the 15 fish in the tank were resting on the bottom of the tank (state C). The respiratory rate of state C fish was, however, no different from that of state B and no eye movement of state C fish could be observed (whereas it could in states A and B). Towards the end of the dark period there was more irregular behaviour: several fish actively swam about, others remained resting on the bottom of the tank and yet others remained stationary in the middle of the tank.

With the onset of light the fish that had been actively swimming around immediately returned to the bottom of the tank (cf. Davis, 1962, who showed that the recovery time from this light shock was dependent on the duration of the dark period) and exhibited a peculiarly distinct behaviour for about 5 min, as follows. The fish were initially still (about 30 s) (cf. Spencer, 1939; John & Haut, 1966) then began to swim backwards in small circles for about 90 s. After this, there was a 1–3 min. rest period during which the colour pattern of the fish changed from one of broad vertical stripes through a bleached stage to one of only one or two horizontal stripes. These changes may be brought about by an interaction of melatonin and serotonin that has recently been shown (Satake, 1979). The fish then exhibited darting behaviour for a minute or two, after which there was intense general swimming activity (more than 45 displacements per min) and strong schooling behaviour of alternating A and B states.

Electrical stimuli given to fish in states A, B and C showed that stimulatory thresholds for the A and B fish were not significantly different but that the threshold for C was significantly higher than for A or B. Furthermore, states A and B fish did not differ with respect to the time delay before the eating of the first bread pellet when fed, whereas during state C there was a significant delay before the onset of feeding.

Over a period of 22 weeks the groups of purely juvenile fish developed a sleep pattern which could not be distinguished from that observed previously in adults. In both groups of juvenile fish not exposed to adults, the order of emergence of individual components of sleep behaviour was identical.

The first observed component was that of behavioural quiescence as defined above, i.e. quiescence of the whole group simultaneously as opposed to individually. This occurred at approximately 5 weeks. Second, after a further 2 weeks, followed the stereotypic posture, viz. all fish 'resting' on the bottom of the tank. Then followed components of the waking behaviour. This evolved in the following order (with duration since the previous step's emergence in parentheses): darting behaviour (1 week), alteration of pigment (3 weeks), initial stillness (2 weeks), backward circling (4 weeks), rest period following circling (1 week). Two weeks after this elevated behavioural thresholds in the fish could be detected by both electrical and feeding stimulation techniques. The last feature to emerge was that of the strong schooling behaviour which occurred after the 'wake-up' routine. Throughout all these stages rapid state reversibility following relatively intense stimulation could be observed.

In the tanks in which adults were introduced at regular intervals, a similar pattern of sleep behaviour was observed to develop. However, there were two distinct differences from the tanks where there were juveniles alone. Firstly, all components occurred at an earlier stage, i.e. the introduction of the adult fish 'allowed the juveniles to mature more rapidly'. For the early components of behaviour this shortening of behavioural development was approximately 2 weeks, but increased so that the final expressions of adult-type behaviour occurred approximately 5 weeks earlier. These results are summarized in Table 1. The second difference was that of the emergence of the strong schooling behaviour following 'waking-up', which appeared last (ninth observable component of behaviour) in the tanks of juveniles but fourth (i.e. after darting, at approximately 8 weeks) in the tanks in which adults were periodically placed.

Discussion

This study adds further support to the few studies documenting sleep in fish (Tauber & Weitzman, 1969; Marshall, 1972; Titkov, 1976).

Table 1. *Features of ontogeny of sleep in groups of* T. mossambica

	Latency of emergence of feature (weeks)	
Behavioural features	Juveniles only	Adults periodically introduced
Behavioural quiescence	5	3
Stereotypic posture	7	5
Darting behaviour	8	6
Alteration of pigment	11	10
Initial stillness	13	11
Backward circling	17	13
Rest period	18	14
Elevated threshold	20	17
Schooling behaviour	22	8

Furthermore we have shown a form of behavioural ontogenetic evolution of sleep in a schooling fish. It should be noted that an ontogenetic component to sleep has been shown in several mammalian species (Jouvet-Mounier, Astic & Lacote, 1969) and in general it has been found that the greater the immaturity of the animal at birth the greater the amount of time spent in REM sleep in the perinatal period. The development of sleep behaviour in *T. mossambica* is influenced by the presence of adults. Whether this influence is of adaptive value in this gregarious species we cannot evaluate at present, but both the changes induced by the presence of adult fish – more rapid development of sleep behaviour in general and specifically earlier development of schooling behaviour – suggest that inter-individual influences may be of adaptive significance. A diurnal variation in the tendency to school in response to an external stimulus has previously been shown (Thines & Vandenbussche, 1966). We consider the observation that 'schooling' as a part of the sleep–waking behaviour is influenced by the adult fish is worthy of further attention.

In referring to the question of sleep in fish Blaxter (1978) has pointed out that schools break up at 'night'. This does much to imply that in studying fish sleep only individuals should be considered, and the behaviour of the group has been emphasized in this study. We consider that the form of sleep observed in fish is neither a 'rest–activity' cycle nor the expression of a sleep system equivalent to that seen in mammals. The behavioural observations support a more definite sleep state than the terminology 'proto-sleep' (Karmanova, Titkov & Popova, 1976) implies. The differences between mammalian and fish brain functions are numerous and obvious and one would not expect sleep to express itself with similar EEG patterns in the two taxa. The aspects of similarity of brain function are sometimes surprising (see Borsook *et al*., this volume), and should give further impetus to the search for common denominators in neurophysiological mechanisms of sleep. One example of such a common denominator may be the nocturnal drop in brain serotonin in certain species of fish (Matty, 1978). This observation, together with those of Voronov *et al.* (1977) and Satake (1979) mentioned above, may lead to a neurophysiological model of sleep in fish comparable to the aminergic models of sleep in mammals (Jouvet, 1967).

References

Adam, K. & Oswald, I. (1977). Sleep is for tissue restoration. *Journal of the Royal College of Physicians*, **11**, 376–88.

Allison, T. & Van Twyver, H. (1970). The evolution of sleep. *Natural History*, **79**, 56–65.

Aschoff, J. (1960). Exogenous and endogenous components in circadian rhythms. *Cold Spring Harbor Symposia on Quantitative Biology*, **25**, 11–28.

Berger, R. J. (1969). Oculomotor control: a possible function of REM sleep. *Psychological Review*, **76**, 144–64.
Beritashvili, I. S. (1972). Phylogeny of memory development in vertebrates. In *Brain and Human Behaviour*, ed. A. G. Karczmar & T. C. Eccles, pp. 341–51. Springer Verlag, New York.
Blaxter, J. H. S. (1978). Summary of symposium on rhythmic activity in fish. In *Rhythmic Activity of Fishes*, ed. J. E. Thorpe, pp. 285–8. Academic Press, New York & London.
Bobillier, P., Sakai, F. & Jouvet, M. (1974). The effect of sleep deprivation upon the in vivo and in vitro incorporation of tritiated amino acids into brain proteins in the rat at three different age levels. *Journal of Neurochemistry*, **22**, 23–31.
Broughton, R. (1972). Phylogenetic evolution of sleep systems. In *The Sleeping Brain*, ed. M. H. Chase, pp. 2–7. Brain Research Institute, Los Angeles.
Davis, R. E. (1962). Daily rhythm in the reaction of fish to light. *Science*, **137**, 430–2.
Empson, J. A. C. & Clarke, P. R. F. (1970). Rapid eye movements and remembering. *Nature, London*, **227**, 287–8.
Fishbein, W. (1970). Interference with conversion of memory from short-term to long-term storage by partial sleep deprivation. *Communications in Behavioural Biology*, **5A**, 171–5.
Flanigan, W. F., Wilcox, R. H. & Rechtschaffen, A. (1973). The EEG and behavioural continuum of the crocodilian, *Caiman sclerops*. *Electroencephalography and Clinical Neurophysiology*, **34**, 521–38.
Fry, F. E. J. & Hochachka, P. W. (1970). Fish. In *Comparative Physiology of Thermoregulation*, vol. I, *Invertebrates and Non-mammalian Vertebrates*, ed. G. C. Whittow, pp. 79–135. Academic Press, New York & London.
Gibson, R. N., Blaxter, J. H. S. & de Groot, S. J. (1978). Developmental changes in the activity rhythms of the plaice (*Pleuronectes platessa* L.). In *Rhythmic Activity of Fishes*, ed. J. E. Thorpe, pp. 169–86. Academic Press, New York & London.
Hart, J. L. (1931). On the daily movements of the coregonine fishes. *Canadian Field Naturalist*, **45**, 8–9.
Hartmann, E. & Stern, W. C. (1972). Desynchronised sleep deprivation: learning deficit and its reversal by increased catecholamines. *Physiology and Behavior*, **8**, 585–7.
Hepburn, H. R., Berman, N. J., Jacobson, H. D. & Fatti, P. (1973). Trends in arthropod defensive secretions, an aquatic predator assay. *Oecologia*, **12**, 373–82.
Hobson, J. A., Goin, O. B. & Goin, C. J. (1968). Electrographic correlates of behaviour in tree frogs. *Nature, London*, **162**, 387–9.
John, K. R. & Haut, M. (1964). Retinomotor cycles and correlated behaviour in the teleost *Astyanax mexicanus* (Fillipi). *Journal of the Fisheries Research Board of Canada*, **21**, 591–5.
Jouvet, M. (1967). Neurophysiology of the states of sleep. *Physiological Reviews*, **47**, 117–77.
Jouvet-Mounier, D., Astic, L. & Lacote, D. (1969). Ontogenesis of the states of sleep in rat, cat and guinea pig during the first postnatal month. *Developmental Psychobiology*, **2**, 216–39.
Karmanova, I. G., Titkov, E. S. & Popova, D. I. (1976). Species

peculiarities of the diurnal rhythm of motor activity in the fishes from the Black Sea. *Journal of Evolutionary Biochemistry and Physiology*, **12**, 486–8.

Lazarev, S. G. (1978*a*). Neurophysiological analysis of the activation spontaneously arising against the background of primary sleep in the frog *Rana temporaria. Journal of Evolutionary Biochemistry and Physiology*, **14**, 507–10.

Lazarev, S. G. (1978*b*). Electrophysiological analysis of wakefulness and primary sleep in the frog *Rana temporaria. Journal of Evolutionary Biochemistry and Physiology*, **14**, 379–84.

Marshall, N. B. (1972). Sleep in fishes. *Proceedings of the Royal Society of Medicine*, **65**, 177–9.

Matty, A. J. (1978). Pineal and some pituitary hormone rhythms in fish. In *Rhythmic Activity of Fishes*, ed. J. E. Thorpe, pp. 1–30. Academic Press, New York & London.

Olla, B. L. & Studholme, A. L. (1978). Comparative aspects of the activity rhythms of Tautog, *Tautoga onitis*, Bluefish, *Pomatomus saltatrix*, and Atlantic mackerel, *Scomber scombrus*, as related to their life habits. In *Rhythmic Activity of Fishes*, ed. J. E. Thorpe, pp. 131–51. Academic Press, New York & London.

Parmeggiani, D. L., Zamboni, G., Cianci, T. & Calasso, M. (1977). Absence of thermoregulatory vasomotor responses during fast wave sleep in cats. *Electroencephalography and Clinical Neurophysiology*, **42**, 373–80.

Peyrethon, J. & Dusan-Peyrethon, S. (1967). Etude polygraphique du cycle veille-sommeil d'un Téléostéen (*Tinca tinca*). *Comptes rendus hebdomadaires des séances de l'Académie des Sciences, Paris*, **161**, 2533–7.

Roffwarg, H. P., Muzio, J. N. & Dement, W. C. (1966). Ontogenetic development of the human sleep-dream cycle. *Science*, **152**, 604–9.

Ruckenbusch, Y. & Toutain, P. L. (1977). La phylogénèse du sommeil. *Confrontations psychiatriques*, **15**, 9–48.

Satake, N. (1979). Melatonin mediation in sedative effect of serotonin in goldfish. *Physiology and Behaviour*, **22**, 817–19.

Schwassmann, H. O. (1971). Biological rhythms. In *Fish Physiology*, ed. W. S. Hoar & D. J. Randall, vol. 6, pp. 371–428. Academic Press, New York & London.

Shapiro, C. M. (1974). A third state of existence. *Leech*, **44**, 13–16.

Shapiro, C. M. & Girdwood, P. (1981). Protein synthesis in rat brain during sleep. *Neuropharmacology*, in press.

Shapiro, C. M. & Hepburn, H. R. (1976). Sleep in a schooling fish, *Tilapia mossambica. Physiology and Behavior*, **16**, 613–15.

Shapiro, C. M., Moore, A. T., Mitchell, D. & Yodaiken, M. I. (1974). How well does man thermoregulate during sleep? *Experientia*, **30**, 1279–81.

Spencer, W. P. (1939). Diurnal activity in fresh-water fishes. *Ohio Journal of Science*, **39**, 119–32.

Tauber, E. S. & Weitzman, E. D. (1969). Eye movements during behavioural inactivity in certain Bermuda reef fish. *Communications in Behavioral Biology*, **3**, 131-5.

Thines, G. & Vandenbussche, E. (1966). The effect of alarm substance on the schooling behaviour of *Rusburg heteromorpha* Duniker in day and night conditions. *Animal Behaviour*, **14**, 296–302.

Titkov, E. S. (1976). Peculiarities of the diurnal rhythm of wakefulness and rest in the catfish *Ictalarus nebulosus*. *Journal of Evolutionary Biochemistry and Physiology*, **12**, 335–40.

Voronov, I. B., Karmanova, I. G. Titkov, E. S. & Rukoyatkina, N. I. (1977). The effect of arecolin on the structure of the rest active wakefulness in the catfish *Ictalurus nebulosus*. *Journal of Evolutionary Biochemistry and Physiology*, **13**, 525–8.

Walker, J. M. & Berger, R. J. (1973). A polygraphic study of the tortoise (*Testudo denticulata*). *Brain, Behavior and Evolution*, **8**, 453–67.

I. G. KARMANOVA, A. I. BELICH &
S. G. LAZAREV

An electrophysiological study of wakefulness and sleep-like states in fish and amphibians

Introduction

Although much is known about sleep, it is not yet possible to say with confidence whether it is a universal, pan-vertebrate phenomenon, inherent in all representatives of this taxonomic group, or whether its emergence is connected with a certain stage of phylogenesis (e.g. with the emergence of poikilothermy, viviparity) (Allison & Van Twyver, 1972). From our point of view two principal factors hinder a solution to this problem. Firstly, it is as yet unknown which periods of vital activity of fish and amphibians are to be considered as functional analogues of homoiotherms' sleep. Secondly, we lack quantitative data about the temporal organization of the sleep–wakefulness cycle in the aforementioned poikilotherms as well as about the neurophysiological peculiarities of states of wakefulness and sleep-like states in these animals. In some cases even the presence of sleep at all in certain poikilotherms is called in question (Hobson, Goin & Goin, 1967). Nevertheless the majority of researchers indicate the existence of torpid or sleep-like states in the behaviour of fish and amphibians (Weber, 1961; Marshall, 1972; Shapiro & Hepburn, 1976). Since research into sleep using objective techniques is difficult in fish (Roschevsky, 1972), most investigations have been carried out using only purely behavioural observations (Pieron, 1913; Weber, 1961, Tauber & Weitzmann, 1969; Marshall, 1972). These authors mainly accentuated the dynamics of locomotory activity in the course of a 24-h period. Attention was also paid to changes in body coloration, eye movements, tactile perception thresholds, etc. Though these observations are unquestionably of interest, they usually do not give an idea of the daily organization of the sleep–wakefulness cycle. We suggest that 24-h continuous monitoring of correlates of sleep and wakefulness is advisable for this purpose, as it gives an opportunity to gain extensive quantitative information about the organization of the sleep-wakefulness cycle in fish and amphibians. In this respect the study of sleep in the teleost *Tinca tinca*, the tench, by Peyrethon (Peyrethon, 1968) deserves to be considered in detail.

The chronic polygraphic recording of several parameters of bioelectric activity was obtained via electrodes implanted in the mid-brain optic tecti and 'neck' muscles. Electrocardiogram (ECG), electrooculogram (EOG) and locomotory activity were monitored as well. Peyrethon was thus able to identify states of wakefulness ('tonic activity state'), relative rest ('non-tonic activity state') and absolute rest. During the absolute resting state fish usually lay motionless on the bottom of the aquarium: the electromyogram (EMG) exhibited the absence of any muscle activity, respiration decelerated and the fish did not react upon the appearance of a novel stimulus in the field of vision. According to Peyrethon's data there was no difference between the heart rates typical of wakefulness and of resting states. The brain bioelectrical activity was also practically identical during rest and wakefulness. This led him to postulate that the EEG criteria of states of wakefulness and rest did not differ because fish lack the forebrain structures that in mammals are responsible for slow-wave synchronization. He also did not notice any signs of active (paradoxical) sleep.

We consider that though Peyrethon began his experiments 2–3 days after surgery (electrode implantation), this time interval was insufficient for the adequate adaptation of his animals; EEG and ECG changes, important for obtaining a clear idea of the organization of fish behaviour were therefore not noticed.

Nevertheless, Peyrethon has made an attempt to study the sleep–wakefulness cycle of tench and demonstrated that active behaviour occurs mainly at night (the fish are active 60% of the night). In the day activity did not occur for more than 20% of the time. The continuous 24-h illumination reduced the duration of the fish's locomotory activity. Finally it must be emphasized that Peyrethon pioneered the identification of three qualitatively different states in the daily vital activity of tench, using behavioural and somatovegetative criteria. The temporal organization of these states had, however, still not been studied sufficiently.

Different views on the problem of the existence of sleep in fish are probably due to the inaccuracy and inadequacy of the methods used. The shortcomings of studies on the sleep–wakefulness cycles of fish, mentioned above, are similar to those found in work on amphibians. Hobson (Hobson, 1967; Hobson *et al*., 1967) should probably be considered as the first researcher into the part played by sleep in the daily cycle of behaviour of amphibians. He made extensive use of polygraphic methods. EEG electrodes were implanted in the forebrain hemispheres and in the midbrain; EMG, ECG and other parameters of bioelectrical activity were monitored day and night. In several species of tree frogs (*Hyla squirella, H. cinerea* and *H. septentrionalis*) the states of alert wakefulness and day-time rest were identified. EEGs of these states differed from each other and the resting state was accompanied by an elevated arousal threshold. However, in the bullfrog (*Rana catesbiana*) behavioural quiescence was not accompanied by any changes in arousal threshold or in the EEG pattern. One must fully agree with Hobson's conclusion that the EEG criteria are not absolute ones. Evidently, further investigations are needed into the possible existence of functional analogues of homoiotherms' sleep in poikilotherms.

The peculiarities of sleep in a caudate amphibian were examined by McGinty (1972) in the tiger salamander, *Ambystoma tigrinum*. Spectral analysis of the forebrain EEG samples suggested certain differences between the power spectra typical of the periods of behavioural activity and quiescence. McGinty believes that the periodically occurring quiescent state of the salamander must be the precursor of homoiotherms' quiet sleep, and emphasizes the necessity for statistical methods of analysis to facilitate the exposure of EEG correlates common to poikilotherms and homoiotherms. We concur with McGinty's (1972) opinion that the criteria applied in mammals must be broadened for the study of amphibian sleep.

According to Broughton (1972) the absence of reticular pontine nuclei, medial or descending vestibular nuclei and of the locus coeruleus in amphibians makes the emergence of sleep stages and/or phases such as are typical of homoiotherms' sleep impossible. However, such a conclusion does not exclude the necessity of a thorough study of the brain bioelectrical activity patterns typical for different states of the amphibians' rest–wakefulness cycle. We believe that the absence of superficial resemblance between EEG patterns of homoiotherms' sleep and those of the resting states of fish and amphibians does not refute the existence of sleep in lower vertebrates.

The lack of comparative-physiological data concerning the organization of the sleep–wakefulness cycle in fish and amphibians must be mentioned here. This has probably resulted from the underestimation of such important physiological correlates as heart rate and phasic vegetative and motor phenomena occurring against the background of the forms of rest studied. According to Peyrethon, the investigation of cardiac activity in tench revealed no correlation between heart rate values and the states of the sleep–wakefulness cycle. But such an effect could be caused by insufficient adaptation of the animals to experimental conditions. Also, a number of researchers express an opinion that mean values of heart rate are less informative than beat-to-beat ones (Brooks *et al*., 1956; Shepovalnikov, 1966; Aldredge & Welch, 1973; Welch & Richardson, 1973; Lisenby, Richardson & Welch, 1976). All this research stimulated our interest in further investigating the peculiarities of wakefulness and sleep-like states in fish and amphibians. Indeed, it seems impossible to understand and interpret correctly the data obtained in the course of research into animal or human sleep without a thorough study of the evolutionary and ecophysiological aspects of this phenomenon. We consider that it is the evolutionary approach that will allow us to obtain an idea about the nature of sleep and to solve the problem of whether homoiotherms' sleep has been a result of an evolutionary transformation of certain ancient forms of sleep or whether it has emerged independently in birds and mammals.

In order to gain insight into the stages in the evolution of sleep, identification of the states in the sleep–wakefulness cycle was undertaken in representatives of all classes of vertebrates on the basis of the following indices: (1) presence or absence of locomotory activity, posture of the animal, character of muscle tone, position of the pupil; (2) heart rate, type of respiration, character of eye movements and of other automatic motor activities; (3) visual and statistical evaluation of the EEGs of different brain structures; (4) levels of arousal thresholds and characteristics of alterations in brain bioelectrical activity and in the heart rate response to an adequate stimulus; (6) quantitative evaluation of the states constituting the sleep–wakefulness cycle and of their temporal organization over the 24-h period; (7) consideration of ecological and seasonal effects. Our study was carried out on fish and amphibians by means of two complementary methods: the 'non-contact', continuous recording of ECG and locomotion in free-moving animals (Belich, 1978), and 'contact' polygraphic recording of EEG, EMG, ECG and EOG via chronically implanted electrodes (Lazarev, 1978*a*, *b*). The method of non-contact monitoring suggested by Rommel (1973) had been used in our laboratory by Titkov (1976) for the investigation of wakefulness and sleep-like states in a catfish, *Ictalurus nebulosus*. This atraumatic method attracted our attention for it permitted the precise identification of the states in the sleep–wakefulness cycles in fish and amphibians using somatovegetative and behavioural criteria (Karmanova, Churnosov & Popova, 1976; Popova & Churnosov, 1976; Karmanova *et al.*, 1977; Shilling, 1979). Nevertheless this technique appeared to be inadequate for the continuous recording of heart rate in cases where the animals stayed in an unfavourable position near the graphite electrodes. Belich (1978) therefore modified the technique to enable continuous beat-to-beat analysis of heart rate for 24-h periods. In this way the peculiarities of heart rate alterations could be studied without disturbing the original pattern of the sleep–wakefulness cycle and with the animals being well adapted to the experimental conditions.

Since we attach great importance to the factor of adaptation, the non-contact experiments usually preceded the polygraphic study of neurophysiological correlates of wakefulness and sleep-like resting states.

Comparative-physiological characteristics of the states constituting the sleep–wakefulness cycle in fish and amphibians

The subjects of our study were catfishes (*Ictalurus nebulosus*) and brown frogs (*Rana temporaria*). All the experiments were carried out during autumn and winter (October–December). Using the beat-to-beat method of continuous 24-h recording of heart rate (Belich, 1978) and subsequent (off-line) computerized analysis of experimental data (Alexandrov *et al*.,

Fig. 1. Postures of the catfish typical of the different forms of sleep-like states that constitute 'primary sleep'. For the SLS with plastic muscle tone (SLS-1) the moment of shaping the animal into an unnatural posture is depicted (*a*). (*b*) The unnatural posture is preserved. An inclination of the body at an angle of about 60° to the bottom of the aquarium is a usual correlate of SLS-3.

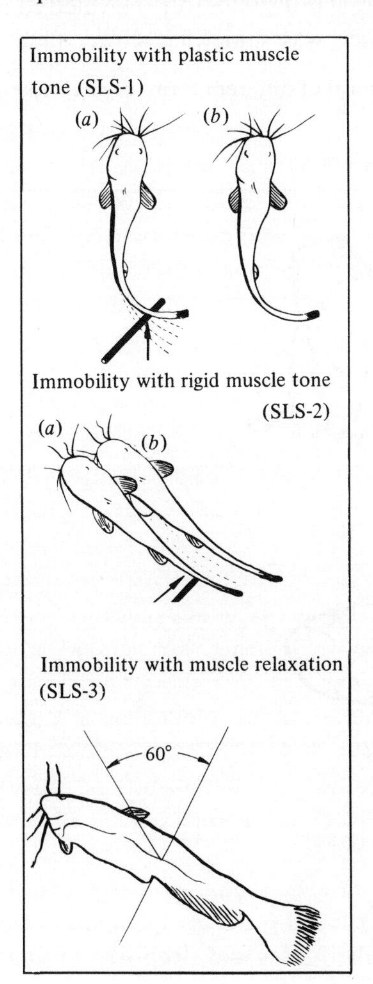

1980) we found five fundamental states in our animals: active and passive wakefulness and three sleep-like states (SLSs). Sleep-like state no. 1 (SLS-1) correlated with a torpid posture of the animals (Figs. 1 and 2). The animals could also, while in this state, be shaped into an unnatural posture, which could be preserved for some time – i.e. there was plastic muscle tone. Arousal thresholds were elevated considerably and the appearance of a novel stimulus caused no avoidance reaction in the animals. Heart rate was relatively low and the reaction to auditory, visual or tactile stimulation was a heart rate deceleration (in catfish) or acceleration (in frog).

Sleep-like state no. 2 (SLS-2) was correlated with a pronounced rigid muscle tone. An attempt to shape the fish or the frog into some unnatural posture resulted in passive displacement of the body or in its turning at an angle of 45–90° (Figs. 1 and 2). Arousal thresholds were elevated compared with those of wakefulness. Heart rate was somewhat slower than during

Fig. 2. Postures of the frog typical of different forms of sleep-like states. For comments, see Fig. 1 and text.

Immobility with plastic muscle tone (SLS-1)

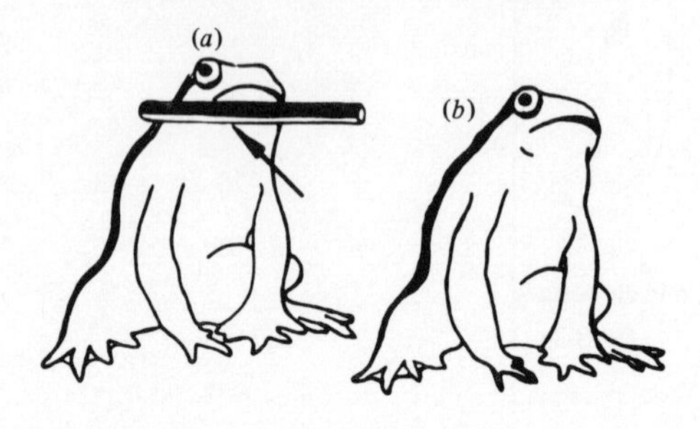

Immobility with rigid muscle tone (SLS-2)

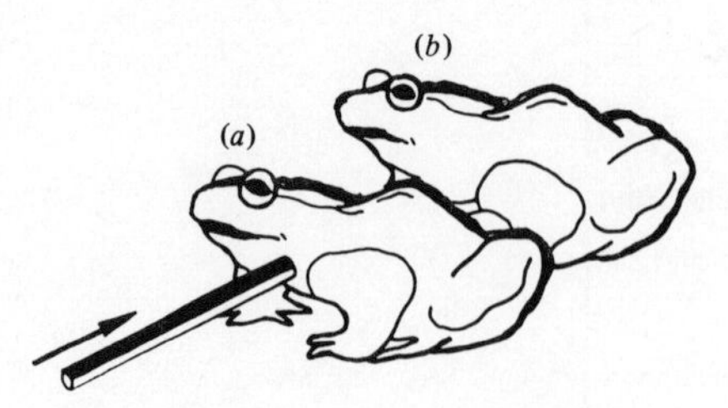

Immobility with muscle relaxation (SLS-3)

Table 1. *Description of the sleep–wakefulness cycle of catfish according to vegetative and behavioural indices (autumn animals)*

Indices[a]	Waking		Sleep-like states		
	Active	Passive	SLS-1	SLS-2	SLS-3
N	98 ± 2.0	80 ± 15.0	50 ± 2.8	63.8 ± 4.2	40 ± 5.1
$T_{R_1R_2}$	0.65 ± 0.04	0.75 ± 0.15	1.2 ± 0.12	0.94 ± 0.07	1.5 ± 0.23
Reaction on tactile stimulation	Locomotion	Locomotion	No locomotion Plastic muscle tone	No locomotion Rigid muscle tone	No locomotion Reduced muscle tone
Reaction on auditory stimulation:					
$\frac{T_{RR_1}}{T_{R_2R_3}}$	0.87 ± 0.04	0.78 ± 0.06	0.6 ± 0.12	0.7 ± 0.05	0.9 ± 0.014

All values mean ± S.D.
[a] N, heart rate (beats per minute);
$T_{R_1R_2}$, interval between two beats (s);
T_{RR_1}, interval between two beats before the auditory stimulation (s);
$T_{R_2R_3}$, interval between two beats against the background of auditory stimulation (s).

Table 2. *Description of the sleep–wakefulness cycle of frog according to vegetative and behavioural indices (autumn animals)*

	Waking		Sleep-like states		
Indices[a]	Active	Passive	SLS–1	SLS–2	SLS–3
N	34±0.4	22±6.0	8.2±3.2	14±1.2	12.2±2.4
$T_{R_1R_2}$	1.8±0.12	2.8±0.98	7.5±2.5	4.1±0.48	4.9±2.1
Reaction on tactile stimulation	Locomotion	Locomotion	No locomotion Plastic muscle tone	No locomotion Rigid muscle tone	No locomotion Reduced muscle tone
Reaction on auditory stimulation $\frac{T_{RR_1}}{T_{R_2R_3}}$	1.0±0.03	1.0±0.9	1.45±0.6	1.6±0.15	1.2±0.1

All values mean ± S.D.
[a] N, heart rate (beats per minute);
$T_{R_1R_2}$, interval between two beats (s);
T_{RR_1}, interval between two beats before the auditory stimulation (s);
$T_{R_2R_3}$, interval between two beats against the background of auditory stimulation (s).

SLS-1. Reactions to auditory, visual or tactile stimulation were represented by a smaller heart rate deceleration (catfish) or acceleration (frog) than during SLS-1.

Sleep-like state no. 3 (SLS-3) was accompanied by muscle relaxation and was correlated with the lowest frequency and least rhythmicity of cardiac activity. In catfish, the body was inclined at an angle of 60°–80° to the horizontal, the tail lying flat and the fins passively stretched. In frogs, muscle relaxation was marked during this state and the eyes were closed (as distinct from SLS-1 and SLS-2 when the eyes were wide open) (Figs. 1 and 2). Arousal thresholds were substantially elevated.

In passive wakefulness the presence of an avoidance reaction was a principal characteristic, an attempt to shape the animal into some unnatural posture resulting in immediate flight. Heart rate exceeded the values obtained for SLS-1, SLS-2 and SLS-3. Dispersion of the R–R intervals of ECG exceeded that of active wakefulness, when R–R interval dispersion approached zero. Thresholds of responses to external stimuli were low.

The quantitative characteristics of states constituting the sleep–wakefulness cycle in catfish and frog (according to somatovegetative and behavioural criteria) are presented in Tables 1 and 2.

Fig. 3. The averaged 24-h pattern of correlation between heart rate values and daily periodicity of sleep–wakefulness cycles in catfish, *Ictalurus nebulosus* (autumn and winter animals). SLS-1, sleep-like state with plastic muscle tone; SLS-2, sleep-like state with rigid muscle tone; SLS-3, sleep-like state with muscle relaxation; W, wakefulness; R–R, interval between two subsequent heart beats (s).

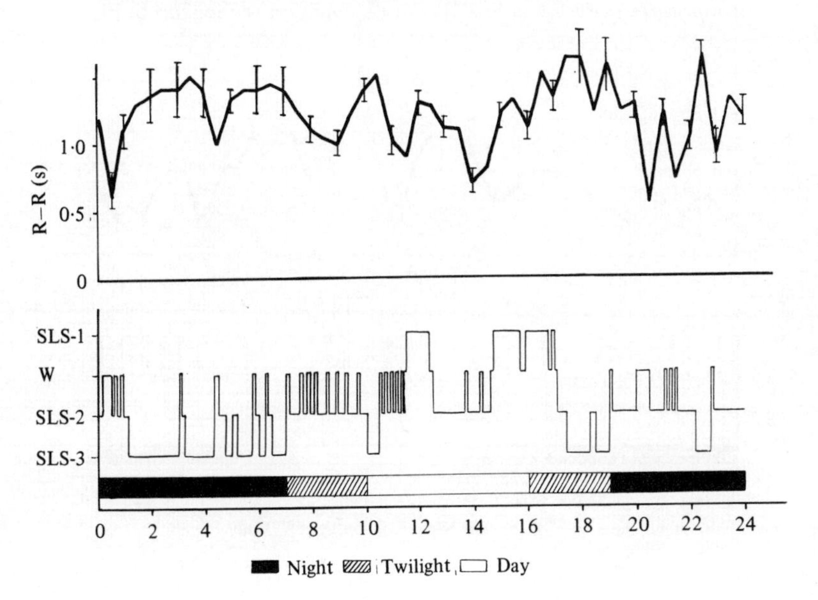

On the basis of these behavioural and physiological data the SLSs identified in the daily cycles of the catfish and frog were amalgamated into a concept we termed 'primary sleep'. It should be emphasized that tonic long-term heart rate deceleration was a common feature of all SLSs. What, then, is the circadian periodicity of these states and what is the correlation between them and the circadian dynamics of heart rate?

Figs. 3 and 4 show a typical relationship between heart rate and the behavioural states of catfish and frog during the autumn–winter season. The polyphasic character of the manifestation of these states in the course of a 24-h period can be seen, as can the predominance of SLS-3 at night and of SLS-1 during the day. SLS-2 is also found mainly at night, but nevertheless occurs during day-time and twilight as well. Analysis of the probability of the behavioural state changing revealed that SLS-2 is an intermediate (transitional) state between wakefulness and SLS-3. But during the day-time wakefulness may lead straight into SLS-1, bypassing SLS-2. Fig. 5 shows the averaged probabilities of transition from one state to another during the day and at night for catfish and frog. These data emphasize that SLS-1 is an adaptive reaction during the day and that SLS-3 is a similar reaction at night. Thus we have a reason to regard SLS-1 as a day-time form of rest and SLS-2 and SLS-3 as night-time ones. SLS-2 may occur as a transitional state during either the day-time or the night-time, but the probability of a transition from wakefulness to SLS-2 increases significantly at night (Fig. 5).

Fig. 4. The averaged 24-h pattern of correlation between heart rate values and daily periodicity of sleep–wakefulness in frog, *Rana temporaria* (autumn animals). For explanation see legend to Fig. 3.

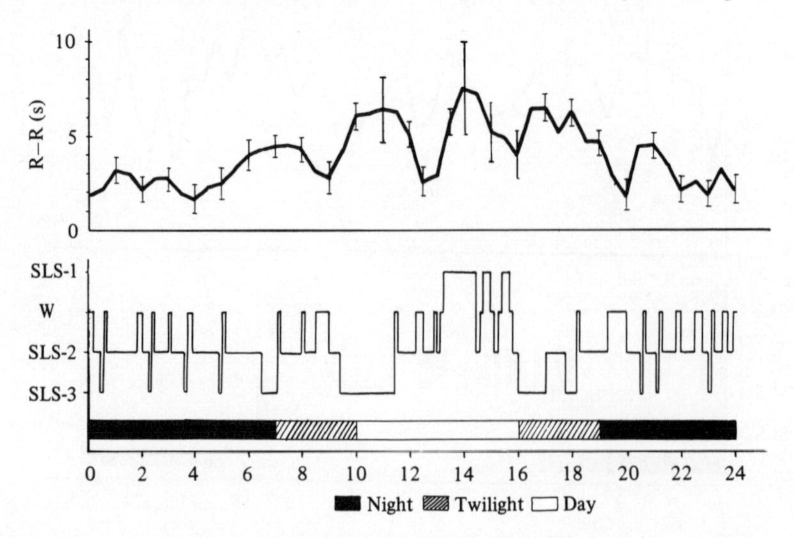

As stated above, the correlations between heart rate values and the states of the daily 'wakefulness–primary sleep' cycle, as well as the dynamics of those states, are polyphasic in both catfish and frog. And in both animals there are different subcircadian cycles: one can distinguish complete cycles of night-type states (e.g. wakefulness–SLS-2–SLS-3–wakefulness) or of the day-type states (e.g. wakefulness–SLS-1–SLS-2–SLS-3–wakefulness). In

Fig. 5. Averaged probabilities of changing for sleep-like states during day (*a*) and at night (*b*) for catfish and frog. The numerals are the probability coefficients of transition from one state to another. W, wakefulness.

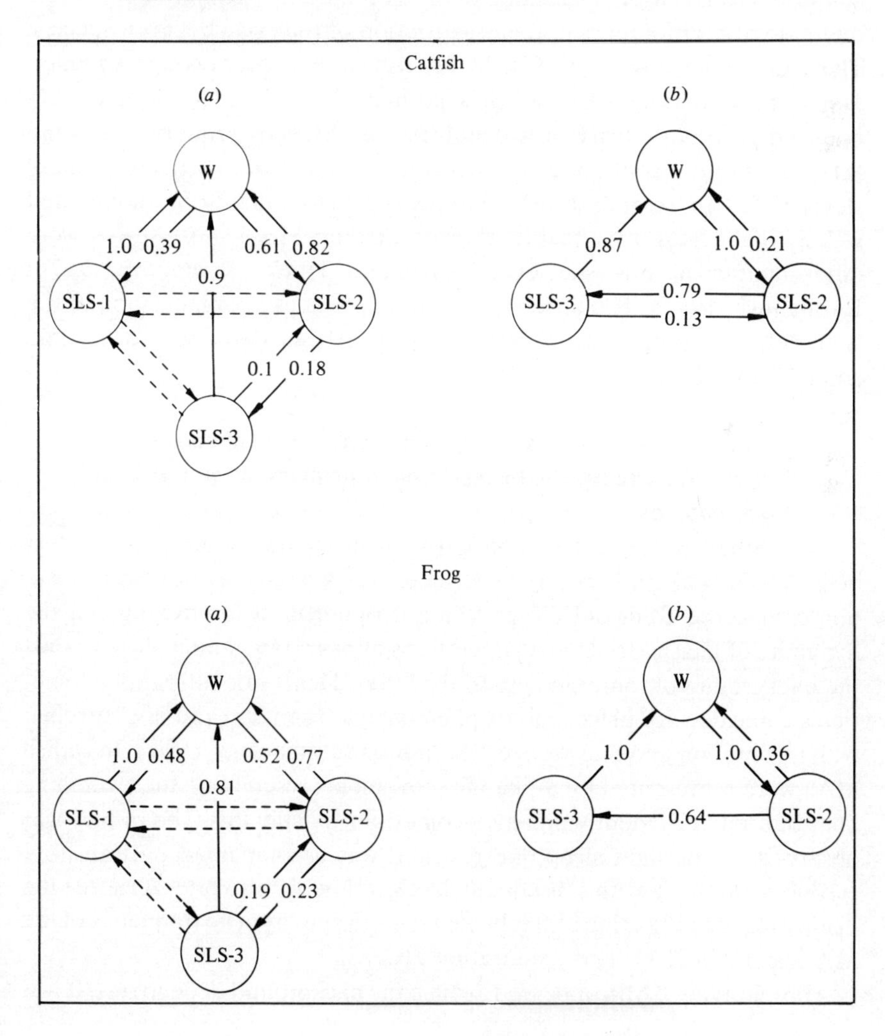

parallel with the complete cycles incomplete ones (subcycles) take place (e.g. wakefulness–SLS-2–wakefulness). It is important to note that the periodicity of the 'wakefulness–primary sleep' cycle nevertheless correlates with alterations in illumination in the course of a 24-h period.

During autumn and winter catfish usually pass 37.9% of the 24-h period in SLS-2, 32.9% in SLS-3 and 12.1% in SLS-1. Active and passive wakefulness occupy up to 17.1% of the time. During the same season frogs pass 57.6% of the time in SLS-2, 20.0% in SLS-3 and 7.6% in SLS-1. Active and passive wakefulness occupy 14.8% of the time (Belich, 1980). In spring the proportion of SLS-1 increases in both catfish and frog, while the proportion of SLS-2 is reduced correspondingly (Titkov, 1976; Karmanova, 1977). In addition to this seasonal variation in the 'wakefulness–primary sleep' cycle there are also differences in the organization of the cycle between species. Thus, among seven species of Black Sea fish studied, only two (grey mullet, *Mugil* sp., and ribbontail, *Dasyatis pastinaca*) exhibited a strongly pronounced 24-h periodicity of wakefulness and SLSs of primary sleep, and *Sciaena umbra*, a continuously active species, showed no signs of primary sleep SLSs. In scorpionfishes (*Scorpena porcus*) the SLS-2-dominated waking behaviour was feebly shown; neither SLS-1 nor SLS-3 were differentiated in this species (Karmanova, Titkov & Popova, 1976; Karmanova, 1977). The differences between species in organization of the 'wakefulness–primary sleep' cycle have also been described for certain amphibians (Karmanova, 1977).

Characteristics of phasic vegetative and motor phenomena occurring against the background of primary sleep in fish and amphibians

Study of the correlation between heart rate values, locomotory activity and 'wakefulness–primary sleep' states by means of beat-to-beat, non-contact recording of ECG provided an opportunity for investigating the dynamics of phasic (transitory) alterations of heart rate taking place against the background of tonic changes in the latter. Heart rate alterations sometimes coincide with phasic motor phenomena. The latter did not correlate with locomotion and thus were called 'motor automatisms' (MA). In catfish MAs were represented by S-like tail movements resembling the swimming ones but not coincident with any locomotion. Usually this type of MA was observed during light sleep-like rest and was accompanied by transitory cardiac arrhythmia with subsequent deceleration of pulse rate. As a rule the deepening of the existing SLS followed such an activity. The frequency of the MA was $1 \pm 0.26\,s^{-1}$ and its duration 20 ± 7 s.

Another type of MA appeared against the background of deep rest. It was characterized by paroxysmal motor phenomena resembling startings or twitching and produced transitory arrythmia and heart rate acceleration. Sometimes the latter resulted in sharp acceleration of pulse rate and in transition from the resting state to wakefulness. The frequency of this MA was about $20 \pm 4\,s^{-1}$, its duration not exceeding 0.2 ± 0.06 s tremor-like activity, according to the EMG data. MA of this type predominated during night-time, mainly against the background of SLS-3.

Rhythmic body movements without locomotion have also been identified against the background of primary sleep in frogs. The first type of MA was observed during a light resting state, and was correlated with cardiac arrhythmia, subsequent tonic deceleration and the transition to a deeper state of rest (Fig. 6 *A* and *B*). The frequency of such an MA did not exceed $0.9 \pm 0.15\,s^{-1}$, its duration being equal to 10 ± 2 s.

Another type of MA observed in frogs in the course of deep primary sleep was represented by spontaneous paroxysmal body startings and twitchings and by rhythmic movements of hindlimbs. After such an MA cardiac arrythmia was usually marked, succeeded by heart rate acceleration and by transition to wakefulness (Fig. 6 *D* and *E*). The frequency of this MA was about $0.8 \pm 1.8\,s^{-1}$, its duration 3.0 ± 1.0 s. MAs of this type predominated at night-time. The total duration of MAs averaged about 4.5% of the total duration of primary sleep.

Activation of the cholinergic brain structures in catfish and in frog caused a substantial increase in MA frequency and duration (Karmanova *et al*., 1977; Voronov *et al*., 1977). It is of interest that different motor phenomena

Fig. 6. ECG patterns and phasic motor phenomena (as 'artifacts') noted against the background of primary sleep SLSs in frog. *A*, SLS-2 and motor automatism (MA), deepening this form of primary sleep; *B*, SLS-3 and MA, deepening this form of primary sleep; *C*, SLS-3 and MA causing heart rate acceleration and elevation of the level of alertness; *D* and *E*, SLS-2 and MA, causing the elevation of level of alertness with subsequent transition to the waking state.

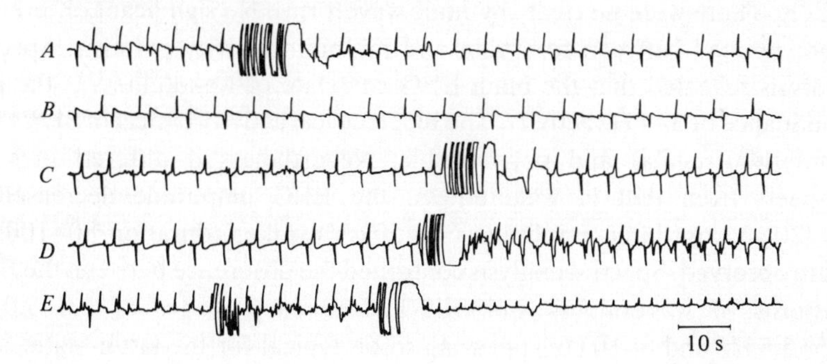

(rhythmic body or head shaking, myoclonic twitches, etc.) are common during sleep in human subjects, particularly infants (Denisova & Figurin, 1926; Shepovalnikov, 1971; Golbin & Stupnitsky, 1976). Intensification of motor phenomena is correlated with certain sleep disorders (Vein, 1974). Thus, the phasic motor and vegetative events described above, when observed against the background of poikilotherms' primary sleep, reveal certain features of the evolution of central sleep-regulating mechanisms. As we have demonstrated, rhythmic MAs promote the deepening of sleep while paroxysmal movements (startings, twitchings, etc.) prevent it. Thus it is clear that primary sleep is not a homogeneous state and that different phasic phenomena arise against its background. These events are probably to be considered as functional analogues of the vegetative and motor phenomena observed in homoiotherms during their slow-wave (SWS) and paradoxical (PS) phases of sleep.

Neurophysiological characteristics of wakefulness and of sleep-like states constituting 'primary sleep'

A thorough study of behavioural and somatovegetative correlates of wakefulness and primary sleep of fish and amphibians has given us an opportunity to investigate the neurophysiology of these states and, specifically, their typical EEG patterns. For this purpose methods of computer analysis (power spectral analysis) were used as well as the conventional visual analysis of EEG. In long-term (up to 24-h) experiments performed on catfishes and frogs it was found that animals with implanted electrodes exhibited the same states of wakefulness and primary sleep that were earlier identified in non-contact experiments. The only exception was SLS-3, whose identification in fish was problematical because of elevated alertness probably caused by surgery. Thus EEG correlates of this state in catfish remain unavailable.

As depicted in Fig. 7, in frog wakefulness is correlated with low-amplitude (15–20 μV in forebrain sites and 25–30 μV in midbrain ones), polymorphic EEGs. There were no clear rhythmic waveforms. No significant differences were noticed between forebrain and midbrain activity patterns. Spectral analysis revealed that the main EEG correlate of wakefulness is the predominance of 3–4 Hz activity. The bioelectrical activity typical of SLS-1 was similarly irregular and polymorphic. Nevertheless it differed in some respects from that in wakefulness: the EEG amplitude decreased to 10–20 μV and large-amplitude (30–50 μV) spikes (duration 80–100 ms) were observed. Spectral analysis confirmed the difference between the EEG patterns of wakefulness and SLS-1, frequency ranges of 2.0–3.0 Hz, 4.5–5.5 Hz and 8–10 Hz appearing to be typical for the latter state.

Analysis of the effect of external stimulation on the EEG of primary sleep was also carried out. As is generally known, in mammals moderate stimulation during slow-wave-sleep cause EEG desynchronization ('arousal reaction'), i.e. emergence of EEG activation without behavioural arousal. Since in lower vertebrates, animals in SLS-1 and SLS-2 have open eyes, we have been able to study the effect of the appearance of a novel stimulus (object) in their field of vision. Such a stimulus caused an increase in the 3–4 Hz activity usually correlated with wakefulness, but there was neither an increase in alertness nor any motor activity, i.e. the usual behavioural pattern of primary sleep was preserved. The emergence of EEG correlates of wakefulness (3–4 Hz activity in this species) against the background of SLS-1 and

Fig. 7. Patterns of frog electrogram typical for SLS-1, SLS-2, SLS-3 and wakefulness (W). 1, EEG of forebrain hippocampal primordium; 2, EEG of midbrain optic tecti; 3, EMG of paravertebral muscles; 4, ECG. Calibration: 50 μV, 1 s. (After Lazarev, 1978*a*.)

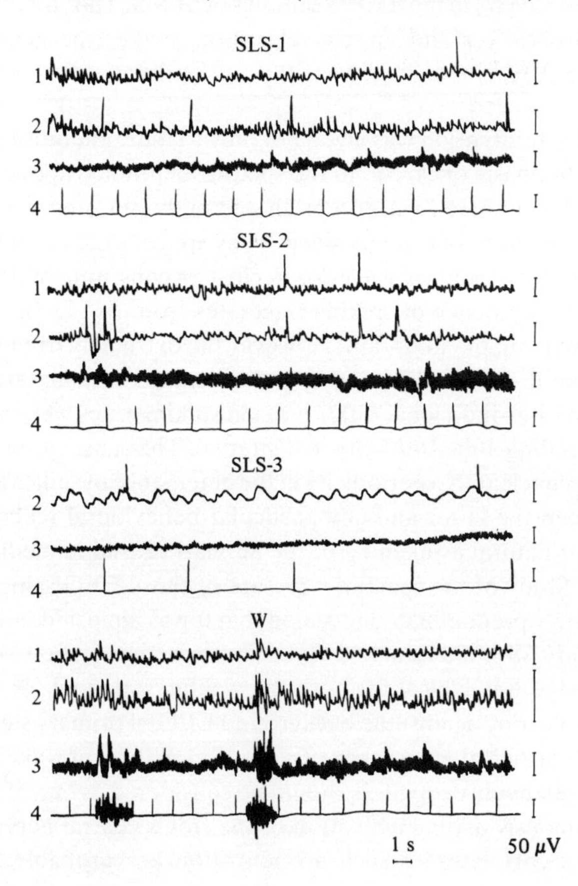

SLS-2 thus enables us to state the existence of a specific 'arousal reaction' in lower vertebrates which differs from that conventionally found in homoiotherms.

Visual analysis shows that in frogs EEG activity during SLS-2 differs only slightly from that of SLS-1, though spectral analysis revealed the domination of a 4–5 Hz frequency range during SLS-2. Obvious and frequent spiking was seen, similar to that of SLS-1. Thus both SLS-1 and SLS-2 differ from the waking state according not only to behavioural and somatovegetative criteria but also to the EEG ones. Computerized quantitative analysis has proved to be a decisive factor in revealing these differences.

Compared to SLS-1 and SLS-2, SLS-3 is easily identified by means of a visual analysis of the EEG. Fig. 7 shows that this state is accompanied by the domination of slow-wave activity in the EEG. Spike-like waves are also present – a common correlate of all sleep-like forms of rest constituting primary sleep. Spectral analysis revealed the domination of a delta-like frequency range (0.5–2.0 Hz) in most EEG samples of SLS-3. This, together with the animals' closed eyes and muscle relaxation, makes the general pattern of frogs' SLS-3 resemble somewhat that of homoiotherms' delta-sleep.

In our study particular attention was given not only to tonic alterations of the EEG but also to to phasic ones, e.g. to the spiking mentioned above. In the frog spikes have been noted from forebrain and midbrain sites against the background of all the SLSs of primary sleep. They appear synchronously or asynchronously, either as solitary waves or in clusters consisting of three to six spikes. Average frequency of spiking fluctuates from 3–15 min^{-1} in forebrain to 10–15 min^{-1} in the midbrain. However the extremely irregular mode of their temporal distribution during the course of primary sleep should be noted. The inter-individual differences in spiking rates were also extremely large, exceeding 300–400% in a 1-h interval. The cause of such a phenomenon remains unclear. No periodicity in the emergence of spikes and no correlation between the latter and any particular behavioural reaction was seen. Artificial or natural awakening of the animals caused immediate cessation of spiking. Study of the spectral structure of those EEG samples where spike-like activity predominated revealed that it was almost identical to the spectral pattern of wakefulness EEG. Thus, predomination of a 3–4 Hz frequency range was typical for both the waking state and for the episodes of spiking occurring against the background EEG of primary sleep. We suggest that this episodical emergence of a spectral pattern of wakefulness in the course of relatively deep SLSs should be considered as one of the correlates of spontaneously arising activation of the frog's central nervous system (CNS). More correlates of such an activation are probably the episodes of hypersynchronized EEG observed against the background of all three forms of SLSs and resembling superficially the artifacts of locomotion. Nevertheless, direct observations revealed that animals were absolutely motionless in 95% of cases during such episodes. Duration of such EEG changes varied from 4 to 20 s, their average frequency being about 1 h^{-1}.

One may therefore assume the existence of three fundamental EEG correlates of CNS activation in the frog. The first of these is the EEG synchronization in the frequency range 3–4 Hz that is typical of wakefulness. Against the background of primary sleep there are, in addition to this, 3–4 Hz activity, spike-like waves and hypersynchronized EEGs that are the signs of spontaneous activation. It is quite probable that though superficially the latter differ from the active (paradoxical) phase of avian and mammalian sleep, their physiological function is similar. We assume that this function is the regulation of the depth of sleep.

Special experiments undertaken in our laboratory in order to study the central mechanisms of spontaneous activation revealed that hypothalamic structures play a considerable role in its realization. In particular it appeared that in frogs, hypothalamic lesions caused reduction of the EEG amplitude and of spiking. The data obtained allowed us to conclude that the hypothalamus has an activating influence on higher brain structures. The following reasoning supports this statement: as mentioned above, spiking had been one of the correlates of spontaneous activation episodically emerging in the course of primary sleep. Thus, a sharp reduction in spike rate may be regarded as an effect of a decrease of hypothalamic influences upon higher structures of the frog's brain. Further, as it is well known that in lower vertebrates activation results not in the decrease of EEG amplitude (as is typical for homoiotherms) but in a considerable increase in the latter, the suppression of EEG after hypothalamic lesions in the frog may be considered as a loss of hypothalamic facilitation of the CNS.

The data obtained thus confirm the existence of a state of primary sleep which was earlier identified by means of non-contact recording and behavioural observations. Moreover, the results of the neurophysiological study confirm the 'heterogeneity' of primary sleep: against the background of all the three forms of SLS that constitute primary sleep spontaneous activation of the CNS emerges episodically, its function probably being similar to that of active sleep in homoiotherms.

It should be mentioned that the results of the neurophysiological study of primary sleep of catfish bore a clear resemblance to those obtained in frog. Fig. 8 illustrates the power spectra of the EEG typical of wakefulness and of SLSs of primary sleep in catfish and in frog. Spectral expression of the reaction of spontaneous activation of CNS is also shown. One cannot but notice certain likenesses between the EEG correlates of the states described in the two species. The differences that exist are probably the result of morphological and functional peculiarities of the brains of fish and amphibians.

Conclusion

Our experimental data can be summarized as follows. In the 24-h periodicity of vital activity of the animals studied, not only wakefulness but also a state of primary sleep exists, consisting of three forms of sleep-like state (SLS-1, SLS-2 and SLS-3). These SLSs differ from each other and from wakefulness according to certain behavioural, vegetative and neurophysiological criteria. Their basic common correlates are the loss of alertness, the specific character of cardiac activity and the presence of a phenomenon of spontaneous CNS activation. Comparison of poikilotherms' primary sleep with types of sleep in birds and mammals gives us an opportunity to surmise the evolutionary relationships between these states.

Thus SLS-3, characterized by closed eyes, muscle relaxation, bradycardia

Fig. 8. Power spectra of EEG of a frog and of a catfish typical for the states of wakefulness (W) and primary sleep (PrS). PrS(Sp), spectra of primary sleep EEG samples during which spike activity predominated (CNS activation). (After Karmanova & Lazarev, 1979.)

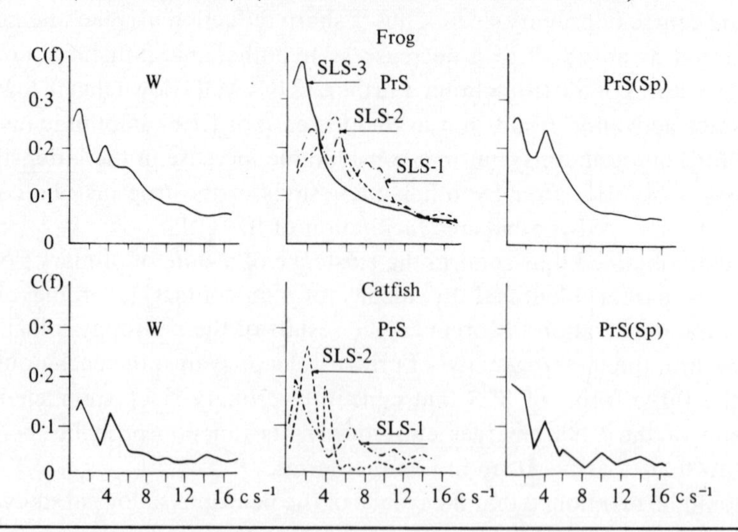

Fig. 9. Schematic depiction of the evolution of sleep-like states in vertebrates. AS, spontaneous CNS activation arising against the background of sleep; SWS, slow-wave sleep; PS, paradoxical sleep. (After Karmanova & Lazarev, 1979.)

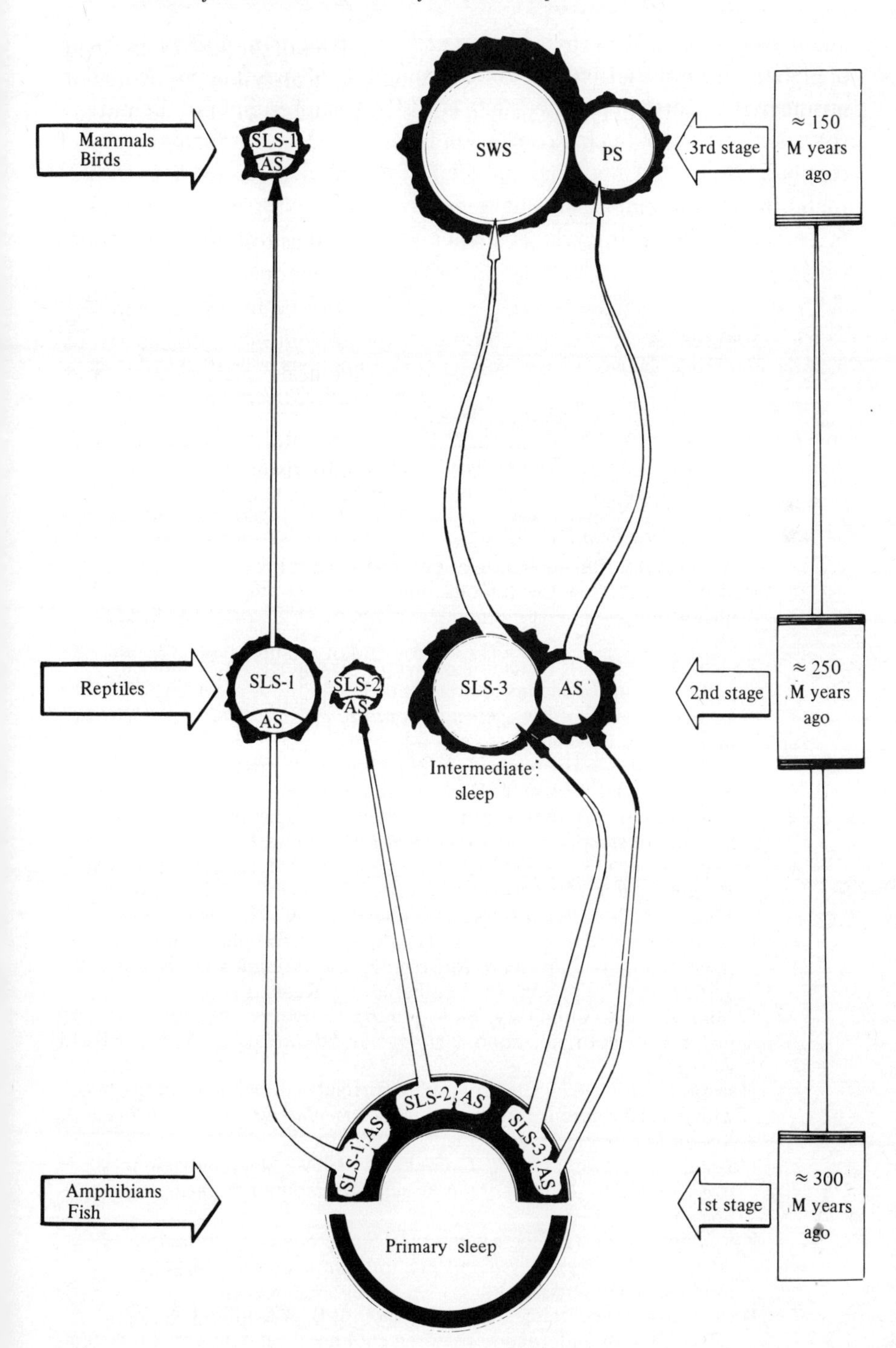
Mammals
Birds
SLS-1
AS
SWS
PS
3rd stage
≈ 150
M years
ago
Reptiles
SLS-1
AS
SLS-2
AS
SLS-3
AS
Intermediate
sleep
2nd stage
≈ 250
M years
ago
SLS-2
AS
SLS-1
AS
SLS-3
AS
Amphibians
Fish
Primary sleep
1st stage
≈ 300
M years
ago

and the slowing of EEG rhythms, is a probable precursor of the delta-sleep of higher vertebrates. Likewise the phenomenon of spontaneous activation arising in the course of primary sleep is probably a forerunner of their active (paradoxical) sleep. Such a conclusion is supported by our further study of reptilian, avian and mammalian sleep. Two other SLSs (SLS-1, SLS-2) found in fish and amphibians have not evolved progressively and are still preserved in the daily cycle of higher vertebrates as rudimentary states (Fig. 9).

References

Aldredge, I. L. & Welch, A. I. (1973). Variations of heart rate during sleep as a function of the sleep cycle. *Electroencephalography and Clinical Neurophysiology*, **35**, 193–8.

Alexandrov, V. V., Belich, A. I., Karmanova, I. G. & Schneiderov, V. S. (1980). Computerized analysis of daily dynamics of heart rate. *Zhurnal evoliutsionnoi biokhimii i fizologii*, **16**, in press. (In Russian.)

Allison, T. & Van Twyver, H. (1972). Electrophysiological studies of the echidna *Tachyglossus aculeatus*. II. Dormancy and hibernation. *Archives italienne de biologie*, **110**, 185–94.

Belich, A. I. (1978). Continuous long-term registration of heart rate by means of non-contact method. *Fiziologicheskii Zhurnal SSSR*, **65**, 1079–82.

Belich, A. I. (1980). Detection of sleep–wakefulness cycle states in poikilotherms by means of non-contact continuous recording of heart rate and body motility. *Zhurnal evoliutsionnoi biokhimii fiziologii*, **16**, in press. (In Russian.)

Brooks, C. M. C., Hoffman, B. F., Suckling, E. E., Kleintjeus, F., Koewig, E. H., Coleman, R. S. & Treumann, H. J. (1956). Sleep and variations in certain functional activities accompanying cyclic changes in depth of sleep. *Journal of Applied Physiology*, **9**, 97–104.

Broughton, R. (1972). Phylogenetic evolution of sleep systems. In *The Sleeping Brain: Perspectives in the Brain Sciences*, vol. 1, ed. M. H. Chase, pp. 2–7. Brain Information Service, UCLA, Los Angeles.

Denisova, M. P. & Figurin, N. L. (1926). Peripheral phenomena during sleep in infants. In *News of Reflexology and Physiology of Nervous System*, vol. 2, pp. 338–45. Leningrad. (In Russian.)

Golbin, A. C. & Stupnitsky, Y. A. (1976). Body motility as an active factor in sleep organization. *Fiziologicheskii Cheloveka*, **3**, 354–61. (In Russian.)

Hobson, J. A. (1967). Electrographic correlates of behavior in the frog, with special reference to sleep. *Electroencephalography and Clinical Neurophysiology*, **22**, 113–21.

Hobson, J. A., Goin, C. J. & Goin, O. B. (1967). Sleep behaviour of frogs. *Quarterly Journal of the Florida Academy of Sciences*, **30**, 184–6.

Karmanova, I. G. (1977). *Evolution of Sleep. Stages of Formation of 'Sleep–Wakefulness' Cycle in Vertebrates*, pp. 1–173. Nauka, Leningrad. (In Russian.)

Karmanova, I. G., Belich, A. I., Voronov, I. B. & Schilling, N. V. (1977). On the interaction between cholinergic and adrenergic systems

during the development of 2 forms of sleep in frog *Rana temporaria* and turtle *Emys orbicularis. Zhurnal evoliutsionnoi biokhimii i fiziologii*, **13**, 706–11.
Karmanova, I. G., Churnosov, E. V. & Popova, D. I. (1976). Peculiarities of day-time rest in a representative of bony fishes, *Ictalurus nebulosus*, and an amphibian, *Rana temporaria*. *Zhurnal evoliutsionnoi biokhimii i fiziologii*, **17**, 572–8. (In Russian.)
Karmanova, I. G. & Lazarev, S. G. (1979). New data on the neurophysiology of sleep in fish and amphibians (on the genesis of slow-wave sleep and paradoxical sleep in homoiotherms). *Doklady Akademii Nauk. SSSR*, **245**, 757–60. (In Russian.)
Karmanova, I. G., Titkov, E. S. & Popova, D. I. (1976). Specific peculiarities of daily periodicity of motor activity and rest in Black Sea fishes. *Zhurnal evoliutsionnoi biokhimii i fiziologii*, **12**, 486–8. (In Russian.)
Lazarev, S. G. (1978*a*). Neurophysiological analysis of wakefulness and primary sleep in frog *Rana temporaria. Zhurnal evoliutsionnoi biokhimii i fiziologii*, **14**, 379–84. (In Russian.)
Lazarev, S. G. (1978*b*). Neurophysiological analysis of activation, spontaneously arising against the background of primary sleep in frog. *Zhurnal evoliutsionnoi biokhimii i fiziologii*, **14**, 507–9. (In Russian.)
Lisenby, M. J., Richardson, P. C. & Welch, A. J. (1976). Detection of cyclic sleep phenomenon using instantaneous heart rate. *Electroencephalography and Clinical Neurophysiology*, **40**, 169–77.
McGinty, D. (1972). Sleep in amphibians. In *The Sleeping Brain: Perspectives in the Brain Sciences*, vol. 1, ed. M. H. Chase, pp. 7–10. Brain Information Service, UCLA, Los Angeles.
Marshall, N. B. (1972). Sleep in fishes. *Proceedings of the Royal Society of Medicine*, **65**, 177–9.
Peyrethon, J. (1968). *Sommeil et évolution. Etude polygraphique des états de sommeil chez les poissons et les reptiles*, pp. 1–102. Tixier, Lyon.
Pieron, H. (1913). *Le problème physiologique du sommeil*, pp. 1–520. Masson, Paris.
Popova, D. I. & Churnosov, E. V. (1976). Daily cycles of wakefulness and rest in frog *Rana temporaria. Zhurnal evoliutsionnoi biokhimii i fiziologii*, **12**, 199–201. (In Russian.)
Rommel, S. A. (1973). A simple method of recording fish heart and operculum beat without the use of implanted electrodes. *Journal of the Fisheries Research Board of Canada*, **19**, 417–22.
Roschevsky, M. P. (1972). *Evolutionary Electrocardiology*, pp. 1–251. Nauka, Leningrad. (In Russian.)
Shapiro, E. M. & Hepburn, H. R. (1976). Sleep in a schooling fish *Tilapia mossambica. Physiology and Behavior*, **16**, 613–15.
Shepovalnikov, A. N. (1966). Heart rate and EEG sleep stages. In *Reports of the 2nd All-Union Conference on the Electrophysiology of Central Nervous System*, pp. 330–1. Tbilisi. (In Russian.)
Shepovalnikov, A. N. (1971). *Activity of the Sleeping Brain*, pp. 1–182. Nauka, Leningrad. (In Russian.)
Shilling, N. V. (1979). On the role of vegetative nervous system in control of primary sleep in frog *Rana temporaria. Zhurnal evoliutsionnoi biokhimii i fiziologii*, **15**, 184–9. (In Russian.)
Tauber, E. S. & Weitzmann, E. D. (1969). Eye movements during

behavioral inactivity in certain Bermuda reef fish. *Communications in Behavioral Biology*, **3A**, 131–5.

Titkov, E. S. (1976). Peculiarities of daily periodicity of wakefulness and rest in catfish *Ictalurus nebulosus. Zhurnal evoliutsionnoi biokhimii i fiziologii*, **12**, 335–40. (In Russian.)

Vein, A. M. (1974). *Disturbances of Sleep and Wakefulness*, pp. 1–186. Medicina, Moscow. (In Russian.)

Voronov, I. B., Karmanova, I. G., Titkov, E. S. & Rukoiatkina, N. I. (1977). Influence of arecolin administration on the structure of rest and active wakefulness in catfish *Ictalurus nebulosus. Zhurnal evoliutsionnoi biokhimii i fiziologii*, **13**, 525–8. (In Russian.)

Wever, E. (1961). Über Ruhelagen von Fischen. *Zeitschrift für Tierpsychologie*, **18**, 517–33.

Welch, A. J. & Richardson, P. C. (1973). Computer sleep stage classification using heart rate data. *Electroencephalography and Clinical Neurophysiology*, **34**, 145–52.

P. R. LAMING

The physiological basis of alert behaviour in fish

Introduction

Observation of the behaviour of any vertebrate leads to the conclusion that the animal is not in a constant state of 'alertness'. Not only does the general level of activity vary, but so does the animal's responsiveness to changes in the environment. Some of the changes of activity in fish are rhythmic in nature, with annual, lunar, circadian or tidal periodicities. Müller (1978) describes seasonal changes in the circadian cycle of activity in a variety of freshwater species, including the minnow (*Phoxinus phoxinus*) and trout (*Salmo trutta*). Some fish, such as eels (*Anguilla anguilla*), show a lunar periodicity in activity, which in this species affects the number of individuals making the seaward migration (Gibson, 1978). Many intertidal species have a tidal rhythm of activity (Gibson, 1978). The most evident activity rhythm, however, is circadian. Fish, like other vertebrates, exhibit periods in which they are relatively inactive and appear to be asleep (Shapiro *et al*., this volume) and particular physiological correlates of this state can be recorded (Karmanova *et al*., this volume). Not only do fish show these changes in activity between 'sleep' and 'wakefulness', but they also exhibit differing degrees of 'alertness' during the waking period. It is these changes with which this paper is concerned.

In this introduction so far, I have used the terms 'alert' and 'active' as if they were interchangeable. This, however, is not the case. The alert fish is one with an increased ability to detect and respond to changes in the environment. It is behaviourally and physiologically aroused. This may not be the case with active fish. Pelagic species, like tuna and mackerel, are constantly active yet it seems unlikely that they are equally constantly aroused. In the context of long-term changes in the behaviour of fish, such as the rhythms already mentioned, it is probable that alert fish are also more active.

During periods of wakefulness in fish changes in alertness occur in response to 'novel' or 'significant' changes in the environmental circumstances. 'Significant' stimuli are those which are biologically important, e.g.

food items, potential mates, competitors or predators. Stimuli of these types cause alerting or orienting responses because they have innately acquired and/or learnt significance to the fish. In contrast to these learnt or innately important stimuli, novel stimuli have significance simply because of their novelty. Responses of fish to stimuli such as food (Peeke & Peeke, 1972) and conspecifics (Peeke & Peeke, 1970) have recently received considerable attention. However, the associated behaviours, like biting at food and aggressive display, are often secondary to orienting or alerting responses, the study of which is thus complicated by the appetitive state of the animal. This paper will therefore only consider the alerting responses of fish to a novel stimulus. Both behavioural and physiological aspects of alertness are assessed in terms of their possible adaptive significance, and are examined under the following headings:

1. Behavioural 'alertness'
2. Autonomic responses
3. Brain regions involved in alertness
4. Seasonal differences in 'alertness'
5. The nature of the EEG and its possible role during changes in responsiveness

Behavioural alertness

Fish normally respond to a novel stimulus in one of two ways: either by an orienting or arousal reaction or by a fright or startle (tail-flip) response. 'Freezing' behaviour can be interpreted in either way depending on whether it is considered as a pause in ongoing behaviour (Goodman & Weinberger, 1973) or as a rigid paralysis induced by extreme fright. In goldfish, Rodgers, Melzack & Segal (1963) described the two types of response to a bang on the side of the aquarium as either 'orienting' or 'tail-flip'. Orienting responses included turning by the fish, moving the fins or head, or otherwise reorienting the body.

Russell (1967) presented guppies (*Poecilia reticularis*) with a shadow stimulus and described the orienting responses in categories of 'jerks', 'single jerks', 'drifting with arched fins' and 'flickering' of fins with no change in position. Later work by Savage (1971) and Laming (unpublished) using goldfish (*Carassius auratus*) has confirmed the impression that orienting responses do not involve significant movement of the fish in the water, but appear to stabilize its position. In goldfish the response consists of erection of the dorsal fin, small dorsoventral flicks of the caudal fin and slight movements of the pectoral fins. An analysis of these behavioural components of the orienting response to a tap on the side of the aquarium was performed in the following experiment on roach (*Rutilus rutilus*) by Laming & Hornby (unpublished). The fish were caught by anglers in the River Bann in Northern Ireland and kept in indoor aquaria for at least 1 week prior to use. Each fish was placed in a 35 × 25 × 25 cm plastic aquarium for 1 h and then observed whilst a tap was delivered to the side of the aquarium by a perspex rod mounted on a solenoid. The activities of the fish were recorded every 5 s for 15-s periods, before the stimulus (B on Fig. 1), after its onset (D), and subsequently (A) (Laming & Hornby, unpublished). The McNemar test was used to indicate whether the changes in fin or body movement occurring at stimulus onset were due to chance (Siegel, 1956). The results indicated that a change in pectoral fin movement is the most consistent and significant ($P < 0.05$) component of the alerting response in roach.

The fright or startle response, unlike the orienting response, is accompanied by rapid movement of the fish. Its main component seems to be what Rogers *et al*. (1963) aptly described as the rapid 'tail-flip'. This response consists of a rapid lateral flexion of the tail which propels the animal forward in the water, and is probably mediated by the giant Mauthner neurons of the medullary reticular formation (Retzlaff, 1957). On repeated presentation of identical stimuli at regular intervals, the tail-flip response becomes weaker

Fig. 1. Fin and body movements in roach before (B), during (D) and after (A) a tap on the side of the aquarium. Black circles indicate significance at $P < 0.05$ (McNemar); $n = 79$.

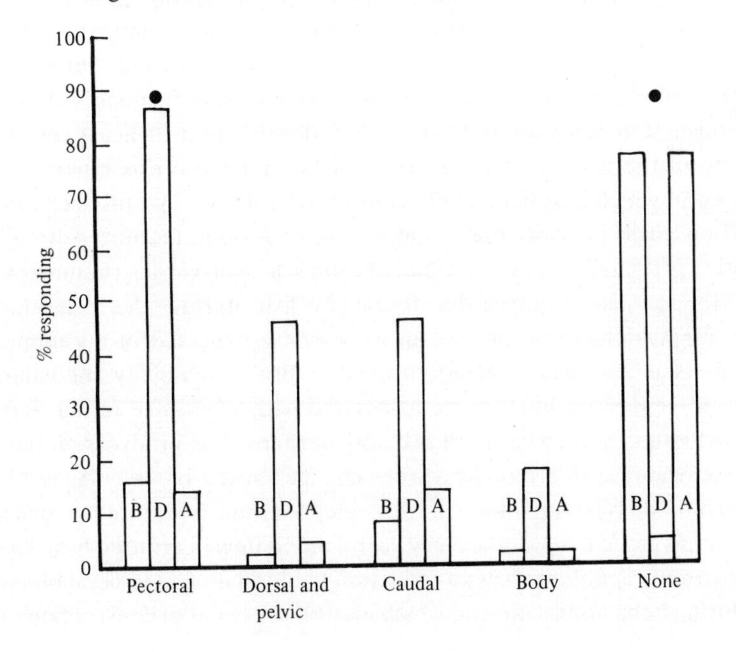

(habituates), to be replaced by an orienting response. Thus, both tail-flip and orienting responses habituate in fish, a phenomenon related to the decreasing novelty of the stimulus. The ability to be less sensitive and responsive to stimuli which have previously occurred is a requisite for the response to novelty, and implies that learning is involved, a topic considered later in this paper.

By the subjective assessment of the responses of animals to stimuli as alerting or orienting responses, we tacitly assume that they prepare the animal to both assess and respond to subsequent environmental changes. The orienting response of fish, as in mammals, is thus seen as an adaptive response to changes in environmental circumstances. In order to justify fully these assumptions about the adaptive functions of the response it is necessary to examine the various ways in which fish respond physiologically to alerting stimuli.

Autonomic responses

As well as responding behaviourally to stimuli, fish also respond physiologically. Heart rate changes have been recorded in response to a variety of environmental changes, including light flashes, mechanical vibration, atmospheric pressure changes, removal from water and touch (Randall, 1966). Otis, Cerf & Thomas (1957) likewise reported changes in ventilation when a light was used as a stimulus, though neither of these quoted reports attempted to correlate the physiological change with behaviour. Recent research has shown that there are clear changes in both heart rate (Alexandrov *et al.*, 1980; Laming & Savage, 1980) and ventilatory rate and amplitude during behavioural orienting or arousal responses of fish.

In goldfish (*Carassius auratus*) there is a deceleration of heart rate to about 70% of the resting level during the 10 s after the presentation of a novel stimulus which has elicited behavioural arousal (Fig. 2). This deceleration or bradycardia is a response common to most vertebrates during orienting reactions, though anuran amphibians show a tachycardia (Laming & Austin, 1981). The bradycardia found in fish during the orienting response might not be the physiological response expected of an animal preparing itself for action. Bradycardias in fish are vagally mediated (cholinergic) and more likely to be associated with vegetative rather than alerting activities. Somewhat paradoxically perhaps, this bradycardia may be a reflex response to a blood pressure change caused by vascular redistribution initiated by sympathetic (adrenergic) neurons. Evidence for such a hypothesis has come from studies on visceral blood flow in aroused fish. The chub (*Leuciscus cephalus*) shows a reduction in blood flow in visceral blood vessels during behavioural arousal which may be the result of constriction of visceral arteries (Laming & Savage, 1978). This vascular redistribution may indirectly be the cause of the bradycardia by effecting a rise in blood pressure in the dorsal aorta. Mott (1957) has demonstrated that a rise in dorsal aortic pressure causes a reflex, vagal bradycardia in fish.

The ventilatory rate in fish during the 10 s after a stimulus which has elicited behavioural arousal, decreases to about 70% of its resting level and its amplitude is decreased by 50% (Laming & Savage, 1980). The physiological mechanisms of this response are unclear, but they may be the result of cardiac and respiratory synchrony described for fish by Satchell (1971).

The behavioural orienting responses of fish to novel stimuli are small, and the muscles which move the fins do not provide significant changes in the

Fig. 2. Cardiac (ECG), ventilatory and electromyograph (EMG) changes during behavioural arousal in goldfish. A, artifact; B, bradycardia; N, normal heart rate.

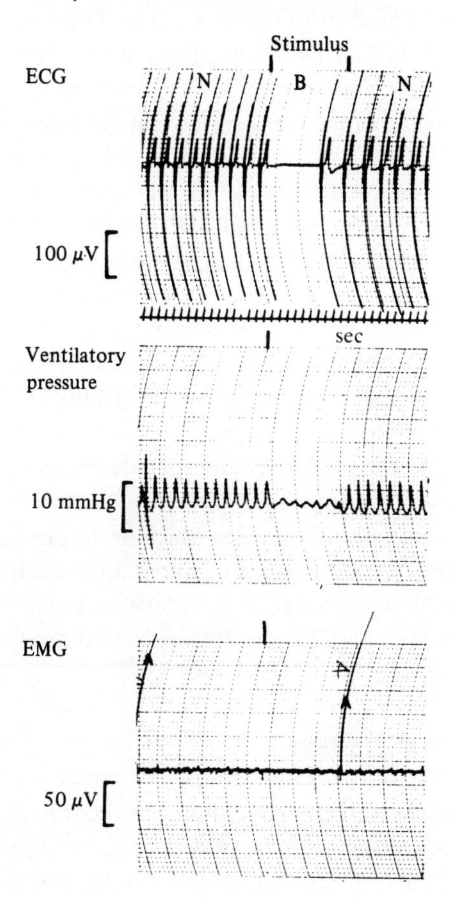

electromyograph as recorded from the dorsal myotomes (Laming & Savage, 1980).

Both the behavioural and autonomic responses of fish are produced by a wide variety of experimental stimuli, including (1) the onset of illumination, (2) a moving shadow over the top of the aquarium, (3) a plunger in the aquarium water, (4) a loud (40 db re microbar^{-1}) 500 Hz sound, (5) a paddle rotating in the water, and (6) a moving black spot of 1 cm diameter passing over the aquarium. The first three stimuli produce evident behavioural and autonomic orienting responses, the plunger even eliciting tail-flip or fright responses on some occasions. Even the three less effective stimuli often elicit autonomic orienting responses, though these are smaller and sometimes occur without the behavioural response, a feature most evident during habituation (Laming & Savage, 1980).

Some stimuli, which might be considered 'strong' in terms of the change in receivable energy they contain, evoke larger orienting responses, and also occasionally tail-flip responses. These latter responses (Fig. 3) are characterized by an initial brief (1–4 s) bradycardia followed by a longer (up to 30 s) tachycardia. Both ventilatory rate and amplitude rise during the 10 s after the start of the stimulus. As the tail of the fish is moved laterally very rapidly in the execution of the tail-flip, there is also an increase in amplitude of the electromyographic potentials recorded from the dorsal myotomes.

Like the orienting response, the fright or tail-flip response can be considered as adaptive in that it causes the removal of the animal from the source of a potentially dangerous stimulus. The often violent behaviour is facilitated by increased cardiac and ventilatory activity. These in turn are probably increased by the direct influence of the sympathetic (adrenergic) nervous elements of the autonomic nervous system and possibly by the direct action of hormonal adrenaline.

Though orienting and fright responses are superficially dissimilar, they may be caused by closely related (adrenergic) physiological mechanisms. Whatever this mechanism, the autonomic changes which occur during fright or arousal are useful indicators of the response of the animal. Both autonomic and behavioural responses are, however, secondary to changes occurring within the central nervous system (CNS) itself, and these changes will be considered in the next section.

Brain regions involved in alertness

There have been comparatively few studies of changes in brain activity during alertness in fish, and these have involved recordings of the electroencephalograph (EEG). Many of these studies have been primarily concerned with changes in the EEG between sleep to wakefulness rather than with changes occurring during wakefulness itself. Karmanova *et al*. (this volume) report on the sleep/wakefulness patterns of the EEG of fish and amphibians.

EEG recordings from fish have usually been made using bipolar silver/silver chloride electrodes, inserted through an opening in the cranium of a clamped or otherwise immobilized animal. This technique has, of course, made the relationship between EEG waveform and behaviour difficult to ascertain. Enger (1957) with cod (*Gadus callarias*) and Laming (1980) with goldfish, chronically implanted EEG electrodes and were able to make some assessment of the behaviour of their unclamped subjects. Laming used stainless steel electrodes to minimize the long-term toxicity effects of silver on the brain (Klemm, 1966) and simultaneously recorded heart and ventilatory changes as indicants of arousal. Table 1 presents some of the

Fig. 3. Cardiac (ECG), ventilatory and electromyograph (EMG) changes during behavioural fright in goldfish. B, bradycardia; N, normal heart rate; T, tachycardia.

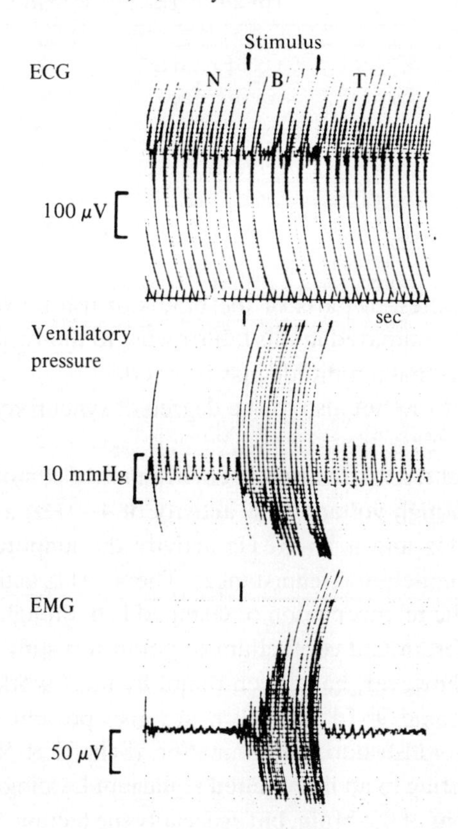

Table 1. *EEGs of teleosts*

Site of recording and reference	Resting frequency ranges (Hz) I	II	III	Frequency in response to stimulus	Species
Telencephalon					
Hara *et al.* (1965)		8–10		—	Salmon
Enger (1957)	4–6	8–13		18–32	Cod
Schadé & Weiler (1959)	4–8	9–14			Goldfish
Laming (1980)	6–9		16–24	16–24	Goldfish
Midbrain					
Hara *et al.* (1965)		9–10			Salmon
Enger (1957)		8–13		18–32	Cod
Schadé & Weiler (1959)		7–14	18–24	18–24	Goldfish
Laming (1980)	6–9		16–24	16–24	Goldfish
Hindbrain					
Enger (1957)			14–32		Cod
Schadé & Weiler (1959)		8–11	25–35		Goldfish
Laming (1980)	6–9		16–24	16–24	Goldfish

frequency ranges found in different parts of the brains of fish by various workers. Amplitudes are not compared in this table owing to the variability of the techniques, and the fact that amplitude may be a reflection not only of the power content of a waveform but also of the degree of synchrony of its components (Elul, 1972).

The EEGs of fish not presented with a discrete environmental stimulus fall roughly into three ranges: a high-voltage slow activity of 4–9 Hz, a lower amplitude activity of 9–14 Hz and a 16–32 Hz activity the amplitude of which varies with the environmental circumstances. The 4–9 Hz activity is most consistently found in the telencephalon of clamped fish, though Laming also found it in the midbrain and cerebellum of goldfish resting in the dark. These latter regions, however, have been found by most workers to have EEG frequencies of some 9–14 Hz, which are also present in the telencephalon of clamped goldfish during illumination (Schadé & Weiler, 1959). In unclamped fish resting in an illuminated aquarium Laming found 16–24 Hz activity in all regions of the brain, but especially the tectum. This is similar to the frequency of 18–24 Hz found by Schadé & Weiler (1959) in the midbrain of goldfish. During the presentation of an environmental stimulus, both Schadé & Weiler (1959) and Laming observed an increase in the amplitude and spread of this 16–24 Hz activity. A similar phenomenon was observed by Enger (1957), who showed the emergence of 18–32 Hz activity in the midbrain and telencephalon of cod presented with a flashing light.

Thus the general characteristics of the EEG of resting fish can be described in terms of a high-amplitude, low-frequency waveform of 4–9 Hz or 9–14 Hz. A 16–32 Hz frequency becomes prevalent when the animal is presented with an environmental stimulus. The apparent variation in the low-frequency component reported by different workers may have its origin in seasonal variation in EEGs of fish, referred to later in this paper.

In goldfish, behavioural arousal is associated with a bradycardia, a reduction in ventilatory rate and an increase in 16–24 Hz activity in the EEG (Laming, 1980). The EEG change is most evident in the midbrain (Schadé & Weiler, 1959; Laming, 1980) where it has the shortest latency after stimulus onset and the longest duration. The functional significance of EEG changes during arousal is unclear, as, until recently, has been their origin. Nevertheless they seem to be closely correlated with behaviour, both in the short-term orienting response reported here and in longer term seasonal changes reported in the next section.

One way of determining which areas of the brain are concerned with arousal is to stimulate electrically and record the electrical threshold of the response – whether it be behavioural (Savage, 1971), or physiological as reported here. Savage (1971) observed the arousing effects of electrical stimulation of the telencephalon and found the lowest thresholds medially, at the level of, and immediately posterior to, the anterior commissure. I have performed similar experiments, also on goldfish, but using chronically implanted stainless steel electrodes with a stimulus frequency of 50 Hz instead of the silver electrodes with a stimulus frequency of 10 or 25 Hz used by Savage (1971). Four presentations of the stimulus were given at intervals of at least 1 min. In the work of Savage (1971) the criterion for arousal was behavioural, whereas in the work reported here it was a 20% decrease in heart or ventilatory rate or a 20% increase in the amplitude of 16–24 Hz activity in the EEG. Irrespective of the differences in stimulating technique and criteria for arousal, both these experiments show the lowest telencephalic thresholds for eliciting arousal (less than 30 μA) to be in the region of the anterior commissure. Fig. 4 illustrates the sites in the telencephalon which were stimulated electrically and indicates the level of stimulus required to produce an arousal response. Stimulation of five out of 24 of the

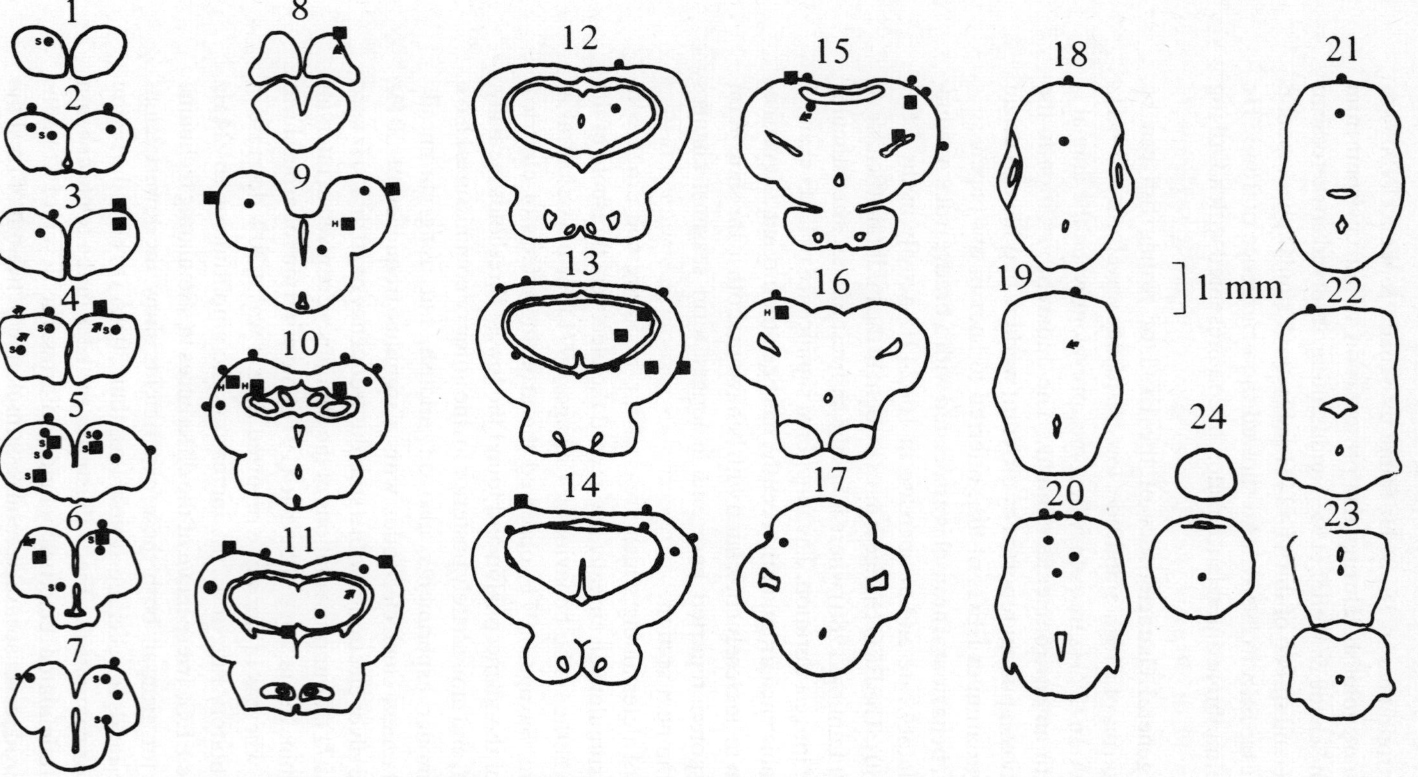

Fig. 4. Implantation sites of electrodes, stimulated electrically to determine thresholds for eliciting arousal. 1–24, transverse sections of the goldfish brain at regular 4% intervals from the front of the telencephalon to the back of the cerebellum. H, sites at which there was no change in threshold and no decrement in the response when stimulated on four separate occasions; S, electrodes implanted by Savage (1971); solid circles, electrode sites giving no response at 30 μA or below; solid squares, electrode sites stimulation of which gave either cardiac, ventilatory or EEG arousal responses, or behavioural arousal for electrodes of Savage (1971), at thresholds below 30 μA; arrows, electrode sites which when stimulated at less than 30 μA produced an EEG in the tectum similar to that of a resting fish.

author's electrode sites and four out of 15 of the sites reported by Savage (1971) gave responses at low electrical thresholds (<30 μA). Apart from one site these were all clustered in the regions either anterior to the anterior commissure (Laming) or posterior to it (Savage, 1971).

Five sites, especially in the region of the anterior commissure, produced a recording which showed a reduction in the frequency of the tectal EEG when stimulated at 30 μA. The slow waves produced (6–8 Hz) were reminiscent of the EEGs of goldfish resting in the dark. No cardiac, ventilatory or behavioural responses accompanied this EEG change.

In the midbrain (Fig. 4, sections 9–16) 18 of the 52 sites elicited arousal at low electrical thresholds, stimulation being most effective in the extreme anterior and posterior regions of the tectum and valvula. Two sites produced an EEG of low frequency when stimulated. A characteristic of tectal sites was that stimulation of these electrodes produced arousal at very low currents.

In the cerebellum only two of the 15 sites produced responses when stimulated. One of these responses was a decrease in the EEG frequency, the other an arousal response.

When the electrical stimuli were repeated, the normal effect was a decline in response or an increment in threshold. This was true of all sites except for three in the anterior tectum and one in the posterior tectum. Thus, the response to stimulation of these sites was more resistant to habituation than it was for other sites.

These results indicate that arousal responses are elicited most readily when the midbrain is stimulated. Few electrodes were implanted in the tegmental regions and so the effect of stimulating these deep midbrain regions cannot be properly assessed. In the tectum, especially the extreme anterior and posterior areas, stimulation elicited low-threshold responses, some of which were apparently resistant to habituation. Such results implicate these, or closely connected regions, in the initiation of arousal responses in fish. The anterior and posterior parts of the tectum have many connections with the anterior tegmentum and telencephalon, and the brainstem reticular formation. Thus, the areas initiating arousal may not be in the tectum itself but in closely connected midbrain or thalamic regions, as in mammals.

In the telencephalon, electrode positions which gave arousal when stimulated at low currents were mainly around the area of the anterior commissure, as were positions which gave a resting EEG response on stimulation. It thus seems that the telencephalon may have both facilitatory and inhibitory effects on arousal in fish. This, of course, is the prime requisite for a modulatory function of the telencephalon on arousal.

One of the characteristics of arousal responses whether they be changes in behaviour, heart or ventilatory rate or the EEG, is that they habituate when the stimulus is repeated. Goldfish presented with a moving shadow stimulus at 2-min intervals habituate these various responses at different rates (Laming, 1980). The order in which they habituate is: cerebellar EEG change, bradycardia, ventilatory rate change, telencephalic EEG change and tectal EEG change.

The accumulated evidence of EEG responses during arousal, the thresholds for electrical stimulation of arousal, and the resistance to habituation of different regional EEG changes, all implicate the midbrain in the origin of the response. On a similar criterion the telencephalon also appears to be involved in arousal, possibly in a modulatory role. Laming & McKee (1981) and Laming & Ennis (unpublished) have found severe deficits in habituation of behavioural and cardiac arousal responses after telencephalic ablation, which would support this contention.

Further, as yet unpublished work by McKee on goldfish and Laming & Hornby on roach (*Rutilus rutilus*) would suggest that small electrolytic lesions in the posterior telencephalon, even when carried out unilaterally, will also impair habituation of arousal. Thus it would appear that in teleosts the midbrain or thalamus are, as in mammals, concerned with the adjustment of the level of alertness of the animal. The telencephalon may exercise a regulatory control, primarily by inhibition, on this arousing mechanism. At this point it is well to say that the analogy with the concept of arousal and its regulation in mammals is a striking one, with the telencephalon of fish operating in a role often attributed to the limbic system of the mammalian brain. Further exploration of the regulation of arousal in teleosts may in the near future provide results the applications of which are not restricted to teleosts alone.

The results discussed here implicate the midbrain in the arousal response to novel stimuli. It appears that high-frequency changes in the EEG are correlated with alertness and activity. One way of testing this impression is to examine the EEGs of fish which show a seasonal decline in alertness or activity. In the experiments reported here, recordings were made in the summer; the next section considers the EEGs of goldfish in winter.

Seasonal differences in alertness

There are few reports of seasonal changes in CNS activity in fish though seasonal changes in behaviour are well documented from both freshwater and marine species. These include minnows (*Phoxinus phoxinus*), trout (*Salmo trutta*) and sand eels (*Ammodytes marinus*) (Winslade, 1974*a*; Müller, 1978). Seasonal increases in behavioural activity are often triggered by increases in water temperature (Winslade, 1974*a*) or illumination (Winslade, 1974*b*). These environmental factors, among many others, with or without an endogenous timing mechanism, presumably trigger the CNS to promote greater physiological and behavioural activity. In the lungfish, *Protopterus annectens*, Godet, Bert & Collomb (1964) have shown a telencephalic 8–13 Hz activity which has its highest amplitude in the wet season, when the frequency becomes modified during behavioural arousal. In the dry season, when the animal aestivates, no such modification occurs. In toads (*Bufo arenarum*) Segura & De Juan (1966) found that the 7–11 Hz activity of the olfactory bulbs was more prevalent in animals examined in summer than in winter. In a related species, *Bufo bufo*, Ewert & Siefert (1974) reported seasonal changes in contrast detection.

Comparable studies on teleosts are lacking. The observations from goldfish which I report here show seasonal differences in the resting characteristics of the EEG and the EEG responses to stimuli. The fish were obtained from stockists in the northern hemisphere during summer, and maintained in large (2 × 1 × 1 mm) outdoor tanks until required. Fifteen days were allowed for temperature acclimation at 15 ± 1 °C in aquaria with a regime of 12 h light/12 h dark. Silver/silver chloride electrodes of 0.5 mm diameter were chronically implanted in the manner described by Laming (1980), as were ECG electrodes and a ventilatory catheter (Laming & Savage, 1980). The fish were revived from the anaesthesia used in the operation and placed in a plastic mesh trough, suspended in a 25 × 25 × 35 cm aquarium of aerated water at 15 ± 1 °C. The aquarium was contained in a black wooden box, the lid of which was provided with a device for automatically presenting a moving shadow to the fish as a stimulus (Laming & Savage, 1980). Each fish was left for 24 h to adapt to this environment and its EEG then recorded whilst it was resting in the illuminated aquarium. The EEGs of fish obtained in this way often showed two frequencies to be present. In this case the frequencies were termed primary and secondary depending on which was most evident, using the criteria described in Laming (1980). The frequencies observed in the EEGs of fish recorded in January to March (winter) and June to July (summer) are compared on Table 2. The primary and secondary frequencies of tectal EEGs of fish from which recordings were made in summer are both significantly higher ($P < 0.05$) than the equivalent recordings from fish in winter. The same is also true of the primary telencephalic frequency. Compared with the two frequency ranges of 6–8 Hz and 16–24 Hz found in the EEGs of fish in summer, the EEGs of fish in winter had frequencies of 3–5 Hz and 7–13 Hz. These are remarkably similar to those of 4–8 Hz and 8–14 Hz found by Hara, Ileda & Gorbrian (1965), Enger (1957) and Schadé & Weiler (1959) but not to those from fish recorded in summer by Laming

Table 2. *Seasonal differences in the EEG frequencies (mean ± S.E.) of goldfish resting in an illuminated aquarium*

	Telencephalon		Tectum		Cerebellum	
	Primary	Secondary	Primary	Secondary	Primary	Secondary
Winter	7.6 ± 0.8	3.7 ± 0.6	8.3 ± 1.9	3.5 ± 0.6	10.7 ± 3.0	—
n	8	3	7	4	6	
Summer	17.5 ± 1.0	6.5 ± 0.7	18.3 ± 1.6	6.3 ± 0.6	7.0 ± 1.0	18.5 ± 1.3
n	6	2	6	3	6	4
P	0.001	—	0.001	0.05	NS	—

(1980). This suggests that at least some of the recordings made by these workers were performed in winter, leading to the disparate results indicated in Table 1. As high frequencies seem to be characteristics of arousal in fish it might appear that fish are less responsive in winter than in summer. This indeed does seem to be the case judging by the minute EEG changes recorded from fish which had behavioural and autonomic responses to a moving edge in winter. The only noticeable change in the EEG during arousal at this time was a transient reduction in amplitude and increase in frequency. This was especially evident in the tectum and telencephalon. These results are in contrast to the large frequency and amplitude changes of EEGs in response to a moving shadow stimulus, of fish in summer.

These experiments were carried out after acclimation of the fish to the experimental temperature for 2 weeks. It would thus seem that a reduction in metabolism due to temperature alone would be an unlikely explanation for the results summarized here. Goldfish show a reduction in brain activity in winter which is correlated with a reduction in responsiveness. This may be an adaptive mechanism to reduce energy expenditure during a season when food is scarce and reproductive activity is at a minimum. The EEG frequency appears to be an indicant of the level of activity and responsiveness of the animal; whether the relationship might be a causal one is open to question. The first stage in the examination of the physiology of behaviour is to establish a correlation. Thus, it has become apparent that a relationship exists between the EEG and behaviour. The next stage is to examine the way in which behaviour and physiology are related. The following section looks at the nature of EEG changes and considers whether they may have an adaptive function, and how such a function might be explored.

The nature of the EEG and its possible role during changes in responsiveness

The grossly recorded electrical activity of the brain, or EEG, has long been known to show characteristic changes coincident with behaviour. This information has largely been derived from studies on mammals. More recently, similar characteristics have been found in the EEGs of other vertebrates. In spite of the correlation between behaviour and EEG pattern, little was known about the physiological basis of EEG patterns until recently, and their function still remains a topic for speculation.

Characteristically, EEGs contain a variety of waveforms which have a long rise time and duration when compared with action potentials. Early EEG investigators thought that combined or massed action potentials formed the basis of the EEG (Adrian & Matthews, 1934), though it has subsequently been found that unit activity and EEG waves are rarely synchronized (Buchwald, Halas & Schramm, 1966). Thus, the slow waves of the EEG are not themselves the massed action potentials of neurons firing in synchrony, though 'spikes' appearing in the EEG may have this origin. These rapid EEG events may be the result of action potentials in sensory or motor pathways caused by either external stimuli or intracranial electrical stimulation. When they are thus elicited they are described as evoked responses. The slow waveforms which have been recorded, from extra- and intracellular as well as surface electrodes, appear to be due to synchronous changes in the membrane potentials of neurons (Klemm, 1969). Intracellular recording has shown regular changes in potential correlated with the grossly recorded EEG (Fujito & Sato, 1964). It would not, however, be possible to record the EEG from surface electrodes if it were simply a change in the membrane potential of individual neurons, and the EEG is therefore considered to be a synchronized membrane potential fluctuation in a population of neurons (Klemm, 1969). The site of the 'generator' of these potential fluctuations within neurons is unclear, but the evidence leans towards a synaptic one according to Elul (1972), who proposes that a number of such generators operating in phase would produce a synchronized EEG. The amplitude of the EEG waveform produced would depend on the amplitude of each individual generator and on their phase relationships.

EEG research has largely been performed with mammals, where the source of the EEG waveform has long been considered to be the cortex. It is now apparent that the synchronizer of the potential generators, necessary to produce a waveform, may be subcortical (Elul, 1972). Certainly in lower vertebrates, like fish, which have no cortex, the synchronizers must be subcortical.

The identification of the EEG waveform as a synchronized fluctuation in membrane potentials leads to speculation on the nature of the 'synchronizer' and the possible adaptive functions of the EEG.

The EEG: a harmonic oscillator?

One interesting feature of the EEG is the regularity of the waveforms which occur in certain situations. 'Alpha' waves of 10 ± 1 Hz recorded from the human cortex are often extremely regular (McClelland, personal communication), as are the 16–24 Hz waveforms found during arousal in fish. Of course, 'perfect' waveforms are rarely found but the fact that they are ever found shows a remarkable degree of synchronization in the activity of neuronal populations. What is more intriguing is the constancy of the frequencies found in any individual animal. Thus one individual fish may have an aroused frequency of 16 Hz and this frequency rarely varies by more than 1 Hz on subsequent arousals of that animal.

Another feature of the EEG becomes apparent when frequencies of the EEGs from individual animals are taken. Close arithmetical relationships are often found between the frequency ranges present at any one time. Thus in the goldfish, if two frequencies are simultaneously evident then the faster is often a multiple, usually 2 to 4 times, of the slower (within the limits of measuring techniques). This suggests that the lower frequency may be a harmonic of the faster one. In 60% of cases where two frequencies are evident simultaneously in the EEG of goldfish, such a harmonic relationship has been found (Laming, unpublished). More research is required to see if such a harmonic relationship between EEG waveforms is a general phenomenon. This will lead to a greater understanding of the control of EEG oscillations in membrane potential.

Membrane potential oscillations: a sensitizing mechanism?

The neuroethologist working with EEG recording on lower vertebrates is tempted to speculate on its possible adaptive functions. When the animal is behaviourally resting the frequency of the EEG oscillations is low. In winter when the animal is similarly unresponsive a similar situation is found. How, therefore, might a faster EEG, correlated with behavioural alertness, aid the animal in perceiving and responding to environmental change? It would be surprising if well synchronized changes in membrane potential did not have an adaptive function of this nature. The assumption will initially be made that the amplitude of the waveforms is constant. The EEG's regular partial depolarization increases in frequency from 6–8 Hz to 16–24 Hz on arousal. The threshold for a sensory input to fire one of the cells involved in the oscillating changes in membrane potential constituting the EEG would be decreased during depolarization. The increase in frequency of such depolarizations would therefore increase the likelihood that short bursts of sensory inputs would coincide with a depolarization, fire a central neuron and therefore be detected. The effect is one of an increase in time-sampling of the input.

The above mechanism would make the brain more capable of rapidly detecting changes in the environment but would not change the overall sensitivity, as we have assumed that the mean level of depolarization/hyperpolarization of central neurons is constant, and that the maximum depolarization is constant also. If the amplitude of the oscillations increases, which overall it does not do significantly (Laming, 1980), then sensitization would be general. A better mechanism for increasing sensitivity, however, would be to partially depolarize all central neurons, i.e. change the 'baseline' for EEG oscillations. Examination of shifts like this in resting potential of large numbers of neurons has ironically been hampered by aspects of the design of electroencephalographic equipment, which has deliberately been made to ignore such slow potential changes. The points made here about the membrane potential fluctuations being able to both increase time-sampling and possibly decrease thresholds of sensory input or motor output imply a facilitatory role of EEG synchronizers. If, however, there were a movement of the base level of membrane potential fluctuations in the direction of hyperpolarization, inhibition would be the result. I would suggest that an important aspect of the future study of arousal in animals will involve a close examination of both EEG waveform relationships and also slower, more durable changes in membrane potential in large populations of CNS neurons.

Conclusion

In a changing environment there are times when it may be adaptively important for animals to be capable of maximally efficient detection of and response to transient changes. In order to reduce overall energy expenditure, maximal efficiency is reserved for special occasions, such as when food or mates are available or when a previously unexperienced change takes place in the environment. This latter case is exemplified by the alerting or arousal response to a novel stimulus. In fish this consists of movements of the fins to stabilize the animal's position in the environment, and is also associated with reductions in heart and ventilatory rate. It is suggested that these physiological changes may be responses to a redistribution of energy supplies, via the circulatory system, to tissues like heart, brain and skeletal muscle, which may need to operate at maximum efficiency.

In the brain itself, alertness is associated with an increase in frequency of the regular oscillations recorded in the EEG. In summer, a regular 6–8 Hz activity is replaced by 16–24 Hz activity which is most evident in the midbrain and telencephalon. Electrical stimulation of the midbrain can also elicit arousal at very low thresholds, implicating this region in the origin of arousal. The telencephalon, on the other hand, can apparently either facilitate or inhibit arousal and may act in a modulatory role. The association between high-frequency waveforms in the EEG and alertness is confirmed by the lack of 16–24 Hz activity in the EEG of goldfish during winter, when the fish are less responsive. Slower 3–4 Hz and 8–14 Hz activity are evident at this time.

The brain is ultimately the origin of adaptive changes which enable animals to function more efficiently during arousal. The relationship between changes in brain activity and this improvement in functional capability is therefore interesting. The EEG waves of regular partial depolarization and repolarization are suggested as being possible mechanisms for increasing the functional capabilities of the CNS during arousal. Detailed research in this area may well prove fruitful for our understanding of the mechanisms of alert behaviour.

References

Adrian, E. D. & Matthews, B. H. C. (1934). The interpretation of potential waves in the cortex. *Journal of Physiology,* **81**, 440–71.

Alexandrov, V. V., Belich, A. I., Karmonova, I. G. & Schneiderov, V. S. (1980). Computerised analysis of daily dynamics of heart rate. *Zhurnal evolutsionnoi biokhimii i fiziologii,* **16**, in press.

Buchwald, J. S., Halas, E. S. & Schramm, S. (1966). Relationships of neuronal spike populations and EEG activity in chronic cats. *Electroencephalography and Clinical Neurophysiology*, **21**, 227–38.

Elul, R. (1972). The genesis of the EEG. In *International Review of Neurobiology,* vol. 15, ed. C. C. Pfeiffer & J. R. Smythies, pp. 228–72. Academic Press, New York & London.

Enger, P. S. (1957). The electroencephalogram of the codfish. *Acta Physiologica Scandinavica,* **39**, 55–72.

Ewert, J. P. & Siefert, G. (1974). Seasonal change in contrast detection in the toad (*Bufo bufo*) visual system. *Journal of Comparative Physiology,* **94**, 177–86.

Fujita, Y. & Sato, T. (1964). Intracellular records from hippocampal pyramidal cells in rabbit during theta rhythm activity. *Journal of Neurophysiology,* **27**, 1011–25.

Gibson, R. N. (1978). Lunar and tidal rhythms in fish. In *Rhythmic Activity of Fishes,* ed. J. E. Thorpe, pp. 201–15. Academic Press, New York & London.

Godet, R., Bert, J. & Collomb, H. (1964). Apparition de la réaction d'éveil telencéphalique chez *Protopterus annectens* et cycle biologique. *Comptes rendus des séances de la Société de Biologie,* **158**, 146–9.

Goodman, D. A. & Weinberger, N. M. (1973). Habituation in 'lower' tetrapod vertebrates. Amphibia as model systems. In *Habituation*, vol. 1, ed. H. V. S. Peeke & M. J. Herz, pp. 85–140. Academic Press, New York & London.

Hara, T. J., Ileda, K. & Gorbrian, A. (1965). Electroencephalographic studies of homing salmon. *Science,* **149**, 884–5.

Klemm, W. R. (1969). *Animal Electroencephalography.* Academic Press, New York & London.

Laming, P. R. (1980). Electroencephalographic studies on arousal in the goldfish (*Carassius auratus*). *Journal of Comparative and Physiological Psychology,* **94**, 238–54.

Laming, P. R. & Austin, M. (1981). Cardiac responses of the anurans, *Bufo bufo* and *Rana pipiens* during behavioural arousal and fright. *Comparative Biochemistry and Physiology*, in press.

Laming, P. R. & McKee, M. (1981). Deficits in the habituation of cardiac arousal responses, incurred by telencephalic ablation in the goldfish *Carassius auratus*: their relation to other telencephalic functions. *Journal of Comparative and Physiological Psychology*, in press.

Laming, P. R. & Savage, G. E. (1978). Flow changes in visceral blood vessels of the chub (*Leuciscus cephalus*) during behavioural arousal. *Comparative Biochemistry and Physiology,* **59A**, 291–3.

Laming, P. R. & Savage, G. E. (1980). Physiological changes observed in the goldfish (*Carassius auratus*) during behavioural arousal and fright. *Behavioural and Neural Biology,* **29**, 255–75.

Mott, J. C. (1957). The cardiovascular system. In *The Physiology of Fishes,* vol. 1, ed. M. E. Browne, pp. 81–108, Academic Press, New York & London.

Müller, K. (1978). Locomotor activity of fish and environmental oscillations. In *Rhythmic Activity of Fishes,* ed. J. E. Thorpe, pp. 1–21. Academic Press, New York & London.

Otis, L. S., Cerf, J. A. & Thomas, G. J. (1957). Conditioned inhibition of respiration and heart rate in the goldfish. *Science.* **126**, 263–4.

Peeke, H. V. S. & Peeke, S. C. (1970). Habituation of aggressive responses in the Siamese fighting fish (*Betta splendens*). *Behavior,* **36**, 232–45.

Peeke, H. V. S. & Peeke, S. C. (1972). Habituation, reinforcement and recovery of predatory responses in two species of fish. (*Carassius auratus* and *Macropodus opercularis*). *Animal Behaviour,* **20**, 268–73.

Randall, D. J. (1966). The nervous control of cardiac activity in the tench (*Tinca tinca*) and the goldfish (*Carassius auratus*). *Physiological Zoology*, **34**, 185–92.

Retzlaff, E. (1957). A mechanism for excitation and inhibition of the Mauthner cells in teleosts. A histological and neurophysiological study. *Journal of Comparative Neurology,* **107**, 209–25.

Rodgers, W. L., Melzack, R. & Segal, J. R. (1963). 'Tail-flip response' in goldfish. *Journal of Comparative and Physiological Psychology,* **56**, 917–23.

Russell, E. M. (1967). Changes in the behaviour of *Lebistes reticularis* upon repeated shadow stimulus. *Animal Behaviour,* **15**, 574–85.

Satchell, G. H. (1971). *Circulation in Fishes.* Cambridge University Press, London.

Savage, G. E. (1971). Behavioural effects of electrical stimulation of the telencephalon of the goldfish (*Carassius auratus*). *Animal Behaviour,* **19**, 661–8.

Schadé, J. P. & Weiler, I. J. (1959). EEG patterns of the goldfish. *Journal of Experimental Biology,* **36**, 435–52.

Segura, E. T. & De Juan, A. (1966). Electroencephalographic studies in toads. *Electroencephalography and Clinical Neurophysiology,* **21**, 373–80.

Siegel, S. (1956). *Nonparametric Statistics for the Behavioural Sciences.* McGraw-Hill, New York.

Winslade, P. (1974*a*). Behavioural studies on the lesser sandeel, *Ammodytes marinus* (Raitt). III. The effect of temperature on activity and the environmental control of the annual cycle of activity. *Journal of Fish Biology,* **6**, 587–600.

Winslade, P. (1974*b*). Behavioural studies on the lesser sandeel, *Ammodytes marinus* (Raitt). II. The effect of light intensity on activity. *Journal of Fish Biology,* **6**, 577–86.

PART IV

Neuronal substrates of appetitive behaviour

LEO S. DEMSKI

Hypothalamic mechanisms of feeding in fishes

Introduction

It has been accepted for some time that hypothalamic areas are involved in regulating food intake in mammals and several detailed hypotheses have emerged to explain the mechanisms involved (Grossman, 1975). In contrast, relatively little is known concerning similar systems in cold-blooded vertebrates. This is especially surprising considering that Herrick pointed out in 1905 that areas in the teleost hypothalamus had anatomical connections indicative of a feeding control centre. Peter (1979) has recently reviewed the present state of general knowledge of neural mechanisms of feeding in fishes and although several aspects of hypothalamic control are discussed, the details of possible pathways and neurophysiological mechanisms were purposely avoided. The present article is intended to augment Peter's review by covering these topics in some detail, including the results of recent microelectrode recording experiments (Demski, unpublished), and to summarize the available information in a single diagrammatic model (Fig. 1). It is hoped that regardless of the model's speculative nature, it will serve to encourage added interest in this area.

Behavioural studies

The first experiments directly concerned with the role of the hypothalamus in feeding in fishes were studies on the behavioural effects of electrical stimulation of this area in free-swimming teleosts. Studies in three species have demonstrated that low-threshold (3–90 μA) complete feeding responses can be evoked from the areas near the lateral recess of the third ventricle in the inferior lobe of the hypothalamus. In sunfish (*Lepomis macrochirus*) stimulation resulted in feeding responses consisting of searching the surface and bottom of the tank, snapping at food and snapping-up gravel (Demski & Knigge, 1971). A similar pattern was observed during stimulation in *Tilapia macrocephala* (Demski, 1973). One of the most consistent responses evoked in *Lepomis* and *Tilapia* was the snapping-up of gravel. Experiments with *Tilapia* demonstrated that neither gravel nor food were necessary for this response to be triggered. In a tank with gravel on one side and bare slate on the other, the fish made the same basic manoeuvre on both sides. The fish pitched forward, snapped at the substrate and made spitting movements. Since a similar response is given by intact control fish upon injection of colourless food extracts into the tank, it was suggested that evoked snapping-up gravel responses were possibly related to activation of an olfactory input into the hypothalamic inferior lobes (Demski, 1973). Studies in goldfish (Savage & Roberts, 1975) are consistent with the results in sunfish and *Tilapia*. Stimulation in the area just dorsal to the lateral recess of the third ventricle and slightly below the nucleus preglomerulosus (area called n. subpreglomerulosus by Savage & Roberts) resulted in the lowest-threshold (20 μA or less) feeding responses. Other less consistent or only partial feeding responses were observed but these were usually evoked at higher thresholds or from more widespread areas of the brain. Thus, low-threshold feeding activity has been evoked from what appear to be equivalent regions of the hypothalamus in sunfish, mouthbreeders and goldfish. In addition, Redgate (1974) has reported that gross electrical activation of the inferior lobe results in feeding responses in carp (*Cyprinus carpio*), and Fiedler (1967), using a variety of species, reported that searching the bottom was evoked from the nearby nucleus prerotundus (n. preglomerulosus of others). It is interesting that stimulation in the inferior lobe in sharks also results in feeding behaviour (Demski, 1977). The combined results suggest that this hypothalamic area may mediate feeding in a wide variety of fish groups.

In addition, feeding has also resulted from stimulation of other brain regions that are likely to have connections with the hypothalamic area from which feeding has been evoked, hereafter referred to as the 'hypothalamic feeding area' (HFA). Grimm (1960) demonstrated that well-developed feeding responses occurred following olfactory tract stimulation consisting of 0.5-s pulses of 75–150 μA at 30–50 Hz. Grimm also reported that feeding responses could be evoked from scattered regions in the dorsal telencephalon. The responses were frequently of low intensity or incomplete compared with those evoked from the olfactory tracts. However, rather low thresholds of 20–60 μA were effective. Several other investigators have reported that inconsistent or partial feeding activity has resulted from stimulation in the telencephalon. Savage (1971) observed diving or darting at the bottom in goldfish as a result of stimulation of two sites in the medial regions of the telencephalon. One site was just dorsal to the area of entry of the olfactory tracts and the other site was dorsal to the anterior commissure. Similarly, feeding responses in sunfish (*Lepomis macrochirus*) were evoked by stimulation in two sites in the ventromedial part of the telencephalon (Demski & Knigge, 1971). In addition, Fiedler (1964, 1967) has suggested that food-related responses such as searching the bottom and the surface, picking-up sand, chewing movements and spitting can be evoked from the telencephalon in various teleosts. Stimulation in the general region of the medial forebrain bundle evoked feeding responses in *Tilapia macrocephala* (Demski, 1973). On the basis of these evoked behavioural responses, as well as anatomical considerations, it was suggested that the medial forebrain bundle connects telencephalic areas involved in feeding with hypothalamic feeding mechanisms (Demski, 1973). Many teleosts have two well-developed gustatory lobes (vagal and facial) in the medulla. Stimulation in the vagal lobe in goldfish has resulted in mouth movements (Demski, unpublished) and weak, incomplete feeding activity (Grimm, 1960). It is conceivable that feeding responses resulting from stimulation of the goldfish vagal lobes were due to activation of the HFA since this diencephalic region apparently receives higher-order gustatory fibres (see below).

Roberts & Savage (1978) have recently studied the effects of hypothalamic lesions on feeding in goldfish. They placed bilateral lesions in several areas and recorded total daily food intake. Feeding in both operant and free-access situations was observed. Their results indicate that the most dramatic reductions in feeding occurred in goldfish with damage to the lateral or inferior lobe area. In these cases cessations of operant feeding of up to 60 days and cessations of feeding in response to manually presented food of up to 35 days were observed. Less severe deficits were noted for fish with anterior-medial hypothalamic lesions and no change occurred in animals with posterior-medial lesions. The authors point out that the lesion results are consistent with the electrical stimulation studies (see above) in that the lateral areas of the inferior lobe including the nucleus subpreglomerulosus are probably critical for normal feeding. This conclusion is also consistent with findings that goldfish with bilateral damage in the inferior lobe near the anterior part of the lateral recess of the third ventricle (n. recessus lateralis according to the terminology of Peter & Gill, 1975) demonstrated retarded growth patterns compared with animals with bilateral lesions in other hypothalamic areas and controls (Peter, 1979). It is noteworthy that the above authors failed to observe increases in food intake or growth following any of the various hypothalamic lesions, as might have been expected if goldfish have an equivalent of the mammalian ventromedial hypothalamic area. In mammals, bilateral destruction of this area may lead to obesity caused by hyperphagia (see review by Grossman, 1975). Also in this regard, Peter, Monckton & McKeown (1976) failed to observe hyperphagia following injection of gold thioglucose into goldfish. This substance is believed to cause obesity in certain mammals by damaging the ventromedial hypothalamic area.

Anatomical studies

Early anatomical studies on the connections of the inferior lobe of the hypothalamus in fishes have been summarized extensively (Herrick, 1905; Ariëns-Kappers, Huber & Crosby, 1936; Crosby & Showers, 1969). For the present purpose it will suffice to mention that Herrick's work on the gustatory connections of the inferior lobe predicted involvement of this area in control of food intake. Recent anatomical studies using more powerful methods have supported and greatly extended many of the older observations. Most of the work has been done on teleosts and is summarized schematically in Fig. 1. Finger (1978, 1980), using degeneration techniques, has demonstrated that second- and third-order gustatory fibres end in the subrotundal area of the inferior lobe of the hypothalamus in catfish and goldfish. This area overlaps the HFA mentioned above. In addition, the large cells in this area in catfish (nucleus of the tractus lobobulbaris) appear to project axons back to the medullary gustatory lobes. This observation is based on retrograde transport of horseradish peroxidase (HRP) from injection into the facial lobes (Finger, 1978). Thus the proposed HFA in teleosts seems to receive at least second-order gustatory information and may have connections involved in the feedback control of gustatory input. Anatomical evidence for olfactory input into the HFA is less direct. On the basis of fibre degeneration studies in catfish, Finger (1975) demonstrated

Fig. 1. Schematic diagram of a sagittal section of a goldfish brain. The 'hypothalamic feeding area' (HFA) represents the general region from which low-threshold feeding responses can be evoked by electrical stimulation and the area in which bilateral lesions cause drastic reductions in feeding (see text for details). Presumed connections of the HFA are indicated, based on results discussed in the text. The experimental set-up for the acute recording studies discussed in the text is also illustrated. R, recording electrode; S, stimulation electrodes; V, motor root of the trigeminal nerve; VII, motor root of the facial nerve.

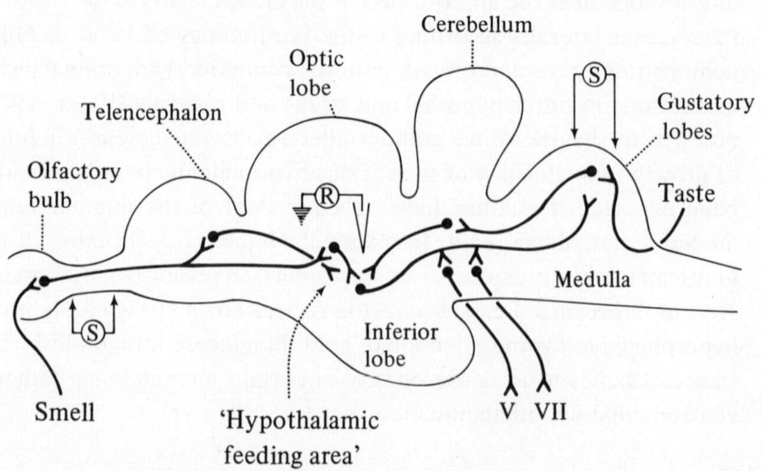

olfactory bulb projections into the posteromedial hypothalamus ('hypothalamic terminal field') where they appear to end in a cell-free zone. This area is, however, just medial to the subrotundus region (subpreglomerulosus of Savage & Roberts, 1975) or HFA and thus the olfactory fibres could perhaps contact long dendrites from the more lateral area or relate to the latter through a hypothalamic interneuron (as suggested in Fig. 1). Davis (1978), using autoradiography following administration of tritium-labelled proline in the olfactory bulb, has found a similar olfactory projection to the hypothalamus in *Macropodus opercularis.* A second and perhaps more important source of olfactory input into the HFA may originate from cells in the olfactory areas of the telencephalic lobes and run into the subrotundal area via fibres in the forebrain bundles (also see suggestions under 'Behavioural studies'). Vanegas & Ebbesson (1976) have demonstrated that several bundles of fibres enter the subrotundal zone and terminate in this and more ventral regions of the inferior lobe following telencephalic resection in the mojarra and squirrel fish. These tracts, observed using degeneration methods, were the lateral forebrain bundle and the strio-lobar bundle. A similar pattern of degeneration has been reported following lateral telencephalic lesions which included the olfactory area in nurse sharks (Ebbesson, 1972). Thus it is likely that at least higher-order olfactory information is relayed to the HFA by perhaps one of several routes in both teleosts and elasmobranchs. Other sensory inputs to the HFA have not yet been reported from studies based on modern anatomical techniques.

With respect to the motor function of the HFA, it is significant that Herrick (1905) pointed out that this area projected to motor cranial nerve nuclei and perhaps the spinal cord via a tractus lobobulbaris. This of course would be consistent with a role in the control of feeding. Recently, Luiten & van der Pers (1977) have demonstrated in carp that HRP injected into the trigeminal and facial motor nuclei was transported back to cell bodies in the nucleus preglomerulosus, n. lateral recessus and n. diffusus, which are in the vicinity of the proposed HFA. The authors point out that these connections are consistent with the earlier findings of Herrick and the idea of the HFA suggested by the stimulation studies (see above). Observations (Demski, 1969, 1978) that isolated mouth movements can frequently be evoked by electrical stimulation in the inferior lobe in sunfish are also consistent with these recent anatomical studies. Luiten & van der Per's results have been included in Fig. 1 on the basis of the likelihood that the cells they found projecting to the trigeminal nucleus are indeed within the zone referred to as the HFA.

Histochemical studies have suggested that the inferior lobe has many catecholamine-containing fibres. Studies in goldfish (Baumgarten & Braak,

1967), sunfish (Parent *et al.,* 1978) and dogfish sharks (Wilson & Dodd, 1973) have all demonstrated fibres containing fluorescent material suggestive of catecholamines in the inferior lobe. At least in sunfish, some of these fibres can be seen running across the subrotundal area here referred to as the HFA. Thus the suggestion may be made that aminergic fibres are part of a feeding system in fishes. This would be consistent with the aminergic feeding system proposed for mammals (Grossman, 1975).

Physiological studies

Detailed studies in this category are lacking. However, some information is available from preliminary experiments concerned with recording

Fig. 2. Responses of HFA units recorded from a single site in goldfish 9. *A–D*, spontaneous activity of a large-amplitude unit (four consecutive sweeps). *E*, response of several units able to follow vagal lobe stimulation at 1 Hz (stimulus artifacts are the large upward spikes). *F*, response of the same units to vagal lobe stimulation at 3 Hz (units unable to follow). *G*, response of the units to injection of 0.02 M acetic acid into the mouth (about 5 ml injected in about 1 s). Injection began about 1–3 s before response which was repeatable. There was no response to control water injections. *H-J*, response of a large-amplitude unit to single vagal lobe stimulations (five consecutive tests). Vagal lobe stimulation 0.1 ms at 0.5 MA; vertical scale 100μV (*A–L*); horizontal scale 1 s (A–*G*), 100 ms (*H–L*).

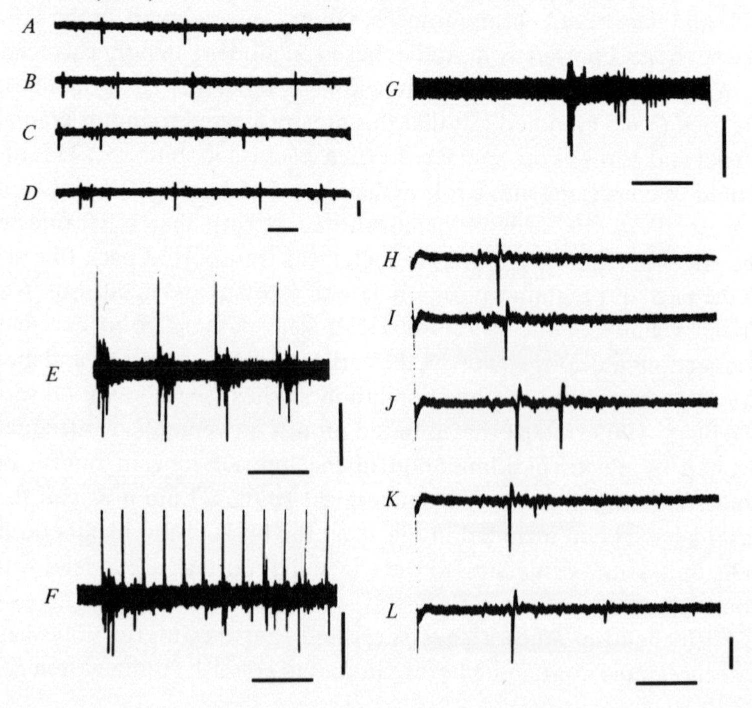

single and multiunit activity from the HFA in goldfish (Demski, unpublished). Animals of approximately 20–25 cm (total length) were first lightly anaesthetized with MS 222 and then immobilized with 1.0–1.5 mg kg^{-1} of D-tubocurarine. The fish were placed in a surgical holder (see description in Demski & Knigge, 1971; Demski, 1977) and all pressure points and incisions infiltrated with xylocaine. Fig. 1 illustrates the experimental scheme. Stimulation electrodes were placed on one each of the vagal lobes (bipolar electrode) and the olfactory tracts (suction electrode) and electrical activity was recorded from the hypothalamus using stainless steel and tungsten microelectrodes (tip diameter of approximately 1 μm). Square-wave stimulation was usually applied to both vagal lobe and olfactory tract when searching for responding units. In some cases various solutions were perfused into the fish's mouth in order to stimulate taste receptors. The activity of HFA units in seven different fish has been sampled. Cells were found that responded to the vagal lobe and/or olfactory tract stimulation by increased firing rate (Figs. 2–4). Whenever both stimuli were tested together the cells either fired in response to both stimuli or failed to respond to either stimulus. Response latencies varied from about 10 to 150 ms (Fig. 2*H–L*, Fig. 3*F–H*). There appeared to be almost a bimodal distribution of latencies with what appeared to be smaller-amplitude units being activated earlier than larger ones. The latencies of the former group are consistent with the anatomical

Fig. 3. Responses of several HFA units recorded from a single site in goldfish 5. *A–B*, response of units to food extracts (5 ml) injected into mouth; *A* is water control and *B* is burst activity in response to food extracts (experiment was repeated three times). *C–E*, response of units to saltwater injection into mouth (5 ml of seawater); *C* and *E* are responses to saltwater, *D* is control freshwater injection. *F–H*, response of units to vagal lobe stimulation (five sweeps per trace); *F* and *H* are stimulation tests, *G* is control (no current passed). Vagal lobe stimulation, 0.3 ms at 1 MA; vertical scale 50 μV (*A–H*); horizontal scale 1 s (*A–E*), 100 ms (*F–H*).

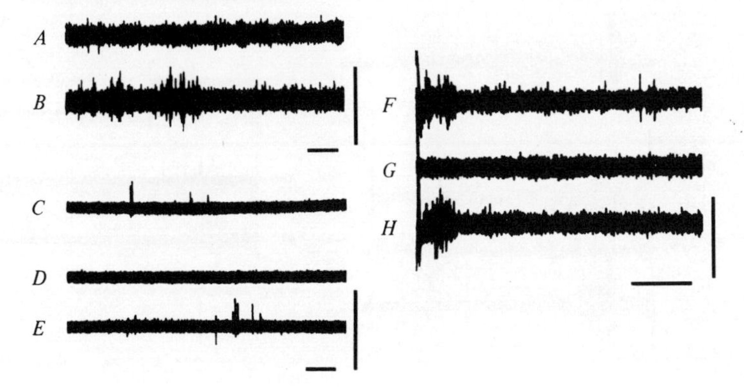

pathways illustrated in Fig. 1. The later responses may represent cells activated by intrahypothalamic connections or more complicated afferent pathways, perhaps involving the reticular formation. Latencies for activation via either stimulus were about the same (Fig. 4*A–G*). Movement of the stimulation electrodes or lowering the stimulation current resulted in loss of the responses (Fig. 4*K–M*). In these cases responses returned on replacement of the electrode or increasing the current. Slight movement of the recording electrode also resulted in loss of the response. In several cases

Fig. 4. Responses of HFA units recorded from a single site in goldfish 1. *A–B*, results of consecutive stimulation tests (*A*, vagal lobe stimulation; *B*, olfactory tract stimulation). *C–D*, repetition of experiment in *A* and *B* (*C*, vagal lobe stimulation; *D*, olfactory tract stimulation). *E–G*, repetition of experiment in *A* and *B* with control test (*F*, no current) between vagal lobe stimulation (*E*) and olfactory tract stimulation (*G*), five sweeps per trace. *H–J*, faster sweep for three consecutive tests of response to olfactory tract stimulation. *K–M*, response of units to increasing strengths of olfactory tract stimulation (*K*, 0.1 MA; *L*, 0.5 MA; *M*, 1 MA), three sweeps per trace. Vagal lobe stimulation, 0.1 ms at 1.5 MA (*A* and *C*), 1 MA (*E*); olfactory tract stimulation 0.1 ms at 0.1 MA (*K*), 0.5 MA (*L*), 1 MA (*B,D,G,H–J,M*); vertical scale 10 μV (*A,M*); horizontal scale 100 ms (*A–M*).

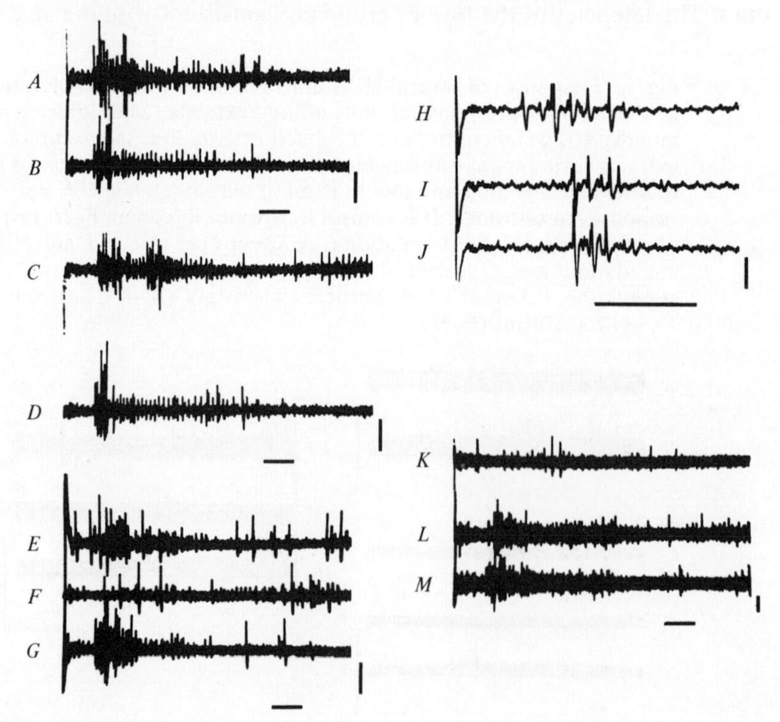

units responding to vagal lobe stimulation were tested for their response to chemical stimulation in the mouth. Food extracts and weak salt and acid solutions (Fig. 2*G*, Fig. 3*A–E*) were effective in increasing the firing rate of these spontaneously slow units (Fig. 2*A–D*) in a manner similar to that described by Hara (this volume). In several cases the recordings were maintained for periods over 3 h at which time the experiment was terminated by moving the recording electrode to be certain of the 'localness' of the response. In a few cases electrolytic lesions were made at the recording site and the brains sectioned in paraffin in order to locate these areas. Fig. 5 illustrates one of the marked sites. The lesion is located in the HFA in the region from which Savage & Roberts (1975) electrically evoked the most consistent and low-threshold feeding responses.

These results support the suggestion based on anatomical grounds (see above) that the HFA in goldfish receives both olfactory and gustatory information, and may help to explain Grimm's (1960) observation that at least partial feeding responses can be evoked by electrical stimulation of the olfactory tracts and vagal lobes in free-swimming goldfish. Perhaps these evoked feeding responses were mediated by activation of cells in the HFA in a manner similar to that observed in these electrophysiological studies. It can also be suggested that the HFA is involved in integrating relevant chemosensory and possibly other types of information and then triggering feeding motor responses via pathways such as the one to the motor trigeminal nucleus (see above). It is noteworthy that visual, auditory and light-touch stimuli did not appear effective in activating the cells in this area of the goldfish hypothalamus. However, it must be pointed out that only crude testing was carried out using light flashes, hand claps, clicks and touch with fine brushes. Electrical stimulation in areas such as the cerebellum and optic lobe was also non-effective, thus giving more support to the specific nature of the chemosensory input to these hypothalamic neurons. The maximum rate at which vagal lobe stimulation could drive cells in the HFA was determined in a few cases and it appeared to be less than 3 Hz (Fig. 2*E* and *F*). This is important since cells in the HFA project to the gustatory area (facial lobe) in, at least, certain catfish (Finger, 1978). The inability of vagal lobe stimulation to drive these cells at high rates and short latencies suggests that a multisynaptic system is involved in the evoked responses rather than antidromic activation of the hypothalamic neurons.

Summary and conclusions

Ideally, most if not all of the following criteria should be met in determining that an area of the brain is involved in the control of feeding behaviour:

Fig. 5. Photomicrograph of a transverse section through the goldfish hypothalamus. The track of a recording electrode (E) can be seen running through the valvula of the cerebellum and the tegmentum of the midbrain. The tip of the electrode (arrows), located by placement of a lesion, was in the dorsal part of the inferior lobe in the subglomerular region of Savage & Roberts (1975) or 'hypothalamic feeding area' as referred to in the text. This site is almost in the same position as some of the lowest-threshold points for eliciting feeding by electrical stimulation in free-swimming goldfish (cf. sites 4, 6 and 7 in Figure 2 of Savage & Roberts, 1975). LR, lateral recess of the third ventricle; OL, optic lobe; NDLI, nucleus diffusus lobi inferiorus; TS, torus semicircularis; VAL, valvula of the cerebellum.

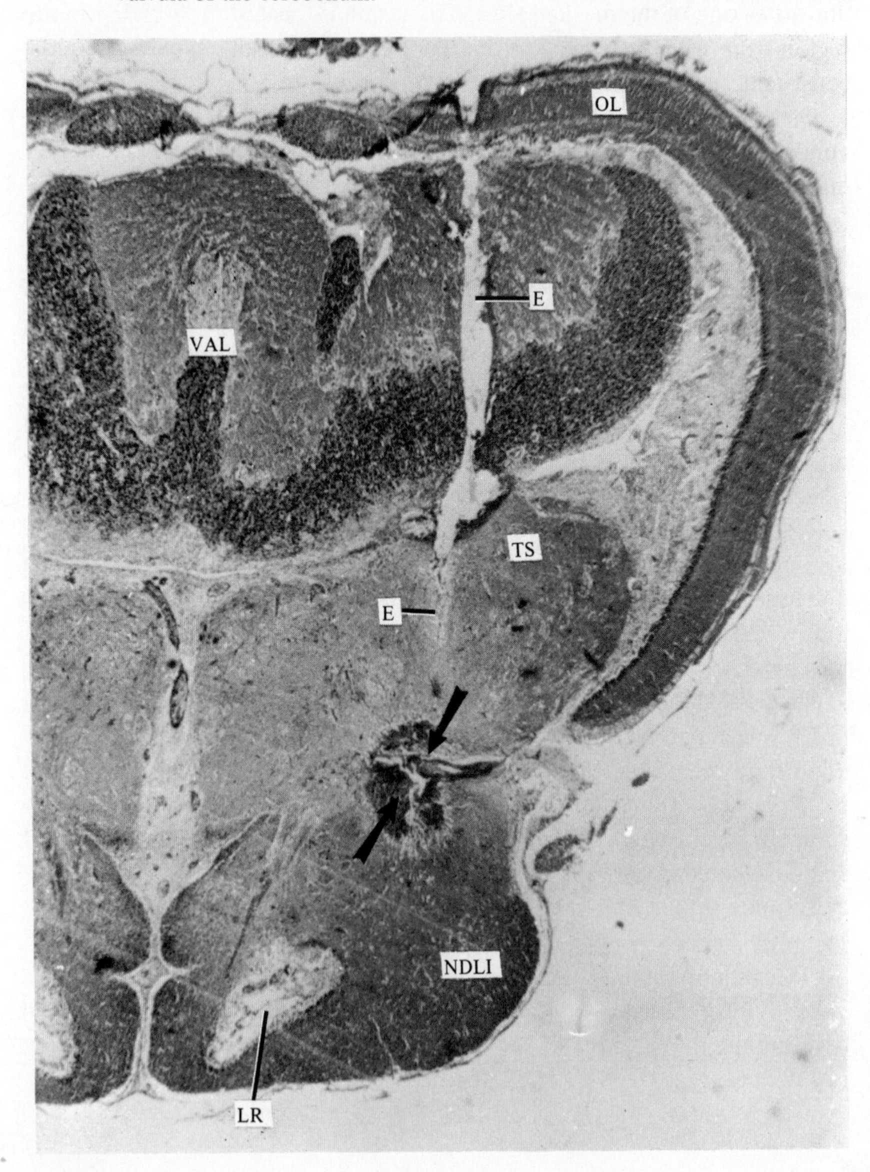

1. Damaging the area bilaterally should result in loss of eating or malfunction of feeding control, e.g. overeating.
2. Artificial activation of the area (usually electrical or chemical) should result in stimulus-bound feeding, disruption or total inhibition of ongoing feeding.
3. The area should have demonstrable functional–anatomical connections with sensory and motor systems involved in the normal feeding behaviour of the species in question.
4. The normal activity of cells in the area should be strongly correlated with certain specific aspects of feeding behaviour.

Criteria 1 and 2 have at least partially been satisfied for the HFA in teleosts (see 'Behavioural studies'). Anatomical and physiological studies in these fishes have indicated that the HFA does appear to have connections with sensory and motor systems consistent with its potential role in the control of feeding and thus criterion 3 has also partially been met. At this time little can be said with regard to the fourth criterion. Studies designed to determine correlations between HFA activity and food deprivation state are needed, as are methods of recording hypothalamic unit activity during feeding in free-swimming fishes.

Although the model presented is indeed tentative and mostly relevant to teleosts which rely heavily on chemosensory mechanisms, enough information is available to begin to outline some of the principal aspects of hypothalamic control of feeding in bony fishes. In contrast, almost nothing is known concerning the central control of feeding behaviour in elasmobranchs. Limited studies on sharks, however, suggest that hypothalamic feeding mechanisms may be similar in elasmobranchs and teleosts.

Author's studies supported in part by National Science Foundation Grant BNS–76–18617 and Biomedical Sciences Research Support Grant RR 07114–08.

References

Ariëns-Kappers, C. U., Huber, G. C. & Crosby, E. C. (1936). *The Comparative Anatomy of the Nervous System of Vertebrates Including Man.* Reprinted 1965 by Hafner, New York.

Baumgarten, H. G. & Braak, H. (1967). Catecholamine im Hypothalamus vom Goldfisch (*Carassius auratus*). *Zeitschrift für Zellforschung,* **80**, 246–63.

Crosby, E. C. & Showers, M. C. (1969). Comparative anatomy of the preoptic and hypothalamic areas. In *The Hypothalamus*, ed. W. Haymaker, E. Anderson & W. J. H. Nauta, pp. 61–135. Charles C. Thomas, Springfield, Ill.

Davis, R. E. (1978). Autoradiographic investigation of the central projections of the olfactory tracts in *Macropodus opercularis* (L.)

(Osteichthyes: Belontidae). *Society for Neuroscience,* abstr. 675.

Demski, L. S. (1969). 'Behavioral effects of electrical stimulation of the brain in free-swimming bluegills (*Lepomis macrochirus*)'. PhD dissertation, University of Rochester, Rochester, NY.

Demski, L. S. (1973). Feeding and aggressive behavior evoked by hypothalamic stimulation in a cichlid fish. *Comparative Biochemistry and Physiology,* **44A**, 685–92.

Demski, L. S. (1977). Electrical stimulation of the shark brain. *American Zoologist,* **17**, 487–500.

Demski, L. S. (1980). Electrical stimulation of the brain. In *Fish Neurobiology and Behavior,* ed. R. G. Northcutt & R. E. Davis. University of Michigan Press, Ann Arbor (in press).

Demski, L. S. & Knigge, K. M. (1971). The telencephalon and hypothalamus of the bluegill (*Lepomis macrochirus*): evoked feeding, aggressive and reproductive behavior with representative frontal sections. *Journal of Comparative Neurology,* **143**, 1–16.

Ebbesson, S. O. E. (1972). New insights into the organization of the shark brain. *Comparative Biochemistry and Physiology,* **42A**, 121–9.

Fiedler, K. (1964). Versuche zur Neuroethologie von Lippfischen und Sonnenbarschen. *Verhandlungen der Deutschen Zoologischen Gesellschaft, Kiel. Zoologisches Anzeiger Supplement* (1965), **28**, 569–80.

Fiedler, K. (1967). Verhaltenswirksame Strukturen im Fischgehirn. *Verhandlungen der Deutschen Zoologischen Gesellschaft, Heidelberg, Zoologischer Anzeiger Supplement* (1968), **31**, 602–16.

Finger, T. E. (1975). The distribution of the olfactory tracts in the bullhead catfish, *Ictalurus nebulosus. Journal of Comparative Neurology,* **161**, 125–42.

Finger, T. E. (1978). Gustatory pathways in the bullhead catfish. II. Facial lobe connections. *Journal of Comparative Neurology,* **180**, 691–706.

Finger, T. E. (1980). The gustatory system in teleost fishes. In *Fish Neurobiology and Behavior,* ed. R. G. Northcutt & R. E. Davis. University of Michigan Press, Ann Arbor (in press).

Grimm, R. J. (1960). Feeding behavior and electrical stimulation of the brain of *Carassius auratus.* Science, **131**, 162–3.

Grossman, S. P. (1975). Role of the hypothalamus in the regulation of food and water intake. *Psychological Review,* **82**, 200–24.

Herrick, C. J. (1905). The central gustatory paths in the brains of bony fishes. *Journal of Comparative Neurology,* **15**, 375–456.

Luiten, P. G. M. & van der Pers, J. N. C. (1977). The connections of the trigeminal and facial motor nuclei in the brain of the carp (*Cyprinus carpio* L.) as revealed by anterograde and retrograde transport of horseradish peroxidase. *Journal of Comparative Neurology,* **174**, 575–90.

Parent, A., Dube, L., Braford, M. R. Jr & Northcutt, R. G. (1978). The organization of monoamine-containing neurons in the brain of the sunfish (*Lepomis gibbosus*) as revealed by fluorescence microscopy. *Journal of Comparative Neurology,* **183**, 495–516.

Peter, R. E. (1979). The brain and feeding behavior. In *Fish Physiology,* vol. 8, ed. W. S. Hoar, D. J. Randall & J. R. Brett, pp. 121–59. Academic Press, New York & London.

Peter, R. E. & Gill, V. E. (1975). A stereotaxic atlas and technique for forebrain nuclei of the goldfish, *Carassius auratus*. *Journal of Comparative Neurology,* **159**, 69–101.
Peter, R. E., Monckton, E. A. & McKeown, B. A. (1976). Effects of gold thioglucose on food intake, growth and forebrain histology in goldfish, *Carassius auratus*. *Physiology and Behavior,* **17**, 303–12.
Redgate, E. S. (1974). Neural control of pituitary adrenal activity in *Cyprinus carpio*. *General and Comparative Endocrinology,* **22**, 35–41.
Roberts, M. G. & Savage, G. E. (1978). Effects of hypothalamic lesions on the food intake of the goldfish (*Carassius auratus*). *Brain, Behavior and Evolution,* **15**, 150–64.
Savage, G. E. (1971). Behavioural effects of electrical stimulation of the telencephalon of the goldfish, *Carassius auratus*. *Animal Behaviour,* **19**, 661–8.
Savage, G. E. & Roberts, M. G. (1975). Behavioural effects of electrical stimulation of the hypothalamus of the goldfish (*Carassius auratus*). *Brain, Behavior Evolution,* **12**, 42–56.
Vanegas, H. & Ebbesson, S. O. E. (1976). Telencephalic projections in two teleost species. *Journal of Comparative Neurology,* **165**, 181–95.
Wilson, J. F. & Dodd, J. M. (1973). Distribution of monoamines in the diencephalon and pituitary of the dogfish, *Scyliorhinus canicula* L. *Zeitschrift für Zellforschung,* **137**, 451–69.

ROGER E. DAVIS, JEFFREY KASSEL &
MIQUEL MARTINEZ

The telencephalon and reproductive behaviour in the teleost *Macropodus opercularis* (L.): effects of lesions on the incidence of spawning and egg cannibalism

Introduction

Bilateral telencephalon ablation has been shown to result in decreased frequencies of male reproductive behaviours in several teleosts (reviews: Segaar, 1965; Aronson & Kaplan, 1968; DeBruin, 1977; Davis & Kassel, 1980). In telencephalon-ablated, or 'diencephalic' males, sexual responses to an intact female and nest building are decreased, and the incidence of egg cannibalism is increased. Diencephalic males occasionally perform partial or even complete sequences of reproductive behaviour, and the behaviour patterns closely resemble those seen in sham-operated or intact males. This supports the interpretation that the telencephalon is necessary for the activation or potentiation of reproduction but not the organization of the motor patterns (Aronson, 1948; Aronson & Kaplan, 1968; Kassel & Davis, 1977).

In *Macropodus*, the paradise fish, social responses to the female, including body displays, approach and chasing behaviour, which occur prior to spawning are only temporarily decreased following telencephalon ablation. These behaviours are not explicitly sexual, though they may be a necessary part of the reproductive cycle (see Davis & Kassel, 1980). When the diencephalic male is allowed to recover for several weeks prior to the first post-operative spawning trial, the frequencies of social behaviours are similar to those of sham-operated and intact males but reproduction is impaired (Kassel, Davis & Schwagmeyer, 1976; Kassel & Davis, 1977). This suggests that the decreased reproductive behaviour in the diencephalic male is a result of damage to brain structures which specifically control reproductive functions.

In teleosts, as in other vertebrates, the preoptic nucleus region contains neurons which concentrate sex steroids, implicating this region of the brain in the control of reproduction (Morrell & Pfaff, 1978). The retention of sex hormones by neurons may be a step in the process by which the hormone alters neuronal function. In *Macropodus*, in addition to the preoptic nucleus region, sex-steroid-concentrating cells also occur immediately anterior and dorsal to the anterior commissure in the area designated by Nieuwenhuys (1963) as the area ventralis pars ventralis, or area Vv (Davis, Morrell & Pfaff, 1977). Removal of the telencephalic hemispheres by the aspiration method employed in behavioural experiments produces a lesion which includes area Vv, the anterior commissure and varying amounts of the preoptic nucleus region as well (Kassel & Davis, 1977). It is possible that the decreased reproduction seen in diencephalic males results from damage done to the preoptic area Vv region.

Reproduction may require perception of specific external stimuli which is impaired in the diencephalic male. Olfaction is necessary for reproductive behaviour in some teleosts but not in others (reviews: Kleerekoper, 1969; Little, 1980). Olfactory deafferentation by bilateral resection of the olfactory tracts results in a decreased incidence of sexual behaviour in the male goldfish, *Carassius auratus* (Partridge, Liley & Stacey, 1976). Male sexual and nest-building behaviour in the blue gourami, *Trichogaster trichopterus*, is decreased following bilateral cauterization of the olfactory mucosa (Pollack, Beeker & Haynes, 1978). In these species, olfactory structures in the telencephalic hemisphere could be a part of the system which potentiates reproductive behaviour. In *Macropodus*, male reproductive behaviour is not limited by olfactory input. The incidence of sexual and nest-building behaviour is unimpaired following bilateral olfactory bulbectomy (Schwagmeyer, Davis & Kassel, 1977). Similar results were reported for male *Gasterosteus aculeatus*, the stickleback (Segaar, 1965). The decrease in reproductive behaviour seen in diencephalic *Macropodus* and *Gasterosteus* may be the result of damage done to other sensory systems. Not surprisingly, removal of both eyes in the male blocks spawning in *Macropodus* (unpublished data). Following removal of one eye latency to spawning is increased, and the male performs the spawning act only when the female approaches from the side with the remaining eye (Reynolds, 1977). Whether telencephalon ablation impairs perception of visual stimuli which elicit reproductive behaviour remains to be demonstrated (Aronson & Kaplan, 1968).

Non-olfactory sensory areas in the pallium, or dorsal zone of the telencephalic hemisphere (Braford & Northcutt, 1978), and neuronal projections from the pallium to the optic tectum and to diencephalic visual system structures have recently been demonstrated in anatomical investigations in various bony fishes (Vanegas & Ebbesson, 1976; Ito & Kishida, 1977, 1978; Ito & Masai, 1978; Bass, 1979; Airhart, 1979). Electrophysiological experiments in the bass, *Micropterus salmoides*, indicate that diencephalic visual areas project to the pallium, and that bioelectrical responses of diencephalic structures to visual stimuli can be modulated by electrical shocks applied to various regions of the hemisphere (Friedlander, 1980). The behavioural functions of visual areas of the hemisphere are unknown.

Few lesion experiments which might elucidate the behavioural functions of non-olfactory structures in the telencephalon have been carried out. Results obtained in the male platyfish, *Xiphophorus maculatus*, suggest that sexual behaviour is decreased by pallial lesions (Kamrin & Aronson, 1954). Bilateral lesion of the pallium or of the pallium and subpallium, or ventral zone of the hemisphere, resulted in decreased copulation. Segaar (1965) reported that in *Gasterosteus* male reproductive behaviour is impaired following the combined removal of the lateral pallium and the anterior region of the pallium and subpallium, but not following removal of either the lateral or the anterior region alone. Superficial lesions administered in the dorsomedial pallium result in altered frequencies of certain components of reproductive behaviour, and social behaviour, in male *Gasterosteus* (Segaar, 1965) and Siamese fighting fish, *Betta splendens* (DeBruin, 1977). Whether the superficial lesions included non-olfactory sensory areas of the hemisphere and whether reproduction was decreased is unclear.

To further investigate the role of the telencephalon in potentiating reproductive behaviour in *Macropodus*, the effects of partial lesions of the hemisphere were examined. Telencephalic cell groups are to be referred to using Nieuwenhuy's (1963) terminology as modified by Northcutt & Braford (1978), in which the groups are identified mainly by their topographic location. Pallial cell masses are denoted as parts of the area dorsalis telencephali, or simply area D. Subpallial cell masses are correspondingly designated as parts of the area ventralis telencephali, or area V. We examined four bilateral lesions: (1) Dl, removal of the lateral part of area D, (2) Dl-Dd-Dm, the combined removal of the lateral, dorsal and medial parts of area D, (3) anterior D + V, the combined removal of the olfactory bulbs and of area D and V anterior to the anterior commissure, (4) OB, olfactory bulbectomy.

The area D lesions included cell groups which receive olfactory bulb fibres. Bulbar afferents terminate extensively in area V and in the basolateral and posterior parts of D (Davis, 1978; review: Northcutt & Braford, 1978). On the basis of data obtained in anatomical and neurophysiological experiments in other species (see above), visual system structures may be located in areas D and V. A goal of the anterior D + V lesion was to damage area Vv, which contains cells which concentrate sex steroids (Davis *et al.*, 1977) and might serve reproductive functions. Schwagmeyer *et al.* (1977) reported that sexual behaviour in *Macropodus* was unimpaired by olfactory bulbectomy but the effect on egg care following spawning was not determined. The possibility that the increased egg-eating seen in diencephalic males (Kassel & Davis, 1977) results from the loss of olfaction was also investigated in the present experiments.

Method

Subjects

These experiments used 35 adult *Macropodus opercularis* (L.) (Class Osteichthyes: Family Belontiidae) males, ranging from 4.5 to 5.3 cm in body length. Prior to a spawning trial the male was kept in an individual home tank as previously described (Davis, Kassel & Schwagmeyer, 1976). The fish were fed once daily, and water temperatures were 25 to 26 °C. Males received bilateral lesions of area Dl ($n=6$), combined areas Dl-Dd-Dm ($n=7$), anterior D + V ($n=8$), the olfactory bulb ($n=6$) or a craniotomy as a sham operation ($n=7$). Three and five weeks following surgery, each male was administered a 5-day spawning trial with an intact female. The spawning tank consisted of a 38-l, clear-glass aquarium with a gravel bottom, a large clay brick and several artificial aquatic plants. The tank was illuminated from 06.00 to 20.00 h daily by an overhead, 20-W fluorescent lamp. Four tanks, separated by opaque partitions, were placed in a row in a darkened room.

Surgery

The fish was anaesthetized in 0.04% Finquel (Ayerst). In the area D lesions, a 2 mm × 3 mm flap of bone in the roof of the cranium over the telencephalon was reflected caudally. The Dl lesion was administered with a Ziegler knife by gently cutting away the lateral one-third of each hemisphere while avoiding the blood vessels located in the external sulcus (Nieuwenhuys, 1963). In the Dl-Dd-Dm lesion, a similar longitudinal cut was made but the plane of the lesion was tilted dorsomedially to remove area Dd and Dm. The anterior D + V lesion was made by cutting away the anterior one-third of the hemisphere followed by aspirating off the hemisphere tissue and the olfactory bulbs. The craniotomy, or sham lesion operation, included removal of a small amount of the loose lipoidal tissue over the telencephalon. The OB lesion was performed as previously described (Schwagmeyer *et al.*, 1977), by aspiration of the bulbs through a 2 mm diameter hole which was cut in the overlying bone. The surgery was completed in several minutes. The bone flap was lowered to close the wound, and the male was returned to the home tank to recover.

Spawning trial

The male was placed in the spawning tank with a female which had 1 day of prior residence. Females which had spawned previously in the laboratory were used, to enhance the probability that they would mate (Kassel & Davis, 1977). The pair was observed frequently throughout the day until the male was seen to perform a 'curve' or 'clasp' response to the female. Thereafter, the frequency of curve and clasp responses were recorded for 10 min at half-hour intervals until spawning was completed or the onset of the daily dark period prevented further observation.

Curve and clasp immediately precede gamete release in the spawning sequence (Kassel & Davis, 1977). Curve, or curve followed by clasp, may occur several to many times before the female first releases eggs. A pair typically spawns several times within a period of 2 to 4 h; then the male chases the female away and cares for the eggs and the foam-nest.

The male was observed frequently following spawning to detect occurrences of egg-eating. Approximately 24 h after spawning the remaining eggs (if any) were placed in culture dishes for 48 h, a period sufficient for hatching (Davis & Kassel, 1975). Virtually every brood contained viable embryos, indicating that the males were fertile. Some males ate all the eggs and others ate varying amounts during the period following spawning. Egg-eating precluded accurate assessment of the effects of the lesions on male fertility.

The OB lesion group was observed during and following spawning to determine whether the male ate eggs. Other aspects of reproductive behaviour in OB-lesioned males have been reported elsewhere (Schwagmeyer *et al.*, 1977).

Lesion verification

The male was sacrificed immediately after the second spawning trial, i.e. approximately 6 weeks post-operatively, and the brain was examined with the aid of a dissecting microscope. The extent of the OB lesions were verified by direct observation. Other lesioned brains were fixed by immersion in 10% formalin, embedded in paraffin and cut in 10-μm thick transverse or horizontal sections. The sections were stained with cresyl violet acetate. Lesions were reconstructed by projecting representative sections on to copies of high-contrast photographs of corresponding sections of a reference brain. The variation in lesion size between individuals in each group was measured by weighing paper cutouts of the lesioned structures.

Results

The proportions of males that spawned in the craniotomy control group and the three lesion groups were not significantly different (Table 1). In trial 2 the proportion of the spawners which ate eggs was increased, relative to the control group, in the anterior D + V and the OB lesion groups but not in the Dl or the Dl-Dd-Dm lesion groups (Table 1). Egg-eating did

Table 1. *The number of males which spawned, and the number of spawners which ate eggs, in the two spawning trials*

Group	*n*	Spawned in trial 1	Spawned in trial 2	Ate eggs in trial 1	Ate eggs in trial 2
Craniotomy	7	6	5	5	0
Dl lesion	6	2	5	2	2
Dl-Dd-Dm lesion	7	3	7	2	4
Anterior D + V lesion	8	5	5	4	4[a]
OB lesion	6	4	6	4	5[a]

[a]Significant at the 2% level; Fisher test.

Fig. 1. (*a*) Transverse sections through the anterior (1), central (2) and posterior (3) regions of the telencephalon of *Macropodus* which were used as reference sections to chart the extent of experimental lesions that are shown in (*b*). The right-hand side of each section is a high-contrast photograph of the cell bodies, which were stained with cresyl violet

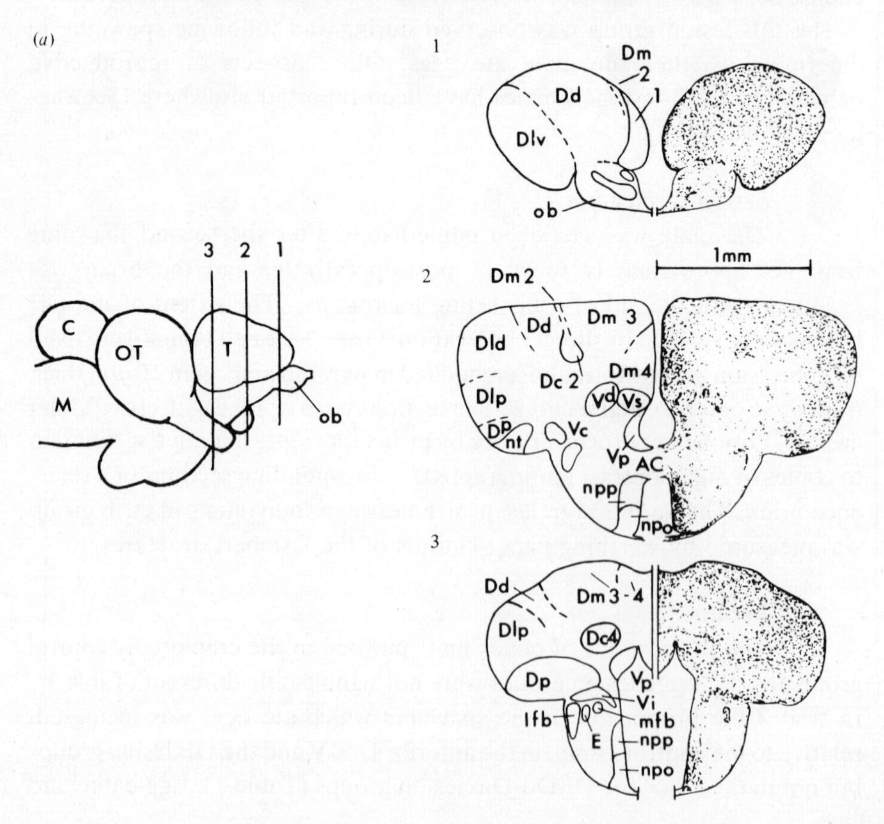

acetate. The left-hand side shows the boundaries of various cell masses and fibre tracts. Abbreviations (Nieuwenhuys, 1963; Northcutt & Braford, 1978): AC, anterior commissure; C, cerebellum; D, area dorsalis telencephali; Dc, central part of D; Dc2,3,4, parts of Dc; Dd, dorsal part of D; Dl, lateral part of D; Dld,p,v, dorsal, posterior and ventral parts of Dl; Dm, medial part of D; Dm1,2,3,4, parts of Dm; Dp, posterior part of D; E, nucleus entopenduncularis; lfb, lateral forebrain bundle; M, medulla; mfb, medial forebrain bundle; npo, nucleus preopticus; npp, nucleus preopticus periventricularis; nt, nucleus taenia; ob, olfactory bulb; OT, optic tectum; T, telencephalic hemisphere; V, area ventralis telencephali; Vc,d,i,l,p,s, central, dorsal intermediate, lateral, post-commissural and supracommissural parts of V. (*b*) Transverse sections through the anterior, central and posterior areas of the telencephalic hemisphere of six male *Macropodus* which were administered lesions of area Dl prior to the spawning trials. The location of each reference section is illustrated in Fig. 1(*a*). The extent of the lesion is denoted by the solid black areas. The number located above each series of sections corresponds to the male number in Table 2.

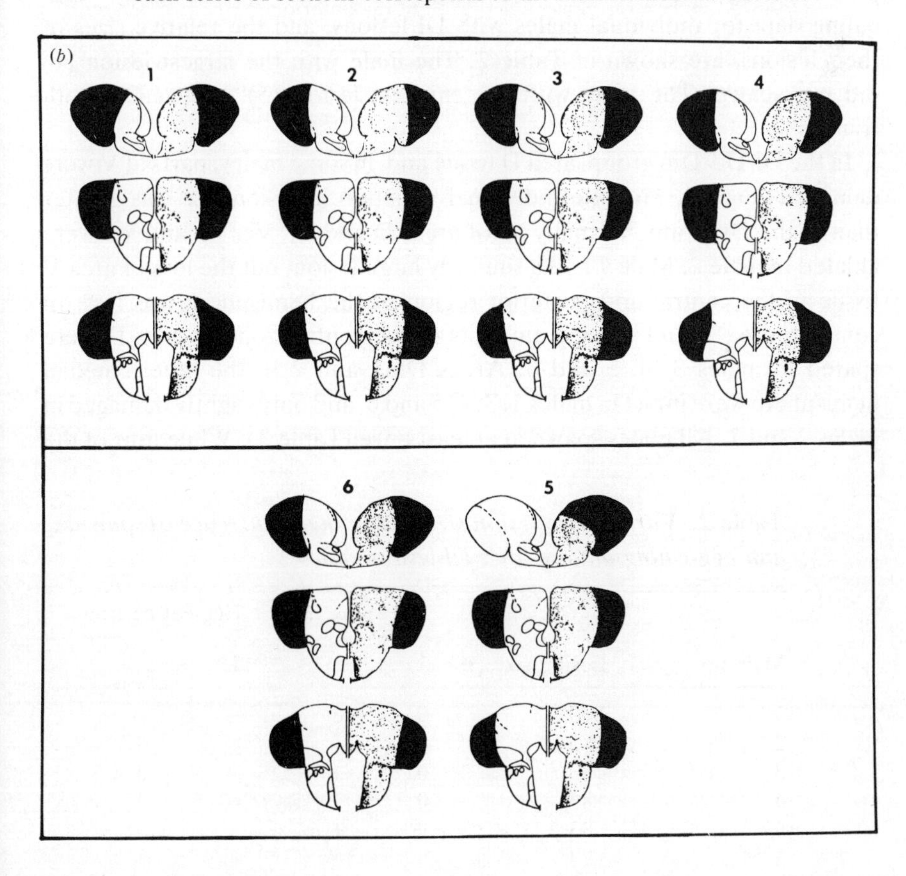

not vary significantly between the groups in trial 1. Spawners and non-spawners all built cohesive foam-nests, except male 1 in the anterior D + V group which did not build a nest or spawn in either trial. The median latency to spawning in each group was 3 days; the latencies of individual males ranged from 2 to 5 days.

The craniotomy and hemispheric-lesion males showed similar frequencies of curve and clasp during spawning. The mean curve frequencies ranged from 7 to 10 per 10 min, and the clasp frequencies ranged from 1.5 to 2.0 per min. The spawning behaviour of the lesioned males was subjectively indistinguishable from that of the control males.

The OB lesions resulted in uniform removal of both olfactory bulbs; the telencephalic hemispheres were left intact. The males of the Dl group sustained relatively uniform lesions in the anterior, central and posterior region of the hemisphere (Fig. 1). The spawning and egg-eating data for individual males with Dl lesions, and the relative sizes of their lesions, are shown in Table 2. The male with the largest lesion (1) did not spawn. The male with the smallest lesion (5) spawned in both trials.

In the Dl-Dd-Dm group, area D tissue and, in some males, parts of V were damaged (Fig. 2). The interindividual variation in lesion size was greater than in the Dl group. Virtually all of areas D and Vi, Vc, Vp and Vd were ablated in male 2. Male 7 had a similarly large lesion, but the loss of area V tissue in the central and posterior region of the hemisphere was less. In contrast, area V and varying amounts of the central zone of area D were spared in males 3, 4, 5 and 6. Areas Dm1 and 2 in the anteriomedial hemisphere were intact in males 1, 3, 4, 5 and 6, and only slightly damaged in males 2 and 7. All males spawned at least once (Table 3). While four of the

Table 2. *Variation in lesion size and in the occurrence of spawning and egg-eating among area Dl lesion males*

		Spawning trial		Egg-eating trial	
Male no.	Lesion size (%)	1	2	1	2
1	100[a]	0	0	–	–
2	83	0	+	–	0
3	83	0	+	–	0
4	83	0	+	–	0
6	83	+	+	+	+
5	73	+	+	+	+

[a] Largest lesion (see Fig. 1*b*).

Table 3. *Variation in lesion size and in the occurrence of spawning and egg-eating among area Dl-Dd-Dm lesion males*

		Spawning trial		Egg-eating trial	
Male no.	Lesion size (%)	1	2	1	2
2	100[a]	0	+	−	+
7	83	+	+	+	+
1	77	0	+	−	+
6	64	+	+	+	+
3	61	0	+	−	0
4	58	0	+	−	0
5	58	+	+	0	0

[a]Largest lesion (see Fig. 2).

Fig. 2. Transverse sections through the anterior, central and posterior areas of the telencephalic hemisphere (see Fig. 1*a*) of seven male *Macropodus* which received bilateral Dl-Dd-Dm lesions. The number located above each series of sections corresponds to the male number in Table 3.

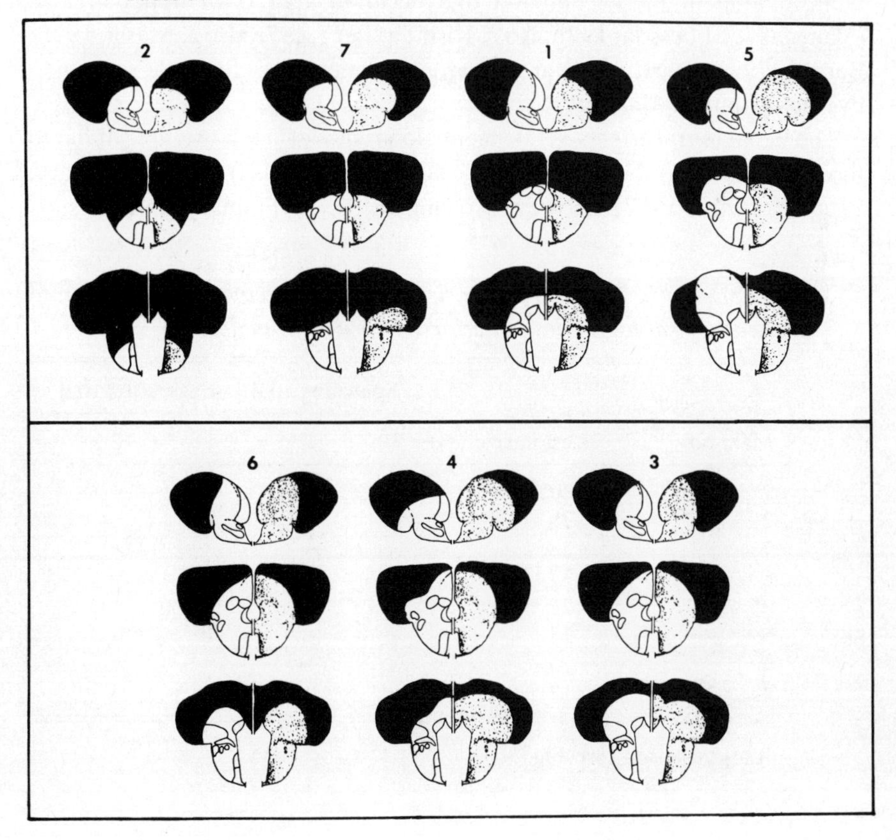

seven males ate eggs in trial 2, the proportion was not significantly different from that observed in the control group (Table 1).

The anterior D + V lesion was of uniform, intermediate size in males 2, 3, 5, 6 and 7, smallest in male 4 and largest in male 1 (Table 4 and Fig. 3). The intermediate lesions resulted in damaging the anterior portions of Dm1, Dm2, Dd, Dlv and Vd. The smallest lesion similarly damaged anterior area D, though more superficially. In addition, areas Vd and Vv appeared to be spared. The largest lesion, in male 1, obliterated area D except for a portion of the central posterior zones, and destroyed Vl and parts of Vd, Vv and Vs. Male 1, as mentioned above, did not spawn, nor was it seen to curve, clasp or nest-build.

Discussion

The incidence of spawning is not significantly decreased following bilateral removal of area D and the posterior and dorsal part of V including Vi, Vp, Vc and Vd (Figs. 1 and 2), whereas, as previously reported (Kassel & Davis, 1977), it is decreased following full hemispherectomy with varying amounts of damage in the preoptic nucleus region. Any connections with the diencephalon or mesencephalon that occur in area D or upper area V (Vanegas & Ebbesson, 1976; Ito & Kishida, 1977, 1978; Ito & Masai, 1978; Bass, 1979; Airhart, 1979) are apparently unnecessary for male reproductive behaviour in *Macropodus*.

The behavioural effects of telencephalon removal in teleosts are similar to those of cortical and limbic system lesions in mammals (review: McCleary, 1966; Pribram, 1967). However, owing to the preliminary state of such

Table 4. *Variation in lesion size and in the occurrence of spawning and egg-eating among anterior D + V lesion males*

		Spawning trial		Egg-eating trial	
Male no.	Lesion size (%)	1	2	1	2
1	100[a]	0	0	−	−
2	38	0	0	−	−
7	38	+	+	+	+
8	37	+	+	+	+
3	25	0	+	−	+
5	25	+	+	0	0
6	25	+	+	+	+
4	17	+	0	+	−

[a]Largest lesion (see Fig. 3*b*).

investigations in fish, in particular studies on the effects of partial hemispheric lesions, direct comparisons with results obtained in mammals are difficult (Davis & Kassel, 1980). Sensorimotor functions are severely impaired in the telencephalon-ablated, or 'thalamic', mammal (Buchwald & Brown, 1973; Grill & Norgren, 1978).

The non-olfactory process(es) which are inferred to potentiate spawning in *Macropodus* may be mediated by integrated functions of diverse parts of the hemisphere, so that to impair the behaviour virtually all of the hemisphere must be removed. The impairment of reproduction may vary with the total amount of the hemisphere that is removed and not the location of the lesion, as Beach (1940) found for neocortical lesions in the rat.

The preoptic nucleus region and the anterior ventral parts of area V, including Vv, Vs and Vl, may contain brain cells which are necessary for male reproductive behaviour in *Macropodus.* Since one hemisphere is sufficient to maintain spawning behaviour (Kassel & Davis, 1977) it is clear that interhemispheric commissural fibre connections are unnecessary. The fibre connections of area V nuclei have unfortunately not been extensively investigated. Olfactory tract afferent fibres terminate in area V (Davis, 1978) in *Macropodus*, and neurophysiological experiments in the bass, *Micropterus*, indicate that areas Vv, Vl and Vd are connected with diencephalic visual areas (Friedlander, 1980). Sex-steroid-sensitive neurons in the preoptic nucleus region and Vv-Vs region (Davis *et al.*, 1977; Kim, Stumpf & Sar, 1978) may potentiate sexual behaviour in teleosts. Behavioural experiments have suggested that the preoptic nucleus area regulates sexual behaviour in some species, but in others the results have been indeterminate or negative. Electrolytic lesion of the preoptic area in *Fundulus heteroclitus*, the mummichog, blocked chemical elicitation of spawning body movement in the male (Macey, Pickford & Peter, 1974; Peter, 1977). In *Lepomis*, sunfish, male sexual behaviour and sperm release were evoked by electrical stimulation of the preoptic region (Demski & Knigge, 1971; Demski, Bauer & Gerald, 1975). However, Kassel & Davis (1977) found that in *Macropodus* the extent of the damage done to the preoptic area in ablating the hemispheres was not at all related to the magnitude of the impairment of reproduction. The behavioural effects in *Macropodus* of localized lesions of the preoptic region and of areas Vv, Vs or Vl need to be investigated.

Kyle & Peter (1979) recently obtained data in *Carassius* which strongly suggest that cells in the Vs-Vv area potentiate male sexual behaviour. Electrolytic lesion of Vs-Vv resulted in a significant decrease in the proportion of males which spawned with sexually active females. In contrast, lesions of the preoptic nucleus region or of Vp did not impair spawning. Thus, the anterior ventral region of area V, and not the preoptic area, may be

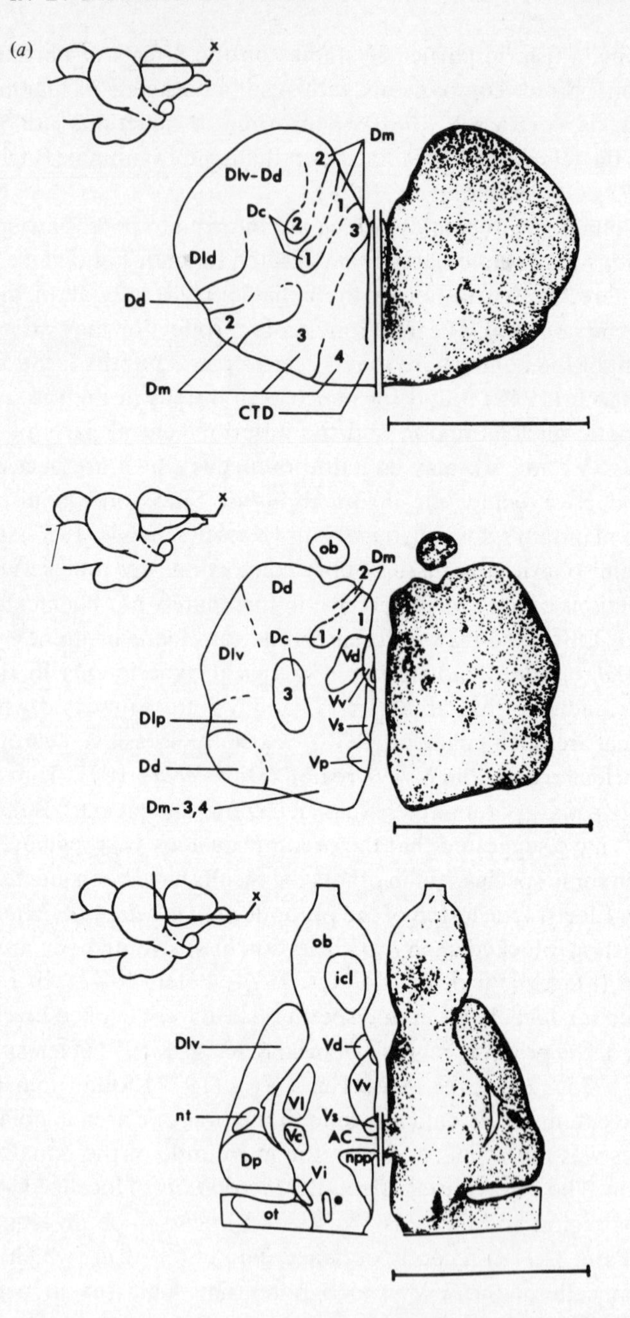

Fig. 3. (*a*) Horizontal reference sections through the dorsal, central and ventral region of the telencephalon which were used to chart the extent of the anterior D + V lesion. The location of each section is illustrated by the lines (×) through the telencephalon on the side-view drawings of the

responsible for activating sexual behaviour in some teleosts. The behavioural effect of the Vv-Vs lesion in *Carassius* might be a result of damage done to sex-steroid-sensitive cells which occur in that area (Kim *et al.*, 1978), but olfactory system damage might also be involved. Olfactory tract fibres terminate in or near the Vv-Vs area in *Carassius* (unpublished data). Inasmuch as olfactory input is necessary for reproduction in this

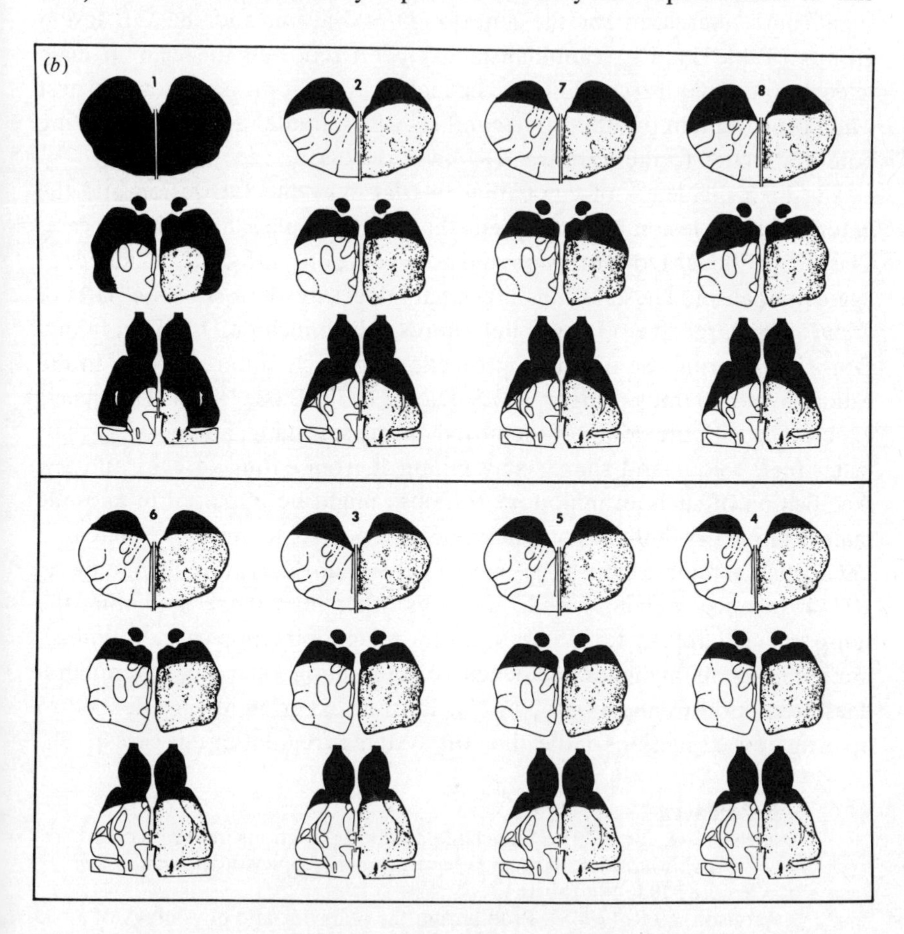

brain on the left-hand side of the figure. The right-hand side of each section is a high-contrast photograph of the tissue, which was stained with cresyl violet acetate. The boundaries of the major cell masses are diagrammed in the outline of the left-hand side of the section. Abbreviations: CTD, commissura telencephali dorsalis; icl, internal cell layer; others as in the legend to Fig. 1. Scale bar = 1 mm. (*b*) The extent of the anterior D + V lesions (solid black areas) which were administered to eight male *Macropodus* prior to the spawning trials. The number above each series of sections corresponds to the male number shown in Table 4.

teleost (Partridge *et al.*, 1976), destruction of olfactory afferents in the Vv-Vs region might also result in decreased sexual behaviour.

Egg-eating was common among both lesion and craniotomy control males in trial 1, but in trial 2 the controls were not seen to eat eggs. Prior reproductive experience and handling the laboratory environment may facilitate egg care. The proportion of males which ate eggs in both trials was significantly increased for the anterior D + V lesion and the OB lesion groups (Table 1). Egg cannibalism has been reported to occur in male teleosts under natural conditions. The male may eat some eggs from the nest when procurement of other food conflicts with parental duties or attracting potential mates to the nest area (Rohwer, 1978).

The high incidence of egg-eating in trial 2 by the OB lesion and the anterior D + V lesion males suggests that olfaction plays a role in egg care. The Dl and the Dl-Dd-Dm lesions did not significantly affect the incidence of egg cannibalism. These lesions presumably destroyed most of the parts of area D which receive olfactory bulb fibres, which include Dlv, Dlp, nt and Dp, while sparing the olfactory projections to nuclei in area V and in the caudal hypothalamus (Finger, 1975; Davis, 1978; Bass, 1979). The male's experience with the odour of the brood during spawning, and possibly their taste, feel, colour and shape, may inhibit it from eating of eggs and fry. Acquisition of such an inhibitory response might be impaired in anosmic males. Olfactory bulb lesions have been reported to result in increased pup-killing in female laboratory mice (Gandelman, Zarrow & Dennenberg, 1971). In mice, olfactory bulb functions other than those mediating the sense of smell appear to be important for maintaining pup care, as bulbectomy results in a higher incidence of pup-killing than does peripheral deafferentation (Vandenberg, 1973). Further experiments are needed to investigate the functions of the olfactory system in regulating egg care in fish.

References

Airhart, M. J. (1979). Telencephalotectal projections in the goldfish, *Carassius auratus*: light and electron-microscopic study. *Anatomical Record*, **193**, 468 (abstr.).

Aronson, L. R. (1948). Problems in the behavior and physiology of a species of African mouthbreeding fish. *Transactions of the New York Academy of Sciences*, **2**, 33–43.

Aronson, L. R. & Kaplan, N. (1968). Functions of the teleostean forebrain. In *The Central Nervous System and Fish Behavior*, ed. D. Ingle, pp. 107–25. University of Chicago Press, Chicago.

Bass, A. (1979). Telencephalic efferents in the channel catfish, *Ictalurus punctatus. Anatomical Record*, **193**, 478 (abstr.).

Beach, F. A. (1940). Effects of cortical lesions upon the copulatory behavior of male rats. *Comparative Psychology*, **29**, 193–245.

Braford, M. R. Jr & Northcutt, R. (1978). Correlation of telencephalic afferents and SDH distribution in the bony fish polypterus. *Brain Research*, **152**, 157–60.

Buchwald, J. S. & Brown, K. A. (1973). Subcortical mechanisms of behavioral plasticity. In *Efferent Organization and Integration of Behavior*, ed. J. D. Naser, pp. 100–36. Academic Press, New York & London.

Davis, R. E. (1978). Autoradiographic investigation of the central projections of the olfactory tracts in *Macropodus opercularis* (L.). *Neuroscience Abstracts*, **4**, 218.

Davis, R. E. & Kassel, J. (1975). The ontogeny of agonistic behavior and the onset of sexual maturation in the paradise fish, *Macropodus opercularis* (Linnaeus). *Behavioral Biology*, **14**, 31–9.

Davis, R. E. & Kassel, J. (1980). Behavioral functions of the telencephalon. In *Fish Neurobiology and Behavior*, ed. R. E. Davis & R. G. Northcutt, vol. 2. University of Michigan Press, Ann Arbor (in press).

Davis, R. E., Kassel, J. & Schwagmeyer, P. (1976). Telencephalic lesions and behavior in the teleost, *Macropodus opercularis*: reproduction, startle reaction, and operant behavior in the male. *Behavioral Biology*, **18**, 165–77.

Davis, R. E., Morrell, J. J. & Pfaff, D. W. (1977). Autoradiographic localization of sex steroid concentrating cells in the brain of the teleost *Macropodus opercularis* (Osteichthyes: Belontiidae). *General and Comparative Endocrinology*, **33**, 496–505.

Davis, R. E., Reynolds, R. C. & Ricks, A. (1978). Suppression behavior increased by telencephalic lesions in the teleost, *Macropodus opercularis*. *Behavioral Biology*, **24**, 32–48.

DeBruin, J. P. C. (1977). 'Telencephalic functions in the behaviour of the Siamese fighting fish, *Betta splendens* Regan (Pisces, Anabantidae)'. Thesis, Central Brain Research Institute, Amsterdam.

Demski, L. S., Bauer, D. J. & Gerald, J. W. (1975). Sperm release evoked by electrical stimulation of the fish brain: a functional–anatomical study. *Journal of Exploratory Zoology*, **191**, 215–32.

Demski, L. S. & Knigge, K. M. (1971). The telencephalon and hypothalamus of the bluegill (*Lepomis macrochirus*): evoked feeding, aggressive and reproductive behavior with representative frontal sections. *Journal of Comparative Neurology*, **143**, 1–16.

Finger, T. (1975). The distribution of the olfactory tracts in the bullhead, *Ictalurus nebulosus*. *Journal of Comparative Neurology*, **161**, 125–42.

Friedlander, M. (1980). The visual prosencephalon of teleosts. In *Fish Neurobiology and Behavior*, ed. R. E. Davis & R. G. Northcutt, vol. 2. University of Michigan Press, Ann Arbor (in press).

Gandelman, R., Zarrow, M. X. & Dennenberg, V. H. (1971). Stimulus control of cannibalism and maternal behavior in anosmic mice. *Physiology and Behavior*, **1**, 583–6.

Grill, H. J. & Norgren, R. (1978). Neurological tests and behavioral deficits in chronic thalamic and chronic decerebrate rats. *Brain Research*, **142**, 299–312.

Ito, H. & Kishida, R. (1977). Tectal afferent neurons identified by the retrograde HRP method in the carp telencephalon. *Brain Research*, **130**, 142–5.

Ito, H. & Kishida, R. (1978). Telencephalic afferent neurons identified by the retrograde HRP method in the carp diencephalon. *Brain Research*, **149**, 211–15.

Ito, H. & Masai, H. (1978). Fiber connections of the carp telencephalon. *International Brain Research Organization Newsletter*, **6**, 33.

Kamrin, R. P. & Aronson, L. R. (1954). The effects of forebrain lesions on mating behavior in the male platyfish, *Xiphophorus maculatus. Zoologica*, **39**, 133–40.

Kassel, J. & Davis, R. E. (1977). Recovery of function following simultaneous and serial ablation in the teleost, *Macropodus opercularis. Behavioral Biology*, **21**, 489–99.

Kassel, J., Davis, R. E. & Schwagmeyer, P. (1976). Telencephalic lesions and behavior in the teleost, *Macropodus opercularis*: further analysis of reproductive and operant behavior in the male. *Behavioral Biology*, **18**, 179–88.

Kim, Y. S., Stumpf, W. E. & Sar, M. (1978). Topography of estrogen target cells in the forebrain of goldfish, *Carassius auratus. Journal of Comparative Neurology*, **182**, 611–20.

Kleerekoper, H. (1969). *Olfaction in Fishes.* Indiana University Press, Bloomington.

Kyle, A. L. & Peter, R. E. (1979). 'Effect of brain lesions on spawning behaviour in the male goldfish'. Paper presented at the Canada West Society for Reproductive Biology, University of Saskatchewan, Canada.

Little, E. E. (1980). Behavioral function of olfaction and taste in fish. In *Fish Neurobiology and Behavior*, ed. R. G. Northcutt & R. E. Davis, vol. 1. University of Michigan Press, Ann Arbor (in press).

McCleary, R. A. (1966). Response-modulating functions of the limbic system: initiation and suppression. *Progress in Physiological Psychology*, **1**, 209–72.

Macey, M. J., Pickford, G. E. & Peter, R. E. (1974). Forebrain localization of the spawning reflex response to the exogenous neurohypophyseal hormones in the killifish, *Fundulus heteroclitus. Journal of Experimental Zoology*, **190**, 269–80.

Morrell, J. I. & Pfaff, D. W. (1978). A neuroendocrine approach to brain function: localization of sex steroid concentrating cells in vertebrate brains. *American Zoologist*, **18**, 447–60.

Nieuwenhuys, R. N. (1963). Further studies on the general structure of the actinopterygian forebrain. *Acta Morphologica Neerlando-Scandinavica*, **6**, 66–79.

Northcutt, R. G. & Braford, M. R. (1978). New observations on the organization and evolution of the telencephalon of actinopterygian fishes. In *Comparative Neurology of the Telencephalon*, ed. S. O. E. Ebbesson, pp. 41–98. Plenum Press, New York.

Partridge, B. L., Liley, N. R. & Stacey, N. E. (1976). The role of pheromones in the sexual behavior of the goldfish. *Animal Behaviour*, **24**, 291–9.

Peter, R. E. (1977). The preoptic nucleus in fishes: a comparative discussion of function–activity relationships. *American Zoologist*, **17**, 775–85.

Pollack, E. I., Beeker, L. R. & Haynes, K. (1978). Sensory control of mating in the blue gourami, *Trichogaster trichopterus* (Pisces, Belontiidae). *Behavioral Biology*, **22**, 92–103.

Pribram, K. H. (1967). The limbic systems: efferent control of neural inhibition and behavior. *Progress in Brain Research*, **27**, 318–36.

Reynolds, R. C. (1977). 'Functional interconnection of the forebrain and midbrain in the paradise fish, *Macropodus opercularis*'. Thesis, Dep. of Natural Resources, University of Michigan, Ann Arbor.

Rohwer, S. (1978). Parent cannibalism of offspring and egg raiding as a courtship strategy. *American Naturalist*, **112**, 429–40.

Schwagmeyer, R., Davis, R. E. & Kassel, J. (1977). Telencephalic lesions and behavior in the teleost, *Macropodus opercularis* (L.): effects of telencephalon and olfactory bulb ablation on spawning and foam-nest building. *Behavioral Biology*, **20**, 463–70.

Segaar, T. (1965). Behavioral aspects of degeneration and regeneration in fish brain: a comparison with higher vertebrates. *Progress in Brain Research*, **14**, 143–231.

Vandenberg, J. G. (1973). Effects of central and peripheral anosmia on reproduction of male mice. *Physiology and Behavior*, **10**, 257–61.

Vanegas, H. & Ebbesson, S. O. E. (1976). Telencephalic projections in two teleost species. *Journal of Comparative Neurology*, **165**, 181–96.

PART V

Learning and memory mechanisms

NANCY BOHAC FLOOD & J. BRUCE OVERMIER

Learning in teleost fish: role of the telencephalon

Allow us to pose a simple question: What is the functional significance of the teleost telencephalon? Is the telencephalon – or forebrain as it is often commonly referred to – there only to receive sensory information from the nose and then to interpret this information in terms of what might be the day's repast? Or is a primary function performed by the telencephalon, one of integrating external environmental stimulation with internal information to regulate the fish's behaviour – to prepare for and execute flight, fight, feast, or further exploration? Does the telencephalon play a role in the *association* of environmental events with physiological and behavioural responses to keep the beast alive, fed, propagating, and responding appropriately and successfully to environmental demands?

These questions have been put forth simplistically; yet they fairly capture the issue at hand. This article discusses the behavioural significance of the teleost fish telencephalon with special emphasis on learning, providing an overview of available psychological data and comment upon various interpretations as to functional mechanisms of the forebrain. The data to be summarized derive largely from ablation studies, the interpretive limitations of which must be clearly recognized (Schoenfeld & Hamilton, 1977). Yet these studies provide us with the broadest array of behavioural and learning data upon which we may base inferences.

On the telencephalon

The teleost telencephalon was long thought to exclusively subserve olfactory functions (e.g. Prosser & Brown, 1961). More recent investigations suggest not only that the telencephalon contributes to more processes than olfaction, but also that olfaction may not even be the primary telencephalic function, and that, perhaps, olfaction evolved from the telencephalon so as to facilitate primary adaptive functions (Riss, Halpern & Scalia, 1969). This would be consistent with Schneirla's general theory of biphasic processes of approach and withdrawal (Piel, 1970), according to which, it is an organism's ability to approach or withdraw appropriately from external stimulation that ensures its survival. Thus, appropriate approach/withdrawal would be the primary function of a rudimentary nervous system. This is the basic function which would need to be expanded and refined by integrating more specialized sensory information into the behaviour-regulating processes.

Olfactory functions continue to be the primary role of the ventral area of the telencephalon where most interconnections are with the olfactory tract (Bernstein, 1970; Scalia & Ebbesson, 1971). And, although the bony fish has not developed any neocortex structures, the cytoarchitectonics of the dorsal area of the telencephalon do suggest progressive development and specialization (Vanegas & Ebbesson, 1976; Nieuwenhuys, 1967). In the more specialized bony fish there is an increase in the thickness of the walls, and an increase in the differentiation of the basic subdivisions with increased diencephalic connections (Bernstein, 1970). This suggests a functional as well as a neuroanatomical change from a basically olfactory structure to one serving additional functions.

In looking for clues about the functions of the fish forebrain, some neuroanatomists have argued that the teleost telencephalon is partly field-homologous with the limbic system of mammals (Riss *et al.*, 1969). As an aside, it is interesting to note that just as the fish telencephalon was thought to subserve only olfactory functions, the mammalian limbic structures, too, were once referred to as the rhinencephalon – the nose brain.

To suggest possible functions of the teleost telencephalon a cursory summary of mammalian limbic behavioural research might be helpful. Both Herrick (1933) and Papez (1937) observed that limbic structures were important to (*a*) memory, (*b*) associative processing and (*c*) emotional experience. Neuropsychological and behavioural research has confirmed, expanded and refined these observations. Limbic ablations and lesions in mammals have little impact upon general sensory-motor functions. There are, however, dramatic effects upon the qualitative nature and patterning of behaviour. Selective limbic lesions impair social behaviours related to threats, aggression and social dominance, and disrupt and disorder parental behaviour patterns. There are also reports of a loss of 'initiative'. In mammalian learning studies, effects of various limbic lesions depend upon the type of learning task. Limbic lesions do not disrupt autonomic classical conditioning nor simple instrumental (goal-seeking) appetitive discriminations. Selective limbic lesions do impair delayed-response learning and reversal learning, augment response perseveration while reducing response flexibility, increase resistance to extinction of simple appetitive tasks, and interfere with the ability to inhibit some responses. Very striking are the effects of limbic lesions upon avoidance behaviours: specific lesions selectively impair active or passive avoidance learning as well as retention of pre-operatively acquired avoidance behaviour. Thorough reviews of these effects are provided by Isaacson (1974) and Iversen (1973).

Effects of teleost telencephalon ablation

With regard to behaviourally assessed sensory functions, taste and temperature processes appear entirely unaffected by telencephalon ablation in the teleost fish (Berwein, 1941). Colour vision is temporarily impaired – perhaps due to surgical trauma – but in any case returns to normal within hours (Bernstein, 1961, 1962). Only olfaction seems to be disrupted by ablation of the telencephalon (Strieck, 1924). That olfaction is disrupted follows from anatomy because the large olfactory tracts from the olfactory bulbs enter the ventral telencephalon. Anatomical and electroencephalography (EEG) studies confirm primary olfactory functions in the ventral telencephalon (Doving & Gemne, 1966). Although processing of olfactory input is located in a portion of the telencephalon – but not exclusively so (Ebbesson & Vanegas, 1976; Vanegas & Ebbesson, 1976) – ablation of the telencephalon does result in permanent loss of olfaction. One consequence of this sensory loss which is confounded with total telencephalic ablations is that all learning experiments on the role of the telencephalon need to incorporate controls for the loss of olfaction – as well as other contributions attributable to olfactory bulb activity (Wenzel, 1974) – if one is to 'localize' any observed disruption as telencephalic in origin. Unfortunately, such controls have not been routinely included.

Motor systems seem unimpaired following telencephalic ablation in teleost fish. No qualitative gross behavioural changes in locomotion, posture or feeding occur post-operatively (Nolte, 1932; Janzen, 1933; Hale, 1956*a*). There are, however, several important *quantitative* changes in behaviour of which the most remarkable is a reduction in response variability. Janzen (1933) reported decreases in the variability of eye movements and selection of illumination, and a general loss of 'initiative', especially when placed in a new environment (Davis, Reynolds & Ricks, 1978). Yet an important recent study by Davis, Kassel & Schwagmeyer (1976) indicated that forebrain ablation results in a hyper-reactive state that may be expressed behaviourally as freezing, submissive posturing or increased startle responses. These alterations in reactivity may have important concomitant effects on both innate and learned behaviours. Important behavioural consequences suggested are altered response latencies and disruption of innate response sequences.

Various species-specific social behaviours *are* disrupted following ablation of the telencephalon. The threshold for the elicitation of aggressive behaviour patterns is increased, although once elicited, the vigour and pattern of the aggressive behaviour is normal (Hale, 1956*b*; Fiedler, 1967, 1968; Karamyan, Malukova & Sergeev, 1967; Kassel & Davis, 1977; Schwagmeyer, Davis & Kassel, 1976, 1977). Nest-building, courtship and parental behaviours are disrupted, but not completely lost following ablations (Noble, 1936, 1939; Noble & Bourne, 1941; Aronson, 1948; Kamrin & Aronson, 1954; Segaar, 1961, 1965; Segaar & Nieuwenhuys, 1963; Overmier & Gross, 1974; de Bruin, 1977). Like the aggressive behaviours, the threshold for elicitation of these reproductive behaviours may be increased, but if the behaviours do occur they are usually appropriate to the eliciting stimulus and are performed in the usual patterns (Aronson, 1948; Fiedler, 1967; Kassel, Davis & Schwagmeyer, 1976; Schwagmeyer *et al.*, 1977; Kassel & Davis, 1977). However, perseveration of components of parental response sequences has been reported (Segaar, 1961).

Telencephalic ablation in teleosts leaves both classical conditioning (Overmier & Savage, 1974), and simple instrumental learning and discrimination (Kholodov, 1960; Savage, 1969b) unimpaired. In other more complex learning tasks, however, there is impairment of learning; telencephalon-ablated fish learn markedly more slowly than normal fish in tasks involving learning of successive reversals of discrimination (Warren, 1961; Frank, Flood & Overmier, 1972), in instrumental discrimination tasks involving delays of reward (Savage, 1969*b*), in avoidance learning (Kaplan & Aronson, 1969; Hainsworth, Overmier & Snowdon, 1967; Savage, 1969*a*), and in tasks involving complex discriminative choices (Warren, 1961). Changes in resistance to extinction follow telencephalic ablations, with increased resistance in appetitive instrumental tasks (Flood & Overmier, 1971) and decreased resistance in avoidance tasks (Hainsworth *et al.*, 1967). (We recognize that the generalizations in this chapter to teleost fish as a group are based upon research with a markedly limited number of species. Nonetheless, to date, whenever different species of teleosts have been subjected to identical experimental operations, the same pattern of results have been obtained.)

A complete, detailed critical review of the relevant learning studies has been provided by Flood, Overmier & Savage (1976). A summary of the behavioural effects of telencephalon ablation in teleosts is presented in Table 1. This table is offered as a gross summarization of the current literature and is not definitive. Only the experiments themselves are definitive and we note that results quoted in Laming (p. 214, this volume) report effects of ablation upon habituation.

Table 1. *Teleost telencephalon ablation: behavioural effects*

Category	Unaffected	Affected
General	Vision	
	Audition	
		Olfaction
	Taste	
	Postural reflexes	
	Gross motor skills	
		Spontaneous activity
		Startle reflex
Species-typical	Feeding	
		Aggression (threshold)
		Reproduction
Learned	Habituation	
	Sensitization	
	Pseudo-conditioning	
	Classical conditioning	
	Escape learning	
	Appetitive learning, asymptote:	Appetitive learning, rate:
	simple	acquisition (somewhat slower)
	choice	extinction (retarded)
	discrimination	reversal learning (retarded)
		delayed reward (disrupted)
		Punishment or passive avoidance (disrupted)
		Avoidance (devastated)

Psychological functions of the teleost telencephalon

From our surveys above we can see that analogous behavioural effects result from telencephalic ablation in fish and from limbic lesions in mammals; indeed, the number of parallels is quite remarkable.

To date, the hypotheses that have been offered to account for the effects of telencephalic ablation in teleost fish have been quite similar to those offered for mammalian limbic functions (e.g. arousal, memory, inhibition, and motivation/reward mechanisms in behaviour).

Four hypothesized functions of the teleost telencephalon will be treated here (although there are several others); these four are the most-promoted hypotheses and focus on (1) non-specific arousal, (2) short-term memory, (3) behavioural inhibition and (4) utilization of conditioned mediational states in response evocation and reinforcement. ('Reinforcer' is a technical psychological term which in essence refers to the event that initiates the process (reinforcement) that stamps in associations between antecedent events. Reinforcers are usually hedonistically important events.) Each hypothesis will be described and then relevant behavioural and learning data presented.

Non-specific arousal

According to the non-specific arousal hypothesis the forebrain is seen as a source of general arousal (Aronson, 1970). That is, the forebrain functions as a system which primes, facilitates or 'energizes the functional activities of lower centers to influence almost every kind of behavior' (Aronson, 1970, pp. 97–8). Thus, the non-specific arousal hypothesis explains disruptions in behaviour or learning by reference to the loss of a facilitating forebrain. The hypothesis is supported by (1) ethological observations that telencephalic ablation results in increased threshold for elicitation of innate responses, (2) behavioural observations that ablation of the telencephalon results in 'loss of initiative' and a decrease in spontaneous activity, and (3) the impaired learning seen in selected tasks.

Arousal has been commonly defined in terms of elicitation thresholds, spontaneous activity levels, and vigour and persistence of responses. Although it is clear that spontaneous activity levels, and perhaps initiative, are reduced after telencephalic ablation, what this arousal hypothesis does not explain are those instances where there is no decrement in behavioural output consequent upon ablation, as has been observed for the startle response (Davis *et al.*, 1976), pseudo-conditioning and sensitization (Overmier & Curnow, 1969), classical conditioning (Karamyan, 1956; Bernstein, 1961, 1962; Overmier & Savage, 1974), and simple appetitive instrumental tasks (Savage, 1969*b*). Additionally, *asymptotic* response latencies have commonly been reported to be *equivalent* for normal and forebrainless fish in both appetitive and aversive tasks (Savage, 1969*b*; Aronson, 1970; Flood & Overmier, 1971; Frank *et al.*, 1972). Finally, the observation that ablated fish are *more* resistant to extinction – that is, continue to respond with *more* vigour for a *longer* time – in appetitive tasks (Flood & Overmier, 1971) seems an especially challenging observation for the arousal hypothesis. The validity of this challenge was recently confirmed by a study in our laboratory; Vaughn & Overmier (unpublished) compared normal and telencephalon-ablated goldfish in their resistance to extinction in a discrete trial, food-rewarded, button-pressing task. They again found that even though the two groups reached equivalent asymptotes, the ablated fish responded with shorter latencies in extinction and did so more persistently than did normal fish (see Fig. 1).

It would seem that the simple non-specific arousal hypothesis, while having some experimental support, is at present insufficiently detailed in both the description of the specific functional properties of the arousal mechanism and the predictions of the behavioural disruptions following ablation.

Short-term memory

A second hypothesis is that the fish forebrain participates in a memory subsystem important for short-term remembering of events such as stimuli and responses. According to this hypothesis there should be no significant learning deficits whenever the signal, response and reinforcer occur contiguously. However, if time should elapse between any of these three elements, the learning and performance of forebrainless fish would be impaired.

Fig. 1. Experiment on response persistence in the absence of reward. Group mean latencies of responding after cued trial onset for normal (solid circles) and forebrain-ablated (open circles) fish during extinction of a button-pushing response previously reinforced with food. The two groups did not differ at the asymptote of acquisition; the mean latency at asymptote was 4.4 s.

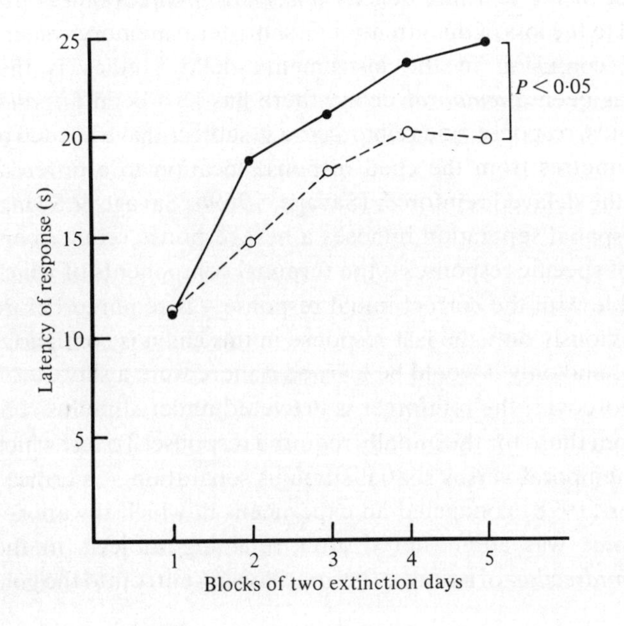

The distinction between short-term memory and long-term memory as two very different kinds of memory mechanisms (bioelectric versus protein-synthetic mechanisms) has been supported by the investigations of Agranoff & Davis (1968), Davis (1968) and Shashoua (1968). Let us first note that long-term memory is not disrupted following telencephalon ablation (Flood, 1975). However, the results of several studies are consistent with the suggestion of deficits in short-term memory following telencephalon ablation (Savage & Swingland, 1969; Savage, 1969*b*). It has been suggested that deficits in short-term memory, a temporary and rapidly fading process, might be overcome by (*a*) preventing memory 'decay' by ensuring the simultaneous occurrence of stimulus, response and reinforcement, and (*b*) massing of training trials (Savage, 1968*b*).

Savage (1968*a*, *b*) found that ablated fish were impaired in an instrumental learning task when reinforcer delivery was delayed a few seconds but were not impaired when such reinforcement was immediate. Thus he inferred that the telencephalon functions as a 'holding' system for event representation. Savage & Swingland (1969) further suggested that perhaps forebrainless fish were unable, after making an operant response, to maintain the excitation of the stimulus 'memory' through several seconds of delay until a reinforcer was delivered. Therefore a delay between response and reinforcer would result in loss of the 'memory' of the signal and/or the correct response through decay of the excitation, and result in no learning. If the reinforcer were received before the 'memory' faded, learning could occur. Savage reasoned that most of the learning deficits observed in forebrainless fish could be attributed to the loss of this attentional, short-term memory system.

One source of confusion in the instrumental-delay studies is that whenever there has been a *temporal* delay, there has also been a *spatial* separation of stimulus, response, and reinforcer: e.g. subjects have needed to swim several centimetres from the cued response location to a different location to obtain the delayed reinforcer (Savage, 1969*b*; Savage & Swingland, 1969). This spatial separation imposes a new response requirement such that a *series* of specific responses – the terminal components of which may be incompatible with the correct initial response – is required before reinforcement. Obviously only the last response in this chain is contiguous with the reinforcer, and only it would be learned if there were a short-term memory deficit. Moreover, the reinforcer is delivered under stimulus conditions different from those for the initially required response. To test which of these factors – temporal versus spatial stimulus separation – is critical, Patten & Overmier (1978) conducted an experiment in which reward for correct colour choice was administered after retaining subjects in the *neutrally coloured midsection* of a runway before allowing entry into the goal box, where the reinforcer (food) was available. Both normal and ablated fish learned this task equally well when the goal boxes were coloured the same as the choice stimuli (see Fig. 2*a*). Thus temporal delay without a concomitant change in stimulus conditions or a change in the response chain did not disrupt learning in forebrain-ablated fish. However, when the goal boxes also were neutral-coloured like the restraining area (see Fig. 2*b*), the ablated fish were markedly more impaired than normal fish (see Fig. 2*c*). This

Fig. 2. Apparatus and results of experiment which demonstrated the effects upon normal and forebrain-ablated fish of receiving rewards under stimulus conditions different from those that cue correct choices. (*a*) and (*b*) Illustrations of apparatus and the stimulus conditions prevailing. (*c*) Mean percentages correct, for the four groups: normal/congruent, normal/irrelevant, ablated/congruent, ablated/irrelevant. The normal and ablated groups did not differ significantly under the congruent conditions but did so under the irrelevant conditions. The '0' point represents pretraining preference levels; all fish were trained *against* their preferred colour. The '0' point was not used in the data analyses but is presented here to show the base rate of choosing the reinforced stimulus.

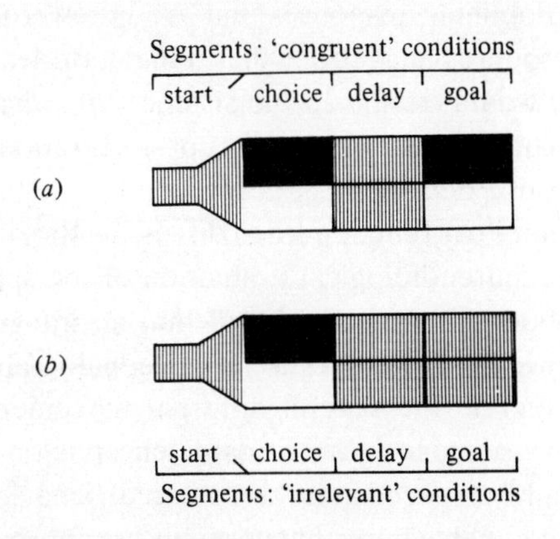

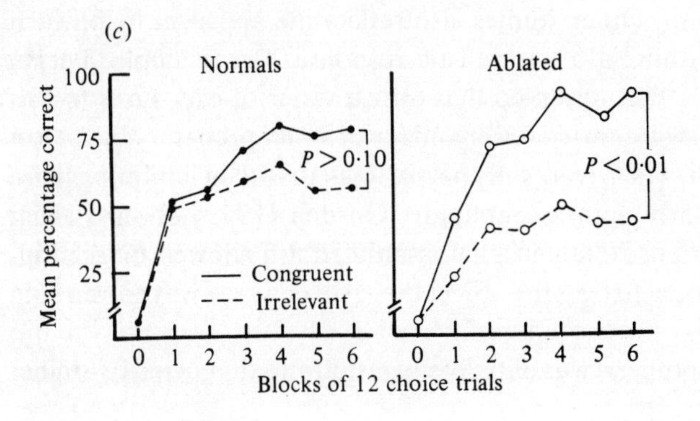

suggests that it was the change in stimulus conditions rather than the delay of reinforcer that was the important causal factor of the impairment observed in Savage's early experiments.

In a more direct test of this memory hypothesis, a comparison was made between normal and forebrainless fish using a time-delay, trace classical conditioning paradigm (Overmier & Savage, 1974). In trace classical conditioning, time intervenes between the termination of signal (CS) and reinforcer (UCS); the subject must 'remember' that the signal has been presented in order to form an association between CS and UCS. This experiment found *no* differences between the two groups of fish in degree of conditioning, thus posing some difficulties for any simple short-term memory hypothesis.

At present the data lead us to be ambivalent about the short-term memory hypothesis. The current data do suggest at least that other functions besides memorial ones are disrupted by telencephalic ablation.

Inhibition of dominant response

An important and viable hypothesis suggests that the fish telencephalon functions to inhibit dominant responses that are unrewarded or punished. Such dominant responses may be either innate or learned. Telencephalic ablation then would result in persistence of whatever response was initially dominant, which would disrupt integration and development of normally adaptive behaviour patterns.

Segaar (1965) invoked an inhibitory function when discussing the roles of the forebrain. On the basis of neuroethological evaluation of the species-specific behaviours of the stickleback, Segaar argued that an integrative function of the forebrain seems apparent because telencephalon-ablated sticklebacks are unable to integrate the several different movements of nest-building into a constructive sequence. After total telencephalon ablation, fish vigorously and incessantly engaged in certain parental behaviours – fanning the nest and digging sand – behaviours that normally are appropriate in a moderate amount. Other studies also reflect the apparent inability of forebrainless fish to inhibit a naturalistic response. For example, Peeke, Peeke & Williston (1972) observed that over a series of exposures to bait behind glass, normal fish emitted fewer and fewer *initial* predatory bites than did forebrainless fish, which were only partially successful in inhibiting these fruitless bites. Similarly, in a recent study, Gordon (1979) observed that over repeated experiences, telencephalon-ablated fish showed marked inability to inhibit approach to a prey object the taste of which was paired with a noxious event.

This inhibition hypothesis was initially suggested and supported by studies of species-specific behaviours, as described, but it also finds support in studies of learned behaviours. As would be predicted by the inhibition hypothesis, Flood & Overmier (1971) reported slower extinction of a learned appetitive response by the forebrainless group. More recently, this markedly slower extinction rate was confirmed by Vaughn & Overmier as discussed earlier (refer to Fig. 1). Additionally, other tasks that require inhibition of a learned response, such as a passive avoidance learning (Overmier & Flood, 1969) and position habit reversal (Warren, 1961; Frank *et al.*, 1972), are learned slowly by forebrain-ablated fish – observations consistent with the idea that telencephalon-ablated fish have great difficulty in inhibiting a dominant learned response.

Regarding the active avoidance learning deficits demonstrated by telencephalon-ablated fish (e.g. Kaplan & Aronson, 1969), the inhibition hypothesis explains the observed learning deficits as failures to inhibit (1) competing, incompatible species-specific defence reactions in avoidance situations (Bolles, 1970), (2) competing exploratory responses (Ingle, 1965), and (3) competing startle reactions (Davis *et al.*, 1976). Competing responses might be freezing, diving or submissive displays (Bolles, 1970). Davis *et al.* were impressed with the degree to which ablated subjects were unable to inhibit startle responses, resulting in a hyper-reactive behavioural state. Such a state impaired the ablated subject's ability to respond appropriately to environmental stimuli, i.e. impaired the 'regulation of responsiveness to external stimuli'.

According to this analysis, (1) increasing the opportunity to engage in competing responses or (2) increasing the degree to which an elicited competing response is incompatible with the required response would result in larger learning deficits in ablated fish. In the case of avoidance learning, Savage (1969*a*) and Aronson & Kaplan (1963, 1968) have both looked at shuttlebox avoidance learning in normal and ablated fish. Both studies found deficits in avoidance learning by the forebrainless fish but the degree of deficit was much larger in Aronson's study. One striking difference between their studies was the apparatus: Aronson's shuttlebox was wide and roomy and the avoidance response was swimming through a small hole in the barrier; Savage's shuttlebox was quite short and narrow and the avoidance response was swimming through a hole almost the width and height of the shuttlebox. Thus, Aronson & Kaplan's shuttlebox permitted a greater number of competing responses to occur, and so the greater deficit in that apparatus is not unexpected under the present hypothesis.

In summary, the inhibition hypothesis sees the role of the telencephalon to be inhibition of dominant responses, innate or learned, thus allowing the development of behaviour adaptive to stimulus conditions and changed feedback, such as reinforcement. This hypothesis correctly accounts for several behavioural disruptions consequent upon telencephalic ablation, for example over-persistence of selected courtship/parental behaviours, persistent unrequited predation, slow passive avoidance learning, slow reversal learning. It offers an account of why ablated fish have slower acquisition rates in those learning tasks affording multiple-response opportunities by reference to the slower elimination of competing responses – although forebrainless fish should, and do, *eventually* reach the same asymptote as normal fish.

There are, however, data that are problematic for the inhibition hypothesis. For example, after a fish has mastered an avoidance task, one might reasonably presume that avoidance response to now be the dominant response. But in point of fact telencephalon-ablated fish show markedly *less* resistance to extinction of avoidance responses than do normal fish (Hainsworth *et al.,* 1967)! Not only does this result seem inconsistent with the inhibition hypothesis but it is also the opposite of the pattern of extinction found in appetitive tasks. These differing results do, however, arise from tasks relying upon different motivation and reward mechanisms. Perhaps this then is a cue to further theoretical analysis of the function(s) of the teleost telencephalon.

Utilization of mediational states

The final hypothesis to be considered here derives from the tenets of the two-process mediational theory of learning (Mowrer, 1947; Rescorla & Solomon, 1967). This is appropriate because the most striking behavioural deficits observed in telencephalon-ablated fish are related to avoidance behaviour, and two-process mediational theory was developed explicitly to account for avoidance behaviour (Mowrer, 1947, 1960). The theory has been extended to instances of appetitive learning as well (Spence, 1956; Trapold & Overmier, 1972). This mediational analysis describes avoidance learning as involving (*a*) the classical conditioning of motivational properties to a signal, which gives that signal power to evoke anticipatorily the instrumental motor response required to prevent shocks, and (*b*) the termination of the signal *and* its conditioned motivational state by this instrumental motor response, which thereby provides reinforcement for the motor response (see Fig. 3). Thus, two learning processes, classical conditioning and instrumental response learning, interact to yield the development and learning of appropriate avoidance behaviours. Telencephalic ablation might selectively impair any of the integral functions: the classical conditioning of motivation; the evocation of the avoidance response by a conditioned state; or the reinforcement of the instrumental response through reduction in conditioned motivation. Since forebrain ablation does not impair classical conditioning, two possibilities remain; these involve the two ways in which the two learning processes interact. Each will be considered in turn.

Response evocation by mediational states. The forebrain is here considered critical for the evocation (or cueing) of instrumental motor responses by classically conditioned mediational states. This has been tested using the transfer-of-control experimental paradigm developed explicitly to test two-process theories. In this test (Overmier & Starkman, 1974) normal and forebrainless fish first were trained in a shuttlebox to avoid shock by swimming across a hurdle whenever a tone warning signal was presented. Although learning by the ablated fish was slow, the task was eventually mastered. In the second step, one group of normal and forebrainless fish

Fig. 3. Outline of the stages of avoidance learning according to two-process theory. The symbol set r_x–s_x represents the unconditioned pain response and its sensory consequences which include 'fear'. The animal initially learns an instrumental motor response to terminate the shock-produced pain and fear. Simultaneously the temporal conjunction of external stimuli (here labelled 'tone') and shock leads to the conditioning of fear to the external stimuli. This conditioned fear endows the stimuli with the capacity to evoke indirectly the previously learned escape response *before* the shock actually occurs, thus removing the fish from danger and the signals for it (e.g. 'tone'). Termination of the fear-eliciting conditions rewards and strengthens the anticipatory escape reaction. The key features are that 'fear' is a state (*a*) which mediates response (R) control, and (*b*) the termination of which provides reinforcement for the response, R.

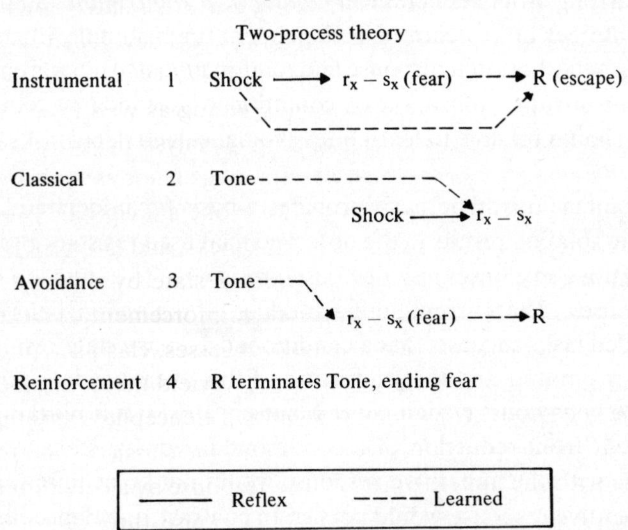

received, in a small chamber, discriminated classical conditioning during which one coloured light was followed by shock while a different light was never followed by shock. The critical question was whether or not the forebrainless fish would be able to integrate these two separate experiences of (*a*) avoidance training to the tone and (*b*) classical conditioning to the light, and thus, without further training, generate in the shuttlebox avoidance responses to the light *previously* paired with shock, even though the light cue had never before occurred in the shuttlebox. Both normal and forebrainless fish promptly performed the avoidance response when the light signal previously paired with shock was presented and did not respond to the control light stimulus. Thus, the forebrain appears *not* integral to the *evocation* of an instrumental response by a classically conditioned stimulus.

Utilization of mediational states as conditioned reinforcers. It is suggested that the forebrain is involved in the learning and utilization of conditioned (secondary) reinforcers. This hypothesis is related to Gloor's (1960) view that mammalian limbic structures provide the motivational mechanisms which *allow learning* of behaviour appropriate to a situation and to Richardson's (1973) view that limbic structures participate in the *utilization* of *reinforcer contingencies* in generating new behaviour.

This hypothesis would seem sufficient to account for the two major learning deficits which occur in avoidance tasks and delay-of-reinforcer tasks. If removal of the forebrain disrupts utilization of conditioned reinforcement, then an avoidance response should be acquired slowly to the extent that learning of the instrumental response is dependent upon reinforcement arising from reduction in *conditioned motivation.* Similarly in delay-of-reinforcer tasks, learning is dependent upon stimuli which bridge the temporal delay between response and reinforcement. Thus, if removal of the forebrain disrupts utilization of conditioned (secondary) reinforcers, ablated fish should be slow to learn both avoidance and delay tasks – which they are.

In like manner, this hypothesis provides a basis for understanding why telencephalic ablation results in the observed increased resistance to extinction of appetitive responses *and* decreased resistance to extinction of avoidance responses. The current understanding of extinction behaviour in food-rewarded tasks suggests that a conditioned aversive state results when the reward is omitted and that reduction of this aversive state reinforces incompatible behaviours (Adelman & Maatsch, 1955). If forebrainless fish cannot benefit from reduction of a conditioned aversive state, responses incompatible with the appetitive response would not be acquired, and the learned appetitive response would persist. In contrast, avoidance responses are thought to be maintained primarily by the response-produced terminations of the conditioned aversive state. However, if this source of reinforcement is weak or ineffective – as hypothesized for ablated fish – the avoidance response should be unstable and less persistent, exactly as has been observed.

Direct empirical support for this hypothesis was obtained in a recent experiment using one of the tests suggested by Flood *et al.* (1976). In this test Farr & Savage (1978) showed that presentation of a classical signal for shock – a conditioned aversive stimulus – contingent upon choice of one side in a T-maze had no effect upon the choice behaviour of forebrainless goldfish; the same treatment caused a significant shift in the choice behaviour of normal fish away from the side of the conditioned aversive stimulus. This differential effect upon choice behaviour is not dependent upon differential activity and was obtained even though both the normal and ablated fish showed equal degrees of classical conditioning as indexed by bradycardia.

While this hypothesis on utilization of mediational states as conditioned reinforcers promises to account for the full array of deficits known in *learning* situations, it seems to leave unaccounted for the disruptions in innate, species-specific behaviour sequences. Hollis & Overmier (1978) have attempted to show how the present hypothesis on utilization of conditioned or secondary reinforcers might be extended to account for the deficits in naturalistic behaviours. Essentially their extension relies upon the suggestion that sequencing and integration of species-specific behaviour patterns is dependent upon the rewarding properties of successive events in the chain. This idea finds support in the work of Thompson (1963), Hogan (1967) and Davis *et al* (1976, and this volume), and others who have demonstrated the reward value of opportunities to engage in selected components of species-specific behaviour chains (e.g. aggressive display, courtship, nest-building). Hollis & Overmier's suggestion is that these species-specific rewarding events are members of a class of weak reinforcers, a class which includes conditioned reinforcers, that require telencephalic processing. Their attempt at extension of the basic hypothesis is quite speculative, but they do propose a programme of research to validate empirically the speculation. (See Hollis (1979) for the first steps in this programme.) Until validation is achieved, however, this last hypothesis, too, can only account for limited portions of the data.

Conclusion

We started with a simple basic question: What is the functional significance of the teleost telencephalon? We end with complicated half-answers. In sharp contrast to the accepted view of only a few years ago, we now know clearly that the teleost telencephalon is integral to the adaptive performance in a broad spectrum of activities, both species-specific and learned. The pattern of deficits seems intriguingly similar to that observed in mammals after selective limbic lesions.

This research on the effects of total telencephalic ablation has successfully identified those classes of tasks in which the teleost telencephalon plays some substantial role. This identification in turn establishes boundary conditions for the nature of the underlying psychological deficits. Moreover, these 'sensitive indicator tasks' provide a basis for an efficient determination of the functional significance of limited, differentiated forebrain regions.

Of the hypotheses about psychological functions which have been suggested, some (e.g. utilization of conditioned reinforcers) seem more promising than others (e.g. arousal). However, none seems sufficient to integrate and 'explain' the full array of known deficits.

One might reasonably question the expectation that a single hypothesis about a single function would prove adequate. (Parsimony is a 'law' of man, not nature.) Indeed, the extent of neurological differentiation in the telencephalon itself suggests multiple functions. Moreover, it is empirically difficult, if not impossible, to design behavioural experiments which definitively test a single hypothetical function against all others. If we accept the idea of multiple functions in the teleost telencephalon, one might suppose that these are localized in the telencephalon at different places. This suggests that now is the time to pursue *more localized* telencephalic lesions, as well as attempting electrical stimulations of the brain and single-unit recordings while the fish are engaged in the identified sensitive indicator tasks. Through task dissociations (see Weiskrantz, 1968) such experiments could help determine where hypothesized functions are localized or at least to which tasks that brain structure contributes.

As we eagerly look to the future, a word of caution seems in order. We should be cautious in ascribing too precise locations to particular functions, because the telencephalon is more likely an integrated set of *interacting* structures than a system of independent structures with independent functions. None the less, the search for identifiable telencephalic functions itself should elucidate the principles of CNS functional organization in ways that anatomy alone cannot.

Preparation of this chapter was supported in part by grants to the Center for Research in Human Learning from NICHD (HD–01136), NSF (BNS77–22075), and the Graduate School of the University of Minnesota.

References

Adelman, H. M. & Maatsch, J. L. (1955). Resistance to extinction as a function of the type of response elicited by frustration. *Journal of Experimental Psychology,* **50**, 61–5.

Agranoff, B. W. & Davis, R. E. (1968). The use of fishes in studies of memory formation. In *The Central Nervous System and Fish Behavior,* ed. D. Ingle, pp. 193–201, University of Chicago Press, Chicago.

Aronson, L. R. (1948). Problems in the behavior and physiology of a species of African mouthbreeding fish. *Transactions of the New York Academy of Sciences,* **11**, 33–42.

Aronson, L. R. (1970). Functional evolution of the forebrain in lower vertebrates. In *Development and Evolution of Behavior,* ed. L. R. Aronson, E. Tobach, D. S. Lehrman & J. Rosenblatt, 75–107. W. H. Freeman, San Francisco.

Aronson, L. R. & Kaplan, H. (1963). Forebrain function in avoidance conditioning. *American Zoologist,* **3**, 483–4.

Aronson, L. R. & Kaplan, H. (1968). Function of the teleostean forebrain. In *The Central Nervous System and Fish Behavior,* ed. D. Ingle, pp. 107–25. University of Chicago Press, Chicago.

Bernstein, J. J. (1961). Brightness discrimination following forebrain ablation in fish. *Experimental Neurology,* **3**, 297–306.

Bernstein, J. J. (1962). Role of the telencephalon in color vision of fish. *Experimental Neurology,* 6, 173–85.

Bernstein, J. J. (1970). Anatomy and physiology of the central nervous system. In *Fish Physiology,* vol. 4, ed. W. S. Hoar & D. J. Randall, pp. 1–90. Academic Press, New York & London.

Berwein, M. (1941) Beobachtungen und Versuche über das gesellige Leben von Elritzen. *Zeitschrift für vergleichende Physiologie,* **28**, 402–20.

Bolles, R. C. (1970). Species-specific defense reactions and avoidance learning. *Psychological Review,* **77**, 32–48.

de Bruin, J. P. C. (1977). 'Telencephalic functions in the behavior of the Siamese fighting fish, *Betta splendens* Regan (Pisces, Anabantidae)'. Published Doctoral Dissertation, University of Amsterdam.

Davis, R. E. (1968). Environmental control of memory fixation in goldfish. *Journal of Comparative Physiological Psychology,* **65**, 72–8.

Davis, R. E., Kassel, J. & Schwagmeyer, P. (1976). Telencephalic lesions and behavior in the teleost. *Macropodus opercularus:* reproduction, startle reaction, and operant behavior in the male. *Behavioral Biology,* **18**, 165–77.

Davis, R. E., Reynolds, R. C. & Ricks, A. (1978). Suppression behavior increased by telencephalic lesions in the teleost, *Macropodus opercularus. Behavioral Biology,* **24**, 32–48.

Doving, K. G. & Gemne, G. (1966). An electrophysiological study of the efferent olfactory system in the barbot. *Journal of Neurophysiology,* **29**, 665–74.

Ebbesson, S. O. S. & Vanegas, H. (1976). Projections of the optic tectum in two teleost species. *Journal of Comparative Neurology,* **165**, 161–80.

Farr, E. J. & Savage, G. E. (1978). First- and second-order conditioning in goldfish and their relation to the telencephalon. *Behavioral Biology,* **22**, 50–9.

Fiedler, K. (1967). Ethologische und neuroanatomische Auswirkungen von Vorderhirnexstirpationen bei Meer brassen (*Diplodus*) and Lippfischen (*Crenilabrus*, Perciformes, Teleostei). *Journal für Hirnforschung,* **9**, 481–563.

Fiedler, K. L. (1968). Verhaltenswirksame Strukturen im Fischgehirn. *Zoologischer Anzeiger,* **31**, 602–16.

Flood, N. C. (1975). Effect of forebrain ablation on long-term retention of a food reinforced shape discrimination. *Psychological Reports,* **36**, 783–6.

Flood, N. C. & Overmier, J. B. (1971). Effects of telencephalic and olfactory lesions on appetitive learning in goldfish. *Physiology and Behavior,* **6**, 35–40.

Flood, N. C., Overmier, J. B. & Savage, G. E. (1976). Teleost telencephalon and learning: an interpretive review of data and hypotheses. *Physiology and Behavior,* **16**, 783–98.

Frank, A. H., Flood, N. C. & Overmier, J. B. (1972). Reversal learning in forebrain ablated and olfactory tract sectioned teleost, *Carassius auratus. Psychonomic Science,* **26**, 149–151.

Gloor, P. (1960). Amygdala. In *Handbook of Physiology*, Sect. 1, *Neurophysiology*, vol. 2, ed. J. Field, H. Magoun & V. Hall, pp. 1395–420. American Physiological Society, Washington, DC.

Gordon, D. (1979). Effects of forebrain ablation on taste aversion in goldfish (*Carassius auratus*). *Experimental Neurology*, **1963**, 356–66.

Hainsworth, F. R. Overmier, J. B. & Snowdon, C. T. (1967). Specific and permanent deficits in instrumental avoidance responding following forebrain ablation in the goldfish. *Journal of Comparative and Physiological Psychology,* **63**, 111–16.

Hale, E. B. (1956*a*). Social facilitation and forebrain function in maze performance of green sunfish, *Lepomis cyanellus. Physiological Zoology,* **29**, 93–106.

Hale, E. B. (1956*b*). Effects of forebrain lesions on the aggressive behavior of green sunfish, *Lepomis cyanellus. Physiological Zoology,* **29**, 107–27.

Herrick, C. J. (1933). The functions of the olfactory parts of the cerebral cortex. *Proceedings of the National Academy of Sciences, USA,* **19**, 7–14.

Hogan, J. A. (1967). Fighting and reinforcement in the Siamese fighting fish (*Betta splendens*). *Journal of Comparative and Physiological Psychology,* **64**, 356–9.

Hollis, K. L. (1979). 'The effect of telencephalon ablation upon the reinforcing and eliciting properties of species-specific events in *Betta splendens*'. Doctoral dissertation, University of Minnesota.

Hollis, K. L. & Overmier, J. B. (1978). The function of the teleost telencephalon in behavior: a reinforcement mediator. In *The Behavior of Fish and Other Aquatic Animals,* ed. D. I. Mostofsky, pp. 139–95. Academic Press, New York & London.

Ingle, D. J. (1965). Behavioral effects of forebrain lesions in goldfish. In *Proceedings of the 73rd Annual Convention of the American Psychological Association,* vol. 1, pp. 143–4.

Issacson, R. L. (1974). *The Limbic System.* Plenum Press, New York.

Iversen, S. D. (1973). Brain lesions and memory in animals. In *Physiological Basis of Memory,* ed. J. A. Deutsch, pp. 305–64. Academic Press, New York & London.

Janzen, W. (1933). Untersuchungen über Grosshirnfunktionen des Goldfisches (*Carassius auratus*). *Zoologische Jahrbucher,* **52**, 591–628.

Kamrin, R. P. & Aronson, L. R. (1954). The effects of forebrain lesions on mating behavior in the male platyfish, *Xiphophorus maculatus. Zoologica, New York,* **39**, 133–40.

Kaplan, H. & Aronson, L. R. (1969). Function of forebrain and cerebellum in learning in the teleost *Tilapia heudelotii macrocephala. Bulletin of the American Museum of Natural History, New York,* **142**, 141–208.

Karamyan, A. I. (1956). Evolutsiya funktsii mozzhechka i bol'shikh polusharii golognogo mozza. (Evolution of the function of the cerebellum and cerebral hemispheres.) Medgiz, Leningrad. (Trans. by National Science Foundation, Washington, DC, 1962, OTS TT 61-31014.)

Karamyan, A. I. Malukova, I. V. & Sergeev, B. F. (1967). Uschastie konechnogo mozgo kostistykh ryb v osushschestvlenii slozhnykh uslovnorej-lektornykh i obshchepovedencheskikh reaktsii. In *Povedenie i Retseptsii Ryb.* Academy of Sciences, USSR, Moscow. (PB 184929 T.)

Kassel, J. & Davis, R. E. (1977). Recovery of function following simultaneous and serial telencephalon ablation. *Behavioral Biology,* **21**, 489–99.

Kassel, J., Davis, R. E. & Schwagmeyer, P. (1976). Telencephalic lesions and behavior in the teleost *Macropodus opercularis:* further analysis of reproductive and operant behavior in the male. *Behavioral Biology,* **18**, 179–88.

Kholodov, Y. A. (1960). Simple and complex food obtaining conditioned reflexes in normal fish and in fish after removal of the forebrain. *Works of the Institute for Higher Nervous Activity, Physiology Series,* **5**, 194–201. (Trans. National Science Foundation, Washington, DC, 1962.)

Mowrer, O. H. (1947). On the dual nature of learning: a reinterpretation of 'conditioning' and 'problem solving' *Harvard Educational Review,* **17**, 102–48.

Mowrer, O. H. (1960). *Learning Theory and Behavior.* Wiley, New York.

Nieuwenhuys, R. (1967). The interpretation of the cell masses in the teleostean forebrain. In *Evolution of the Forebrain,* ed. R. Hassler & H. Stephan, pp. 32–40. Plenum Press, New York.

Noble, G. K. (1936). The function of the corpus striatum in the social behavior of fishes. *Anatomical Record (Suppl.),* **64**, 34.

Noble, G. K. (1939). Neural basis of social behavior in vertebrates. *Collecting Net,* **14**, 121–4.

Noble, G. K. & Borne, R. (1941). The effect of forebrain lesions on the sexual and fighting behavior of *Betta splendens* and other fishes. *Anatomical Record (Suppl.),* **79**, 49.

Nolte, W. (1932). Experimentelle Untersuchungen zum Problem der Lokalisation des Assoziationsvermogens im Fischgehirn. *Zeitschrift für vergleichende Physiologie,* **18**, 255–79.

Overmier, J. B. & Curnow, P. F. (1969). Classical conditioning, pseudoconditioning, and sensitization in 'normal' and forebrainless goldfish. *Journal of Comparative and Physiological Psychology,* **68**, 193–8.

Overmier, J. B. & Flood, N. C. (1969). Passive avoidance in forebrain ablated teleost fish, *Carassius auratus. Physiology and Behavior,* **4**, 791–4.

Overmier, J. B. & Gross, D. (1974). Effects of telencephalic ablation upon nest-building and avoidance behaviors in East African mouth breeding fish, *Tilapia mossambica. Behavioral Biology,* **12**, 211–22.

Overmier, J. B. & Savage, G. E. (1974). Effects of telencephalic ablation on trace classical conditioning of heart rate in goldfish. *Experimental Neurology,* **42**, 339–46.

Overmier, J. B. & Starkman, N. (1974). Transfer of control avoidance behavior in normal and telencephalon ablated goldfish (*Carassius auratus*). *Physiology and Behavior,* **12**, 605–8.

Papez, J. W. (1937). A proposed mechanism of emotion. *Archives of Neurology and Psychiatry,* **38**, 725–43.

Patten, R. & Overmier, J. B. (1978). 'Teleost telencephalon and secondary reinforcement'. Paper presented at Midwestern Psychological Association.

Peeke, H. V., Peeke, S. C. & Williston, J. S. (1972). Long-term memory deficits for habituation of predatory behavior in the forebrain ablated goldfish (*Carassius auratus*). *Experimental Neurology,* **36**, 288–94.

Piel, G. (1970). The comparative psychology of T. C. Schneirla. In *Development and Evolution of Behavior,* ed. L. R. Aronson, E. Tobach, D. S. Lehrman & J. S. Rosenblatt, pp. 1–17. W. H. Freeman, San Francisco.

Prosser, C. L. & Brown, F. A. (1961). *Comparative Animal Physiology,* p. 638. Saunders, Philadelphia.

Rescorla, R. A. & Solomon, R. L. (1967). Two-process learning theory: relationships between Pavlovian conditioning and instrumental learning. *Psychological Review,* **74**, 151–82.

Richardson, J. S. (1973). The amygdala: historical and functional analysis. *Acta Neurobiologiae Experimentalis,* **33**, 623–48.

Riss, W., Halpern, M. & Scalia, F. (1969). Anatomical aspects of the evolution of the limbic and olfactory systems and their potential significance for behavior. *Annals of the New York Academy of Sciences,* **159**, 1096–111.

Savage, G. E. (1968*a*). Function of the forebrain in the memory system of the fish. In *The Central Nervous System and Fish Behavior*, ed. D. Ingle, pp. 127–38. University of Chicago Press, Chicago.

Savage, G. E. (1968*b*). Temporal factors in avoidance learning in normal and forebrainless fish (*Carassius auratus*). *Nature, London,* **218**, 1168–9.

Savage, G. E. (1969*a*). Telencephalic lesions and avoidance behavior in the goldfish (*Carassius auratus*). *Animal Behaviour,* **17**, 363–73.

Savage, G. E. (1969*b*). Some preliminary observations on the role of the telencephalon in food-reinforced behavior in the goldfish, *Carassius auratus. Animal Behaviour,* **17**, 760–72.

Savage, G. E. & Swingland, I. R. (1969). Positively reinforced behavior and the forebrain in goldfish. *Nature, London,* **221**, 878–9.

Scalia, F. & Ebbesson, S. O. S. (1971). The central projections of the olfactory bulb in a teleost, (*Gymnothorax funebris*). *Brain, Behavior and Evolution,* **4**, 376–99.

Schoenfeld, T. A. & Hamilton, L. W. (1977). Secondary brain changes

following lesions: a new paradigm for lesion experimentation. *Physiology and Behavior,* **18**, 951–67.
Schwagmeyer, P., Davis, R. E. & Kassel, J. (1976). Telencephalic lesions and behavior in the teleost *Macropodus opercularis:* further analysis of reproductive and operant behavior in the male. *Behavioral Biology,* **18**, 179–88.
Schwagmeyer, P., Davis, R. E. & Kassel, J. (1977). Telencephalic lesions and behavior in the teleost *Macropodus opercularis:* effects of telencephalon and olfactory bulb ablation on spawning and foam-nest building. *Behavioral Biology,* **20**, 463–70.
Segaar, J. (1961). Telencephalon and behavior in *Gasterosteus aculeatus. Behaviour,* **18**, 256–87.
Segaar, J. (1965). Behavioural aspects of degeneration and regeneration in fish brain: a comparison with higher vertebrates. *Progress in Brain Research,* **4**, 143–231.
Segaar, J. & Nieuwenhuys, R. (1963). New etho-physiological experiments with male *Gasterosteus aculeatus,* with anatomical comment. *Animal Behaviour,* **11**, 331–44.
Shashoua, V. E. (1968). The relation of RNA metabolism in the brain to learning in the goldfish. In *The Central Nervous System and Fish Behavior,* ed. D. Ingle, pp. 203–13. University of Chicago Press, Chicago.
Spence, K. W. (1956). *Behavior Theory and Conditioning.* Yale University Press, New Haven.
Strieck, F. (1924). Untersuchungen über den Geruchs und Geschmackssinn der Ellritze (*Phoxinus laevis*). *Zeitschrift für vergleichende Physiologie,* **2**, 122–54.
Thompson, T. (1963). Visual reinforcement in Siamese fighting fish. *Science,* **141**, 55–7.
Trapold, M. A. & Overmier, J. B. (1972). The second learning process in instrumental training. In *Classical Conditioning,* vol. 2, *Current Theory and Research,* ed. A. H. Black & W. F. Prokasy, pp. 427–52. Appleton-Century-Crofts, New York.
Vanegas, H. & Ebbesson, S. O. S. (1976). Telencephalic projection in two teleost species. *Journal of Comparative Neurology,* **165**, 181–95.
Warren, J. M. (1961). The effect of telencephalic injuries on learning by paradise fish, *Macropodus opercularis. Journal of Comparative and Physiological Psychology,* **54**, 130–2.
Weiskrantz, L. (1968). Treatments, inferences, and brain functions. In *Analysis of Behavioral Change,* ed. L. Weiskrantz, pp. 400–14. Harper, New York.
Wenzel, B. M. (1974). The olfactory system and behavior. In *Limbic and Automatic Nervous Systems Research,* ed. L. V. DiCara, pp. 1–40. Plenum Press, New York.

G. E. SAVAGE & D. E. WRIGHT

Electrical stimulation of the telencephalon and learning in a teleost fish

Introduction

There is considerable evidence to show that the telencephalon of bony fish is involved in the formation and maintenance of certain types of learned association. For instance, fish with telencephalic damage are inferior to sham-operated fish in learning to avoid stimuli whose onset signals subsequent delivery of an electrical shock (Kaplan & Aronson, 1967; Hainsworth, Overmier & Snowdon, 1967; Savage, 1968). Surprisingly, if the same signal and shock are used in a classical Pavlovian paradigm, where the response is not a skilled movement, but general activity or a reflex slowing of the heart, learning is apparently unimpaired by telencephalic ablation (Overmier & Curnow, 1969; Overmier & Savage, 1974; Farr & Savage, 1978). The inconsistency in results between the two forms of training cannot be attributed to a failure of motor skill, for it has been shown that telencephalon-damaged fish will rapidly learn shape and colour discrimination problems when required to swim to one or other of the discriminanda from a start box (Janzen, 1933; Nolte, 1933; Savage, 1969).

It has been suggested (Flood, Overmier & Savage, 1976) that a failure to process correctly information about second-order reinforcement may lead to the difficulties observed in learning. Hence a first-order association between light and shock can be established, but should the experiment require a further association between, for instance, sound and light, a fish without a telencephalon cannot learn this, since the conditioned fear induced by the light does not act as a reinforcer when associated with the neutral sound stimulus. An experiment by Farr & Savage (1978) has provided evidence for such a failure of telencephalon-ablated fish to use secondary reinforcement.

Although this hypothesis can explain the majority of acquisition deficits associated with teleost telencephalic lesions, it is less powerful in predicting the types of retention deficit to be expected following telencephalic operations. Routtenberg (1968) distinguished between reticular and limbic arousal mechanisms and Savage (1980) suggested that a similar distinction may be made between conventional arousal and reinforcement-linked arousal when considering the teleost telencephalon. Thus although the measurement of various indices of arousal in the orientation reaction yields inconclusive results when control and telencephalon-ablated fish are considered (Savage, 1980), this does not mean that Routtenberg's type II arousal need be unaffected. Such an arousal failure may cause the total and immediate deficits observed in retention of avoidance associations following telencephalic ablation.

Since almost all current knowledge of the function of the telencephalon stems from brain lesion studies, it was decided to investigate the effects of using telencephalic stimulation as reward in similar avoidance and classical conditioning situations to those used in ablation experiments.

Observations on brain stimulation effects in fish, especially those relating to the telencephalon, are few. Stimulation in the anterior region, the olfactory tracts and their immediate termination, can elicit feeding responses (Grimm, 1960; Fiedler, 1964, 1968; Demski & Knigge, 1971). Only one investigation has examined the effects of telencephalic stimulation on retention. Malyukova (1964) trained fish to respond differentially to three shapes to obtain food, then observed the behaviour of the animals when receiving brain stimulation. Strong aversive reactions were seen, which obscured any effects of stimulation on retention. This finding is not surprising, for Savage (1971) observed that fish given a free choice would seek stimulation if stimulated in the anterior telencephalon, whereas Boyd & Gardner (1962) and Savage (1971) both reported aversive reactions when the posterior part of the telencephalon was stimulated. The stimulation used by Malyukova (1964) provided negative reinforcement whilst the fish were approaching food, thus probably eliciting conflict behaviour. Any investigation of the effects of telencephalic stimulation must make allowance for reinforcement effects, quite apart from any effects on acquisition or retention. In the studies by Boyd & Gardner (1962) and Savage (1971) no attempt was made to compare the rates of learning or the retention levels of the brain-stimulated fish with conventionally rewarded fish, and it was therefore of interest to make such comparisons.

Methods

Animals

The animals used were goldfish (*Carassius auratus*) 15–17.5 cm in length from snout to base of tail. They were obtained from a standard source, and were maintained in well-aerated and filtered stock tanks for 14 days before receiving surgery. The animals were fed daily with a mixture of chopped sheep's heart and Tetramin dried fish food.

Pilot studies using the classical conditioning regime outlined below demonstrated that acquisition and retention were temperature-dependent. At 6–10 °C learning was poor, and extinction correspondingly rapid, compared with that shown by animals trained at 16–18 °C; above this, no significant variations in response with temperature were observed. Accordingly all experimental tanks were maintained at 16–18 °C, a heating and aerating system being introduced to maintain the temperature where necessary.

Electrodes

Electrodes for telencephalic stimulation were made from electrolytically sharpened 22 gauge tungsten wire. The nominal tip diameter of these electrodes was 5 μm. Once the electrodes had been varnished and the tip broken electrically, pairs of electrodes were cemented together, so that a tip separation of 1 mm was achieved. A Grass SD5 stimulator was used for stimulation at 30 Hz, and the waveform across an accurate 10-Ω resistor was monitored on a Tektronix 502A oscilloscope. Electrodes for stimulation of the dorsal body musculature in the free-swimming experiments were made in a similar fashion. Electrodes for recording the electrocardiogram (ECG) were prepared as described in a previous publication (Roberts, Wright & Savage, 1973).

Surgery

All surgery was performed with the animals under MS 222 (methane tricaine sulphonate) anaesthesia, at a concentration of 1:5000.

Animals to be used for the free-swimming instrumental training experiments were anaesthetized and then placed in a Perspex clamp. A disc of bone was trephined from the skull of those fish to be used for telencephalic stimulation studies, then the electrodes were mounted in a Prior micromanipulator and placed in position in the brain. A suitable notch was cut in the skull disc to allow for the electrodes, then the disc was replaced, and it and the electrode assembly were sealed to the skull with a mixture of Eastman 910 adhesive and Simplex Rapid dental cement. The electrode wires were led to a suture on the back of the fish, where a Radiospares miniature two-pin socket was attached. Those fish to be used for free-swimming experiments with stimulation of the body sides as negative reinforcement were first anaesthetized then a suture sewn into the dorsal musculature and the electrode tips placed in the muscles. The electrode leads were joined to a Radiospares miniature two-pin socket just dorsal to the suture.

Cardiac electrodes for the classical conditioning studies were prepared and implanted using the method described by Roberts *et al.* (1973).

Training

Instrumental conditioning. A 30.5 × 30.5 × 61 cm tank was used in these experiments. One half of the tank was painted black, the other end, white. The tank was covered with foil so that no extraneous light reached the fish. When light was provided, the observer could note the position of the animals by peering through a slit along the side of the aquarium. During experiments illumination was provided by a 25 cm 20 W striplight, fixed centrally 46 cm above the water. Each day for 5 days fish were introduced singly into the tank, connected to the electrode lead, and left for 30 min to habituate. Then the light was switched on for 4 min, and the unrestricted movements of the fish into the black or white end of the tank were recorded on a Rustrak 92 event recorder. For the next 5 or 8 days, telencephalic or muscle electrodes were connected to a Grass SD5 stimulator by a long flexible counterweighted lead and a Radiospares two-pin miniature plug, and after the 30-min habituation period, when the light came on, animals were stimulated when they entered the previously preferred end. These trials lasted 4 min. A final series of five 4-min trials was conducted under extinction conditions, that is without stimulation.

Classical conditioning. All fish used for classical conditioning experiments were implanted with cardiac electrodes as outlined above. Two days after this operation they were placed in a Perspex sponge-rubber-lined clamp, with an oral tube delivering a steady flow of aerated water to the gills. Animals receiving 'conventional' body-wall shock (B) had plastic-coated wire laid along their sides, with 5 mm bare wire at the ends, lying just dorsal to the lateral line. For animals to receive telencephalic stimulation (T) a different procedure was employed, to avoid possible anaesthetic effects. Two days before training, these animals were anaesthetized and a disc of bone trephined from the skull. The disc was immediately sutured back, but was removed on the day of training merely by severing the suture. Bipolar electrodes mounted on a Prior micromanipulator were then placed in the telencephalon. Trials were conducted in a darkened box.

All fish had 30 min to habituate in the clamp, and were then given ten 10-s presentations of 10-Hz flashes from a Grass PS2 stroboscope to habituate the bradycardiac response. Next the animals received 20 pairings of light (10 s) and shock (0.25 s) under delay conditioning conditions; finally there were up to 60 unreinforced presentations of the light, sufficient to extinguish the conditioned response to a criterion of two successive trials of zero bradycardia. An intertrial interval of 90 ± 30 s was used in all these experiments. The magnitude of the bradycardiac response was calculated as $(A - B)/A \times 100\%$, where A was the number of beats in the 10-s period before the light was presented and B the number of beats in the 10-s period of light presentation.

Results

When fish are shocked there is a dramatic unconditioned slowing of both cardiac and respiratory activity (Otis, Cerf & Thomas, 1957), and it was decided to use the first of these responses as a measure of the strength of stimulus delivered to a fish, whether by telencephalic (T) or body-wall (B) electrodes. In each experiment, whether instrumental or classical types of association were to be trained, tests were conducted prior to training to ascertain the voltage at which a mean 25% bradycardiac response over 10 s could be elicited. The response to muscular shock remained constant over periods far in excess of those used. With regard to telencephalic stimulation, it was found that electrode placements in the posterior dorsal telencephalon caused the lowest-threshold responses; bradycardias were seen, but no change was observed in the shape of the ECG. (It is interesting to note that

Fig. 1. Cardiac decelerations elicited by stimulation of one hemisphere of the posterior dorsal telencephalon (open circles) and of the centre of one hemisphere of the optic tectum (solid circles). Stimulation was delivered at constant voltage for 5 s each minute for 120 min for the telencephalon, and for 5 s each minute for two 20-min periods, separated by an 80-min period without stimulation, for the optic tectum. Six animals were used for each experiment, and were unanaesthetized.

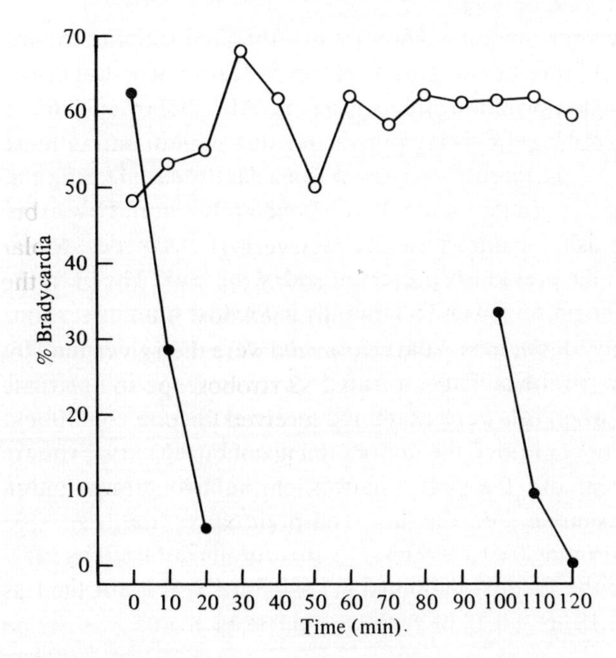

similar effects have been found in amphibians, by Segura (1969) studying toads.) The response was very consistent, and showed none of the rapid waning demonstrated by responses to similar stimulation of the optic tectum (Fig. 1). Despite this invariance, the bradycardiac response initiated as a result of telencephalic stimulation was very sensitive to the effects of MS 222 anaesthesia. Concentrations of 1:40 000 of the anaesthetic in water produced abolition of the responses to stimulation; these levels of anaesthetic are so low that they do not necessarily produce any behavioural signs of anaesthesia. Conversely, stimulation in the vagal lobes of the medulla, that is at the root of the cardiac branch of the vagus, produced consistent decelerations at these anaesthetic concentrations, and higher levels of anaesthetic (1:20 000) were necessary even to reduce the level of response. In view of the extreme anaesthetic sensitivity of the telencephalon the measures outlined above were introduced to avoid anaesthetizing animals less than 24 hr before training commenced. Tests showed that no observable trace of an anaesthetic effect on heart rate responses was present by this time.

In spite of the telencephalon's considerable influence on heart rate when stimulated, it appears that it exerts no tonic effect. The mean heart rates of seven fish were determined each day for a week. The telencephalon was then removed and heart rates noted daily, again for a week. No significant changes in rate were observed following the operation.

Instrumental conditioning

Simple preference reversal. The first instrumental training experiment compared the abilities of two groups of fish to learn and extinguish a new end-preference when swimming freely in a tank. All fish showed definite contrast preferences during the 5-day preference test period, and in most instances the degree of preference increased from day to day. During the second 5 days of the experiment animals received stimulation, by way of telencephalic (seven fish) or intramuscular (seven fish) electrodes, for as long as they stayed in the previously preferred end of the tank. The percentage of time spent in the preferred end fell rapidly in the first training session, and decreased steadily for the next 4 days, to a value of 5–10%. Despite this considerable change of behaviour, a rapid reversion to the original preference occurred when fish were examined under extinction conditions. Statistical comparisons confirmed the impression given by the curves shown in Fig. 2; at no point did the performances of the two groups differ significantly. (For example, for the last end-preference trial, $t = 1.36$, $P > 0.1$; for the first training trial, $t = 0.61$, $P > 0.1$; for the last training trial, $t = 1.19$, $P > 0.1$; for the first extinction trial, $t = 0.38$, $P > 0.1$; for the last extinction trial, $t = 0.12$, $P > 0.1$. In each case d.f. = 12.)

In view of the known peculiarities of fish with telencephalic damage, for instance a decrease in activity under some circumstances (Savage, 1968) and the marked inter-session forgetting followed by rapid intra-session acquisition seen by Peeke, Peeke & Williston (1972), the results were analysed in two other ways. Firstly, to demonstrate any difference in activity, comparisons were made of the mean number of crossings of the white/black boundary, but no significant differences were detected at any of the three stages of the experiment. A second possible difference was that although the performances of the groups in each day's session might be comparable, the intra-session distribution of errors might be different. With this in mind, mean scores for the first minute of training were compared with mean scores for the last minute of training, and a similar comparison was made for the extinction trials. These tests showed that the average within-trial performance of the two groups of fish was almost identical, and certainly did not support the hypothesis that the fish receiving telencephalic stimulation were starting each day's training at a poor level of performance, and then improving rapidly, as compared with the fish receiving conventional reinforcement. (For instance, comparing the two groups in the first minute of training,

Fig. 2. End-preference responses of two groups of fish given 5 days of preference trials, 5 days of receiving electrical stimulation for entering the previously preferred end of the tank, and 5 days of extinction trials. Open circles, seven fish receiving electrical stimulation of the body musculature; solid circles, seven fish receiving electrical stimulation of the telencephalon.

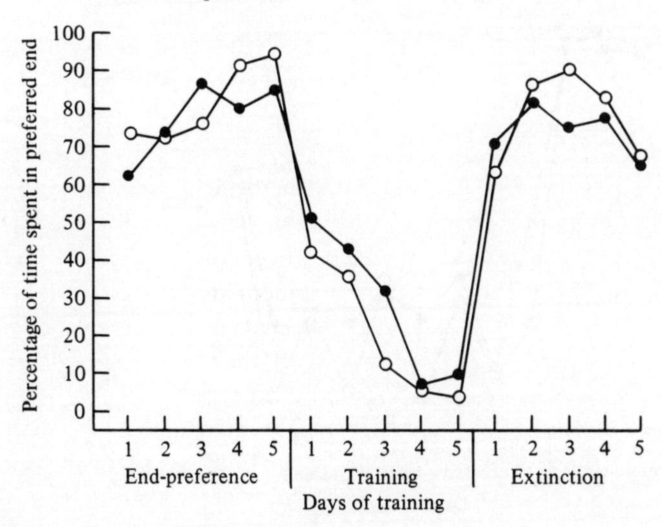

$t = 0.48$, d.f. $= 12$, $P > 0.1$; for the last minute of training, $t = 1.46$, d.f. $= 12$, $P > 0.1$.)

Successive preference reversal. Previous work by Warren (1961) has suggested that although interference with the telencephalon may cause little deterioration of the ability to learn successfully the sort of problem just outlined, more complex manipulations of the same paradigm may demonstrate significant deficits, particularly if the lesioned animals are required to reverse successively their preference. In the next experiment, therefore, after fish had been subjected to five daily sessions of testing for end-preference they were then trained over 8 days with daily reversal of the end of the tank in which stimulation was administered. Finally, 5 days of extinction trials were given. A group of seven fish was subjected to muscle stimulation as reinforcement; the corresponding telencephalic stimulation group consisted of eight fish. The results obtained are shown in Fig. 3.

The performance of the two groups of fish during the end-preference trials closely paralleled that seen in the previous experiment in that a very definite preference was manifest on the first day and increased during subsequent trials. Differences did appear between the two groups of fish when successive reversal training was introduced. For the first 3 days the performances of the

Fig. 3. End-preference responses of two groups of fish given 5 days of preference trials, 8 days of training to avoid the end-preference established each previous day, and 5 days of extinction trials. Open circles, seven fish receiving electrical stimulation of the body musculature; closed circles, eight fish receiving electrical stimulation of the telencephalon.

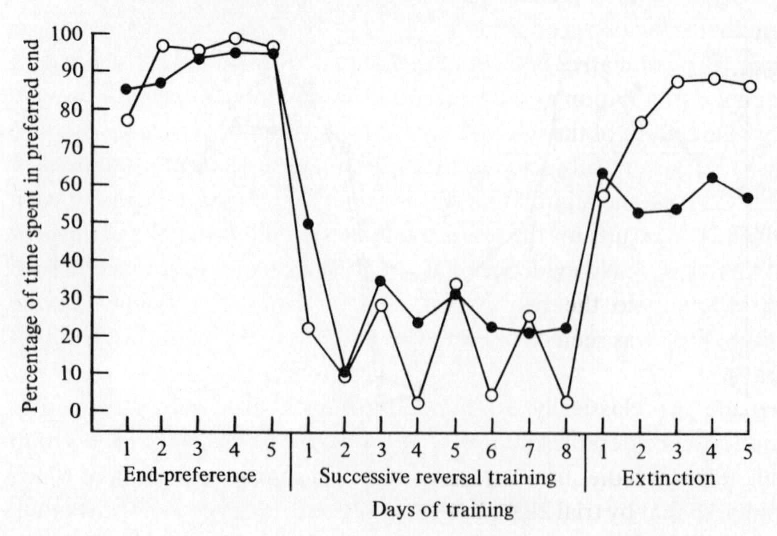

two groups were statistically similar, but for the training sessions 4, 6 and 8 the performance of the conventionally reinforced group (B) was indicative of greater levels of avoidance of stimulation than was that of the group receiving telencephalic stimulation (T) (comparing the groups, training session 4, $t = 3.11$, d.f. = 13, $P < 0.01$; session 6, $t = 2.21$, d.f. = 13, $P > 0.05 < 0.02$; session 8, $t = 2.68$, d.f. = 13, $P ⌓ 0.02$). Thus the B group were avoiding stimulation more successfully when stimulated for entering their original non-preferred end, and less successfully when stimulated for entering their original preferred end, hence retaining some evidence of the original end-preference. The animals receiving telencephalic stimulation appeared to achieve similar scores whether the training was in favour of or against their original preference. During extinction, the performance of the two groups was statistically similar for the first 2 days, but thereafter a significant preference difference developed, with the B group animals reverting more strongly to their original preference than did the animals of the T group. This difference persisted until the final day of training (for instance comparing the groups on day 3 of extinction, $t = 3.48$, d.f. = 13, $P > 0.01$).

Classical conditioning. The cardiac orientation response shown by naïve fish to the first presentation of the light conditioned stimulus (CS) was generally in the order of 10–15% bradycardia; this response habituated rapidly within 10 trials. Subsequently, fish trained with stimulation of the telencephalon (seven fish) or body musculature (nine fish) as unconditioned stimulus (UCS) manifested a conditioned response (CR) of around 50% bradycardia on trials 10 and 20. The magnitudes of response on the final trial were similar for the two groups ($t = 0.38$, d.f. = 14, $P > 0.1$). Since nothing is known of non-associative behavioural changes contingent upon the use of telencephalic stimulation as UCS, pseudoconditioning control groups were used for stimulation of the telencephalon (four fish) and of the musculature (four fish). Fish were subjected to random non-contingent presentations of the light 'CS' and the stimulation 'UCS', such that equal numbers of each were given. The results for the tenth and twentieth presentations of the light are shown in Fig. 4. No evidence of any 'CR' was obtained. (In fact 50 such trials were given to the two groups, and no significant increase in the response to light was seen compared with that observed on the last habituation trial.)

When the two classically conditioned groups of fish were subjected to extinction, a considerable difference was observed (Fig. 4). The T group receiving telencephalic stimulation showed cardiac responses to light which fell rapidly, so that by trial 20 no bradycardias were observable. The B group of fish, with stimulation of the musculature, showed gradual waning of the conditioned response over some 60 trials. Comparing the actual numbers of trials needed to satisfy an extinction criterion of two successive responses to the CS being within the intertrial range of heart rate values, a highly significant difference was noted between the groups ($t = 9.37$, d.f. = 14, $P < 0.001$).

Thus despite the close comparability of the conditioned responses just outlined, and the initial attempt, described above, to equate unconditioned response (UCR) magnitudes in the two groups, a large difference in extinction performance was seen. In order to investigate the phenomenon further, a group of five fish was subjected to a regime identical to that just described,

Fig. 4. Bradycardiac responses of fish trained under classical conditioning (circles) or pseudoconditioning (triangles) regimes. All animals received 10 10-s presentations of 10-Hz light, to habituate initial cardiac orientation responses. Two groups of fish were then subjected to classical conditioning regimes (CS = 10-Hz light for 10 s, UCS = 0.25-s stimulation) for 20 trials, followed by 60 unreinforced (extinction) presentations of the light. The intertrial interval throughout was 90 ± 30 s. Open circles, nine fish conditioned with stimulation of the body musculature as UCS; solid circles, seven fish conditioned with stimulation of the telencephalon as UCS. Two groups of fish were given random unassociated presentations of light and stimulation. Open triangles, four fish pseudoconditioned with stimulation of the body musculature; solid triangles, four fish pseudoconditioned with stimulation of the telencephalon.

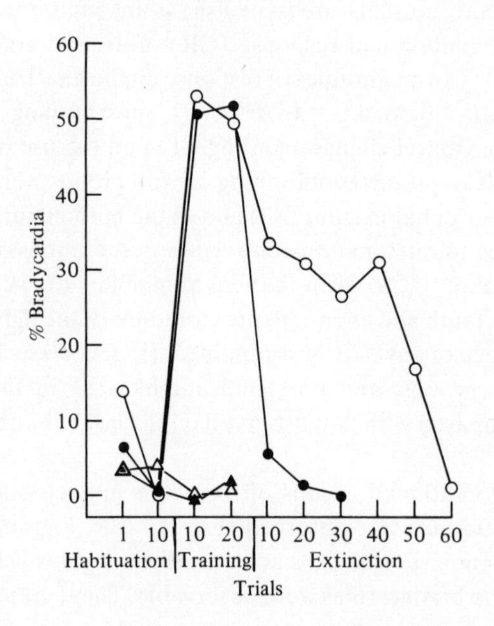

except that the UCS was double (B + T), that is there was simultaneous stimulation of the musculature and the telencephalon, both at voltages which reliably produced a minimum of 25% unconditioned bradycardiac response. It was hoped to determine whether the accelerated extinction observed in the initial classical conditioning experiment was due to some inadequacy of reinforcement delivered via the telencephalic electrodes or whether, despite increasing the level of reinforcement, the information acquired was rendered more labile by telencephalic stimulation. The animals used in this experiment showed conditioned responses lower than those for the B or T groups, at a mean of 35%, but nevertheless statistically comparable with these two groups (for trial 20, B + T versus B, $t = 1.38$, d.f. = 12, $P > 0.1$; B + T versus T, $t = 2.05$, d.f. = 10, $P > 0.05$). In extinction, these fish showed a rapid decrease of CR, reaching zero in an average of 14.4 trials. In this, they resembled the telencephalic stimulation group (B + T versus T, $t = 0.70$, d.f. = 10, $P > 0.1$) but differed considerably from the muscle stimulation group (B + T versus B, $t = 8.64$, d.f. = 12, $P <$

Fig. 5. Bradycardiac responses of three fish subjected to two classical conditioning regimes with a separation of 30 min. The animals received 10 10-s presentations of 10-Hz light, then 20 10-s presentations of light as CS, each followed by 0.25-s stimulation of the body musculature as UCS. Finally 60 10-s unreinforced presentations of the light were given. An intertrial interval of 90 ± 30 s was used.

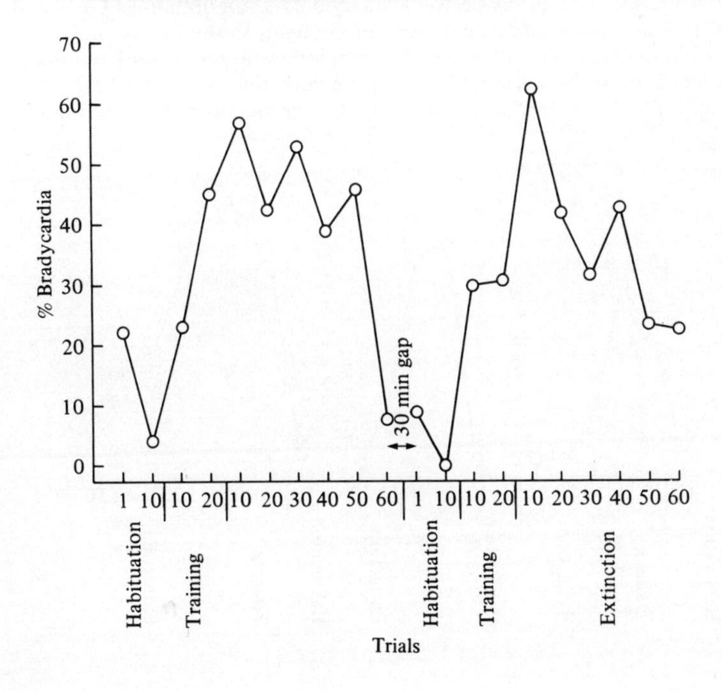

0.001). Thus paired reinforcement did not enhance acquisition, nor did it prevent rapid extinction, as compared with animals receiving a telencephalic stimulation UCS.

It could be argued that brain stimulation *per se*, rather than the activation of a specific area, was the cause of the extinction effect seen in previous experiments. To investigate this possibility a second double UCS experiment was performed, in which body muscle stimulation was paired with cerebellar stimulation. (Cerebellar stimulation was found to be aversive, and to give decelerative effects of repeatable magnitude.) Two fish were subjected to this regime. They showed a mean conditioned bradycardia of 59.5% on trial 20, and took a mean of 55 trials to extinguish the response; thus the extinction times closely resembled those for conventionally reinforced animals.

The next experiments were concerned with investigating the possibility that the effects of telencephalic stimulation, expressed as interference with extinction, might outlast the immediate period of training investigated so far. Fig. 5 shows results for a group of three fish conditioned with stimulation of the lateral musculature as reinforcement, allowed to extinguish their

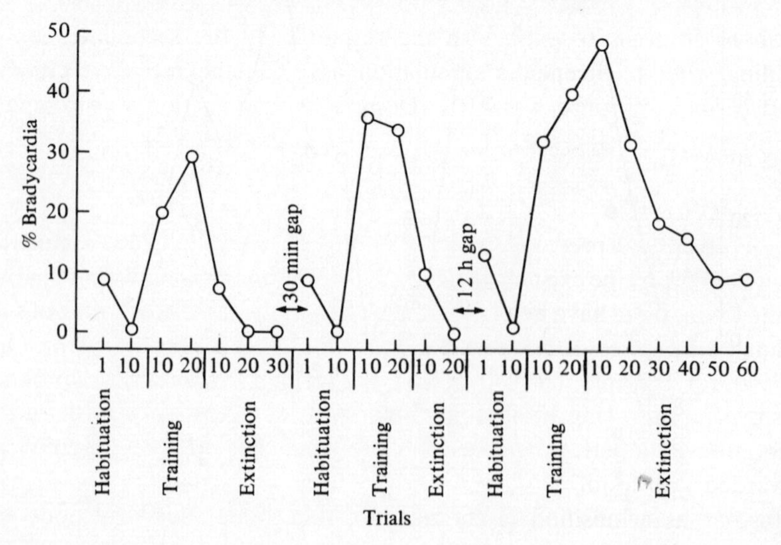

Fig. 6. Bradycardiac responses of six fish subjected to three classical conditioning regimes with separations of 30 min and 12 h respectively. The animals received 10 10-s presentations of 10-Hz light, then 20 10-s presentations of light as CS followed by 0.25-s stimulation as UCS, and 60 10-s unreinforced presentations of the light. In the first training session the UCS was telencephalic stimulation; in the second and third sessions the UCS was stimulation of the body musculature. The telencephalic electrodes remained in place throughout the experiment.

response, and then subjected to an identical regime of training 30 min later. Although the mean conditioned deceleration at trial 20 was greater (46%) for the first training session than for the second (31%), extinction took an average of 56.7 trials in the former case and an average of more than 60 trials in the latter. Thus it appears that closely repeated training does not promote rapid extinction if fish are reinforced conventionally.

Fig. 6 shows the results of a similar experiment. Six fish were first conditioned with telencephalic stimulation (T) as UCS; 30 min later they were subjected to the same regime but with muscle stimulation (B1) as UCS, and 12 h later the last regime (B2) was repeated. In the second training session (B1), 30 min after the first, the bradycardiac responses in trial 20 were similar to those seen in the first muscle stimulation UCS group described (Fig. 4) at the beginning of this section ($t = 0.28$, d.f. $= 13$, $P > 0.1$). Despite this, the animals described in Fig. 6 extinguished their conditioned responses in an average of 14 trials (B1). When the same animals were trained after 12 h (B2), the magnitude of conditioned bradycardia in trial 20 was again comparable with that observed in the original control group ($t = 0.64$, d.f. $= 13$, $P > 0.1$). The differences between the second and third training sessions (B1 and B2) were in the numbers of trials needed to extinguish the response; after the second session 14 trials were needed, while after the third session 60 trials were needed.

A final group of five fish was trained with conventional muscle stimulation reinforcement (B), given 30 min rest, and then reconditioned with telencephalic stimulation (T) as UCS. As can be seen in Fig. 7, the first training session (B) produced a trial 20 score of 46% bradycardia, similar to the 49% mean shown by the first B group of fish (shown in Fig. 4). The fish took a mean of 54 trials to extinguish the response. In the second session of training, with telencephalic stimulation as UCS, the mean conditioned bradycardiac response was 60%. Despite this, extinction was complete within a mean of 4–6 trials.

Discussion

Although previous studies of the effects of telencephalic stimulation have established the existence of negatively reinforcing areas within this part of the brain, there have been no comparisons of the rates of learning of such animals with those of fish receiving more peripheral reinforcement. Olds (1956), for example, has shown that rats trained to run a maze to obtain positively reinforcing brain stimulation will learn and extinguish such a response with a performance which closely matches that of a group of rats rewarded with food.

Insofar as acquisition is concerned, brain stimulation and body-wall stimulation used as UCS were equally effective in promoting the rapid development of a bradycardiac classically conditioned response, and similar magnitudes of such response were observed. In addition, the presence of inert telencephalic electrodes had no effect on the formation of a conditioned response when body-wall stimulation was used as UCS. All animals were tested to obtain *c.* 25% unconditioned bradycardiac responses, and it was shown that over a period greater than that used in the experiment, such responses to a constant level of shock remained constant. Thus any subsequent differences in acquisition behaviour are unlikely to have been due to waning UCS efficiency in promoting UCR behaviour; however no such differences were manifested. Since previous work (Savage, 1971) has shown that behaviour indicative of negative reinforcement and behavioural arousal can be elicited from adjacent telencephalic sites, it is possible that stimu-

Fig. 7. Bradycardiac responses of five fish subjected to two classical conditioning regimes with a separation of 30 min. The animals received 10 10-s presentations of 10-Hz light, then 20 10-s presentations of light as CS, each followed by 0.25-s stimulation as UCS, and then 60 10-s unreinforced presentations of the light. In the first training session the UCS was stimulation of the body musculature, in the second session the UCS was telencephalic stimulation. Telencephalic electrodes were in place throughout the experiment.

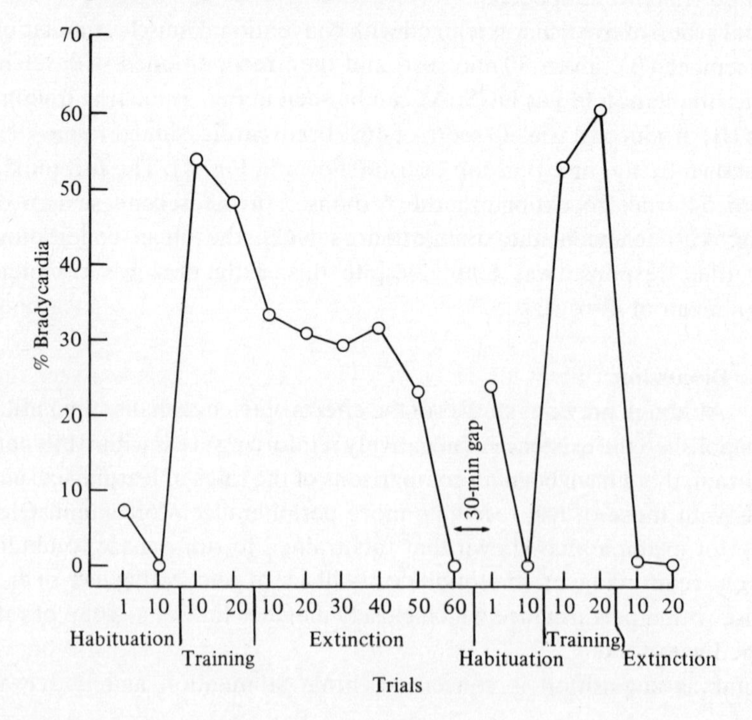

lation of the telencephalon produces hyper-arousal, which in turn causes animals to manifest a larger than usual cardiac component of the orientating response when presented with a neutral stimulus. This possibility can be discounted on the basis of the pseudoconditioning experiments, since neither B nor T groups gave any evidence of enhanced cardiac responses to light as a result of random stimulation and light presentations.

Studies of the acquisition of classically conditioned responses by fish with total telencephalic ablation have been consistent in their findings that no deficits are observed, whether the paradigm used be delay or trace conditioning (see, for example, Overmier & Curnow, 1969; Farr & Savage, 1978). Thus given that telencephalic stimulation is negatively reinforcing, no further effects on acquisition of a classically conditioned response would have been predicted from the lesion experiments.

The position with regard to the acquisition of instrumental responses by animals with telencephalic ablation is less clear, and hence so are the results expected from any study of acquisition of such responses utilizing brain stimulation reinforcement. There appears to be little impairment in acquisition as a result of removal of the telencephalon when such tasks as runway learning (Flood & Overmier, 1971), barrier crossing (Savage, 1969), colour discrimination learning (Janzen, 1933) or shape discrimination learning (Savage, 1969) are considered. However, instrumental avoidance studies have shown the telencephalon to be implicated in the learning of such tasks (see, for example, Kaplan & Aronson, 1967). Despite this, Dewsbury & Bernstein (1969) have demonstrated that fish lacking the telencephalon will successfully learn such problems if the barrier between the start and goal boxes is small. Thus when considering the probable effect of telencephalic stimulation on the acquisition of mastery of the end-preference task used in the present experiments, it is possible to argue either that the telencephalon is known to be involved in learning such tasks and that stimulation should interfere with learning, or that since no barriers were present, learning should be unimpaired.

In the actual experiment, there was a total lack of difference between the B and T groups of fish confronted with the simple preference reversal task. The habituation performances of the two groups were similar; thus the slightly different arrangements of wires just external to the body had no detectably different effect on their patterns of swimming. When stimulation was applied, there were no differences observed between fish subjected to the two types of stimulation as reinforcement. Thus as in the case of the classical conditioning experiments, telencephalic stimulation as UCS was as effective in promoting the formation of an association as was the body-musculature stimulation.

With regard to the successive reversal experiment, a difference in acquisition was detected; fish receiving brain stimulation reward showed a more consistent level of response than did fish receiving more conventionally delivered reward. The oscillating performance of the B group appears to indicate that these animals retained sufficient memory of their original preference for it always to be easier to learn to avoid the non-preferred end than to avoid the preferred end. The T group showed a similar oscillation initially, but the performance of these animals then remained more stable, and approximated to that of the B group when its members were learning to avoid the preferred end. Such a performance could be indicative of rapid forgetting in the T group; this is unlikely for two reasons. First, the extinction curves for B and T groups in the simple reversal experiment are practically identical. Secondly, if forgetting had occurred in the T group, learning should have been easier in the absence of an opposing engram. Thus the performances of the T group should have resembled those of the B group fish when they were avoiding the non-preferred end. Since this was not the case, it is probable that the animals of the T group were not showing inter-session forgetting. The extinction performance of the T group shows that the original end-preference had been lessened as compared with that shown by the B group. This could be taken once more as indicating that the original end performance of the B group remained, and that the animals of the T group, with strengthening of both performances, were less biased in their final choice.

In the classical conditioning experiments, extinction performances of the initial B and T groups were dramatically different. Even when a paired UCS (B + T) was used, telencephalic stimulation might promote high levels of conditioned bradycardia, but there was rapid waning of the conditioned response when extinction trials were given. Since a similar pairing of B with cerebellum stimulation yielded acquisition and extinction results similar to those of the initial B group, the extinction effect was peculiar to the experiments involving telencephalic stimulation.

The accelerated extinction observed could have been due to decay of information between trials or in consolidation, or to an increased labile state of the memory which would only cause decay if actual extinction trials were given. If intertrial information decay was occurring, inferior acquisition results as compared with the B group would have been expected. This was not observed, nor was such a deficit seen when the two types of UCS were administered simultaneously. Similarly, if consolidation failure was the cause of the effect, this should have appeared most obviously in the day-to-day results for the single reversal operant T group, but no evidence was obtained for overnight waning of performance as compared with the B group. For these reasons, it is suggested that the effect observed is one of increased susceptibility to extinction, not of spontaneous decay.

The question arises as to whether the effect was due to the telencephalic stimulation or merely to the presence of the electrodes. A control experiment showed that it was possible to perform two training and extinction sequences within 30 min of one another, without any noticeable performance decrement, if the UCS was muscular stimulation. In a similar experiment with T followed by B stimulation as UCS, equal conditioned responses were obtained, and rapid extinction was observed, when the telencephalic electrodes remained in place throughout the experiment. When the same animals were trained 12 h later, still with the telencephalic electrodes in place, no accelerated extinction was observed. In an experiment reversing this order, animals trained with muscle stimulation as UCS, but with inert telencephalic electrodes in place, showed slow extinction of the conditioned response, and showed accelerated extinction 30 min later when trained with telencephalic stimulation as UCS. Thus the presence of telencephalic electrodes appears not to interfere with the acquisition of a conditioned response trained with muscle stimulation UCS, nor to accelerate the process of extinction if unreinforced presentations of the CS are made. (Since no such effects were apparent during the extinction trials of the instrumental single reversal T group, it appears that the effect of telencephalic stimulation on promoting accelerated extinction is seen only shortly after training has ceased, and not if extinction trials are given 24 h later.) It is therefore not possible to argue that the effect reported here is due to the presence of the electrodes alone, as has been shown in rats with hippocampal electrodes (Landfield, Tusa & McGaugh, 1973). The after-effect of telencephalic stimulation lasted at least 30 min, and since it was possible to train a fish with B stimulation 30 min after training with T stimulation, it argues against the effect being due to some inadequacy of telencephalic stimulation as UCS.

In a recent review (Savage, 1980) it has been suggested that although the proposal that the telencephalon mediates secondary reinforcement (Flood *et al.,* 1976) has much to support it, the effects on retention of experiments using lesions can only be adequately explained if loss of some arousal state is also involved. Routtenberg (1968) has suggested that there may be separate reticular and limbic arousal mechanisms, and that the latter are specifically involved with reinforcement-related arousal. The telencephalic stimulation may cause such a state to be set up, which facilitates extinction. The most easily accomplished test of the hypothesis that the fish telencephalon is serving a function of this sort would be to train animals with telencephalic stimulation and then to conduct extinction trials immediately or at intervals of $\frac{1}{2}$, 1 and 2 h after training. If the hypothesized excitatory state had decayed by 2 h after training, extinctions should be less rapid than if examined immediately after training. In addition, tests need to be made by administering stimulation after conventional training, to see if retention is thus enhanced.

In summary, the results reported here are open to interpretation in two ways: first that telencephalic stimulation may interfere with retention, second that it may not interfere with retention *per se* but may render the stored information more susceptible to active extinction. From the evidence available, it seems more likely that the second hypothesis is correct, but experiments of the type outlined above are necessary to investigate this.

In the Introduction to this chapter it was noted that current theories of telencephalic function in teleost fish incline to the view that since the formation of first-order or classical associations is apparently unaffected by telencephalic damage, the dramatic avoidance learning deficits seen in such fish must be due to the involvement of the telencephalon in second-order information processing. Since the experiments reported here have demonstrated that telencephalic stimulation can affect the stability of a classically conditioned association, it seems that such an absolute distinction is too sweeping, and that some state induced in the telencephalon by stimulation is able to influence lower levels in the brain where presumably storage of simple associations must occur. Such observations as these suggest that it is unwise to construct a model of telencephalic function based solely on ablation studies. It appears that although simple associations are not stored in the telencephalon, its action may be to increase access to the memory system. So far, interference has acted to enhance extinction; it remains to be seen whether the driving of telencephalic rhythms can enhance retention.

The work described here was supported by the Science Research Council, whose assistance is gratefully acknowledged.

References

Boyd, E. S. & Gardner, L. C. (1962). Positive and negative reinforcement from intracranial stimulation of a teleost. *Science,* **136**, 648–59.

Demski, L. S. & Knigge, K. M. (1971). The telencephalon and hypothalamus of the Bluegill (*Lepomis macrochirus*): evoked feeding, aggressive and reproductive behavior with representative frontal sections. *Journal of Comparative Neurology,* **143**, 1–16.

Dewsbury, D. A. & Bernstein, J. J. (1969). Role of the telencephalon in performance of conditioned avoidance responses by goldfish. *Experimental Neurology,* **23**, 445–56.

Farr, E. J. & Savage, G. E. (1978). First- and second-order conditioning in goldfish and their relation to the telencephalon. *Behavioral Biology,* **22**, 50–9.

Fiedler, K. (1964). Versuche zur Neuroethologie von Lippfischen und Sonnenbarschen. *Verhandlungen der Deutschen Zoologischen Gesellschaft, Kiel,* **28**, 569–80.

Fiedler, K. (1968). Verhaltenswirksame Strukturen in Fischgehirn. *Verhandlungen der Deutschen Zoologischen Gesellschaft, Heidelberg,* **31**, 602–16.

Flood, N. B. & Overmier, J. B. (1971). Effects of telencephalic and olfactory lesions on appetitive learning in goldfish. *Physiology and Behavior,* **6**, 35–40.

Flood, N. B., Overmier, J. B. & Savage, G. E. (1976). Teleost telencephalon and learning: an interpretive review of data and hypotheses. *Physiology and Behavior,* **16**, 783–98.

Grimm, R. J. (1960). Feeding behavior and electrical stimulation of the brain of *Carassius auratus. Science,* **131**, 162–3.

Hainsworth, F. R., Overmier, J. B. & Snowdon, C. T. (1967). Specific and permanent deficits in instrumental avoidance responding following forebrain ablation in the goldfish. *Journal of Comparative and Physiological Psychology,* **63**, 111–16.

Janzen, W. (1933). Untersuchungen über Grosshirn funktionen des Goldfisches (*Carassius auratus*). *Zoologische Jahrbucher,* **52**, 591–628.

Kaplan, H. & Aronson, L. R. (1967). Effect of forebrain ablation on the performance of a conditioned response in the teleost fish, *Tilapia macrocephala. Animal Behaviour,* **15**, 438–56.

Landfield, P. W., Tusa, R. J. & McGaugh, J. L. (1973). Effects of post-trial hippocampal stimulation on memory storage and EEG activity. *Behavioral Biology,* **8**, 485–505.

Malyukova, I. V. (1964). Effect of electrical stimulation and partial coagulation of the forebrain (prosencephalon) and valvula cerebelli on the food-obtaining conditional reflexes in fishes. *Zhurnal Vysshei Nervnoi Devatelnosti imeni I.P. Pavlova*, **14**, 895–903.

Nolte, W. (1933). Experimentelle Untersuchungen zum Problem der Lokalisation des Associationsvermögens im Fischgehirn. *Zeitschrift für Vergleichende Physiologie,* **18**, 255–79.

Olds, J. (1956). Runway and maze behavior controlled by basomedial forebrain stimulation in the rat. *Journal of Comparative and Physiological Psychology,* **49**, 507–12.

Otis, L. S., Cerf, J. A. & Thomas, G. J. (1957). Conditioned inhibition of respiration and heart rate in the goldfish. *Science,* **126**, 263–4.

Overmier, J. B. & Curnow, P. F. (1969). Classical conditioning, pseudoconditioning and sensitization in 'normal' and forebrainless goldfish. *Journal of Comparative and Physiological Psychology,* **68**, 193–8.

Overmier, J. B. & Savage, G. E. (1974). Effects of telencephalic ablation on trace classical conditioning of heart rate in goldfish. *Experimental Neurology,* **42**, 339–46.

Peeke, H. V., Peeke, S. C. & Williston, J. S. (1972). Long-term memory deficits for habituation of predatory behavior in the forebrain ablated goldfish. *Experimental Neurology,* **36**, 288–94.

Roberts, M. G., Wright, D. E. & Savage, G. E. (1973). A technique for obtaining the ECG of fish. *Comparative Biochemistry and Physiology,* **44A**, 665–8.

Routtenberg, A. (1968). The two-arousal hypothesis: reticular formation and limbic system. *Psychological Review,* **75**, 51–80.

Savage, G. E. (1968). Function of the forebrain in the memory system of fish. In *The Central Nervous System and Fish Behavior,* ed. D. J. Ingle, pp. 127–38. University of Chicago Press, Chicago.

Savage, G. E. (1969). Some preliminary observations on the role of the telencephalon in food-reinforced behaviour in the goldfish, *Carassius auratus. Animal Behaviour,* **17**, 760–72.

Savage, G. E. (1971). Behavioural effects of electrical stimulation of the telencephalon of the goldfish, *Carassius auratus. Animal Behaviour,* **19**, 661–8.

Savage, G. E. (1980). The fish telencephalon and its relation to learning. In *Comparative Neurology of the Telencephalon,* ed. S. O. E. Ebbesson, pp. 129–74. Plenum Press, New York.

Segura, E. T. (1969). Effect of forebrain stimulation on blood pressure, heart rate and ST–T complex in toads. *American Journal of Physiology,* **217**, 1149–52.

Warren, J. M. (1961). The effect of telencephalic injuries on learning by paradise fish, *Macropodus opercularis. Journal of Comparative and Physiological Psychology,* **54**, 130–2.

DAVID BORSOOK, CLIFFORD J. WOOLF,
DALE VELLET, IAN S. ABRAMSON &
COLIN M. SHAPIRO

Pituitary peptides and memory in fish

Introduction

A large number of different functionally active neuropeptides are present within the nervous system (Elde & Hokfelt, 1979). Some of these peptides are involved in the control of the pituitary, but others seem to function as neurotransmitters or neuromodulators in a variety of sites which are believed to influence different behavioural states (Barchas, *et al.,* 1978). Of great interest is the apparent capacity of the posterior pituitary peptide hormones vasopressin and oxytocin to alter the adaptive behaviour of rats in performing learning tasks (De Wied, Bohus & Van Wimersma Griedanus, 1974). This effect on memory and learning is, moreover, independent of the hormonal actions of these peptides (Greven & De Wied, 1975; Bohus *et al*., 1978). In addition to this effect of posterior pituitary peptides, ACTHα and βMSH also alter specific behavioural responses in rats, by increasing the resistance to extinction of conditioned active and passive avoidance learning tasks and by enhancing the acquisition of certain learning tasks (Bohus, Gispen & De Wied, 1973; De Wied *et al*., 1974; Van Wimersma Griedanus & De Wied, 1975). Hypophysectomy impairs the performance of rats during conditional avoidance tasks and this impairment can be reversed by the administration of pituitary extracts or synthetic peptides (Bohus *et al.,* 1973). Antisera against peptides inhibit memory expression when administered into the cerebrospinal fluid of rats (Van Wimersma Griedanus, Dogterom & De Wied, 1975).

There is therefore strong evidence that exogenous and endogenous peptides significantly influence the acquisition and retention of information (Barker, 1975). Almost all of these studies have, however, only been performed on rats, using either conditioned active or passive avoidance learning tasks. We have now investigated whether the pituitary peptide modifies an adaptive motor response in fish. The model we have chosen to use was first described by Shashoua in 1968, when he demonstrated that goldfish have the ability to learn to swim upright following the attachment of floats to their ventral surfaces. This learning task includes a memory component that can be inhibited by puromycin (Shashoua, 1968) or actinomycin D and cyclohexamide (Woolf & Willies, unpublished). Changes in brain RNA (Shashoua, 1968; Kaplan, Dyer & Sirlin, 1973; Woolf *et al.*, 1974) and protein synthesis (Shashoua, 1976) have been found using this model. A neurochemical correlate of the learning task has also shown that the changes in RNA synthesis are time-dependent during the consolidation phase (Woolf *et al.*, 1974).

Methods

Animals and training procedure

Carp fingerlings (*Cyprinus carpio* L.) 7–8 cm in length were used in all experiments. The fish were maintained in large storage tanks with free running water and fed regularly. Four days prior to the acquisition trial the fish were transferred to 80-l aquaria equipped with aerators. These tanks were maintained at a room temperature of 21–22 °C. Twenty-four hours before the acquisition trial the fish were weighed and a permanent suture was made on the ventral surface, 1 mm caudal to the first pair of lateral fins. For identification purposes a small aluminium tag was secured around the suture. Polystyrene floats, 8.75% of the mass of a fish in 600 ml of water, were made. This formula was empirically derived to present the fish with a polystyrene float of such a density and size that the learning performance agreed with previously reported trials (Kaplan *et al.*, 1973). A suture was made through one side of the float and a small hook was used to couple this to the suture on the fish.

At the start of the experimental run the floats were attached to the sutures in eight fish and the fish placed in a tank (25 × 25 × 40 cm) equipped with an aerator and containing 30 l of water. The fish were observed at 15-min intervals and their progress scored following the technique of Kaplan *et al.* (1973). Three arbitrary stages of training were assigned: stage I, the fish swam upside down or were pulled to the surface; stage II, the fish swam continuously at a downward angle of 45°; stage III, the fish had acquired the skill to swim normally with the float. The following formula (Kaplan *et al.*, 1973) was used to calculate the percentage trained at each scoring interval for a group of eight fish:

$$\text{Percentage trained} = \frac{N\text{II}/2 + N\text{III}}{N_{\text{T}}} \times 100$$

where NII = number of fish at stage II;
NIII = number of fish at stage III;
N_{T} = total number of fish.

In addition to this scoring technique, the time taken for each fish to reach stage III was noted.

Two observers scored independently of each other and the results were compared. In most cases the results were practically identical. In order to measure the decay of the learnt information with time, the fish were subjected to three trials each. Trial 1 (day 1), the initial or acquisition trial, lasted 180 min; trial 2 (day 2), the second trial, took place 24 h after trial I and lasted 90 m; trial 3 (day 6), 4 days after the second trial, presented the fish with the same challenge again, for 120 min. All fish acquired stage III within the times allocated for each trial. The same float was used for a particular fish in all trials.

The fish were fed 24 h before a trial and all trials were performed at the same time of day (evening) and in the same season (summer).

Drug administration

Peptides. Drugs were administered intracranially by injections which were given through the anterior cranium 5 mm caudal to the eyes, using a microsyringe adapted to prevent intracerebral penetration. The peptides ACTH (ProActon, Propan), lysine-vasopressin (Ciba) and oxytocin (Syntocinon, Sandoz) were administered in volumes of 3 μl containing 0.006 IU. Control fish received the same volumes of saline. The fish were injected 1 h prior to the first trial and no further injections were given.

Denatured ACTH. ACTH was denatured by boiling for 5 min. The denatured peptide was administered in the same dose and volume as was the peptide.

Handling-stress control

In order to determine that the improved performance of the fish on repeated exposure to the learning task was the result of retention of specifically acquired information (i.e. memory) and not the result of extraneous sensory stimulation, for example handling, we performed the following experiment. Eight fish were presented with a float the volume of which greatly exceeded the capacity for which the fish could learn to compensate. These fish remained in stage I for the entire 180 min of their first trial. Twenty-four hours later these fish were given normal-sized floats and their performance assessed as if it were trial 1 of naïve fish. Trials 2 and trial 3 were also performed.

Results

Typical learning curves for the three trials in controls (i.e. fish with no intracranial injections) are shown in Fig. 1. These learning curves are similar to those reported by Kaplan *et al.* (1973) using goldfish. The mean times ± S.E. the fish took to learn to swim horizontally (stage III) were: 105 ± 11.8 min for trial 1, 40.7 ± 7.1 min for trial 2, and 62.1 ± 11.1 min for trial 3. The ability of the fish to adapt to the floats improved significantly when tested 24 h after acquisition ($P > 0.001$). The learned skill was still retained 4 days after the second trial because the fish in trial 3 reached stage III significantly faster than they did during trial I ($P > 0.05$). There is, nevertheless, some loss of memory, because the results for trial 3 were slower than those for trial 2. The fish thus demonstrated that during acquisition of the adaptive motor response information was stored and that this information decayed with time. Intracranial injections of saline had no significant effect on the performance of the fish.

The results of the stressed controls indicate that there is indeed a memory component associated with the learning task. Attaching large floats to the fish 24 h prior to a trial in which they had normal-sized floats attached did not improve the performance of these fish compared with naïve control fish. The modification of performance on repeated training in control fish is, there-

Fig. 1. Learning and retention scores for a group of eight control fish which were challenged with floats in three trials. The first trial (T1) lasted 180 min on day 1 and tested the acquisition of the new swimming skill; the second trial (T2) lasted 90 min on day 2; and the third trial (T3) lasted 180 min on day 6 (these trials assessed the retention by the fish of the new swimming skill).

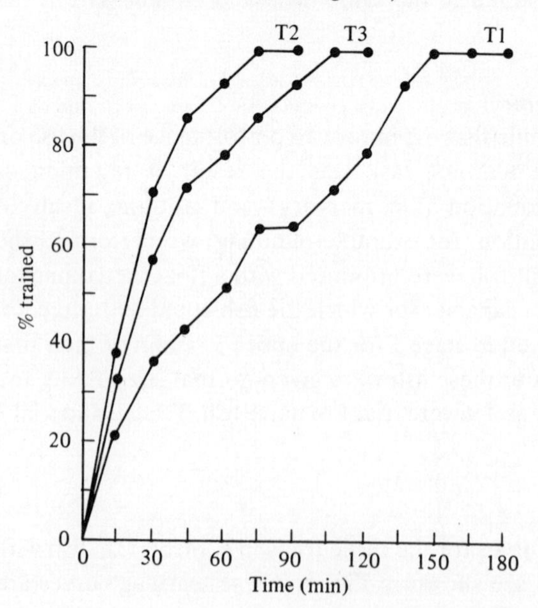

fore, the result of acquiring a new skill and not dependent on other extraneous factors.

Intracranial administration of ACTH, lysine-vasopressin and oxytocin markedly altered the performance of the fish (Fig. 2). All the peptides significantly enhanced the acquisition of the adaptive motor response during trial 1. The fish demonstrated a greatly increased capacity to overcome the disturbance produced by the float. Table 1 gives the mean time ± S.E. taken for each group of eight fish to reach 100% trained in the three trials.

Table 1. *The effects of intracranial injections of peptide on the acquisition of a float-impeded swimming task*

	Drug administration			
Trial	Control	ACTH	Vasopressin	Oxytocin
I	94.4 ± 16.6	11.2 ± 7.4 ($P < 0.001$)	45.0 ± 8.0 ($P < 0.01$)	50.6 ± 2.7 ($P < 0.02$)
II	45.0 ± 7.4	6.9 ± 0.9 ($P < 0.001$)	18.7 ± 3.3 ($P < 0.01$)	48.7 ± 5.4 (NS)
III	61.9 ± 9.8	7.5 ± 1.6 ($P < 0.001$)	54.4 ± 5.2 (NS)	52.5 ± 6.3 (NS)

The mean times ± S.E. for the control and peptide-injected fish ($n = 8$ per group) to reach 100% trained are given. The P values in parentheses represent the significant difference between the controls and the experimental fish (Student's t-test).

Fig. 2. Learning and retention scores for fish injected intracranially with peptides. Eight fish in each group were given 0.006 IU in 3 μl of one of the three peptides; controls received 3 μl of 0.7% saline. The fish were challenged with polystyrene floats in three trials as described in Fig. 1. ●——●, control; ■——■, ACTH; ▲ – – – – ▲, vasopressin; ○·········○, oxytocin.

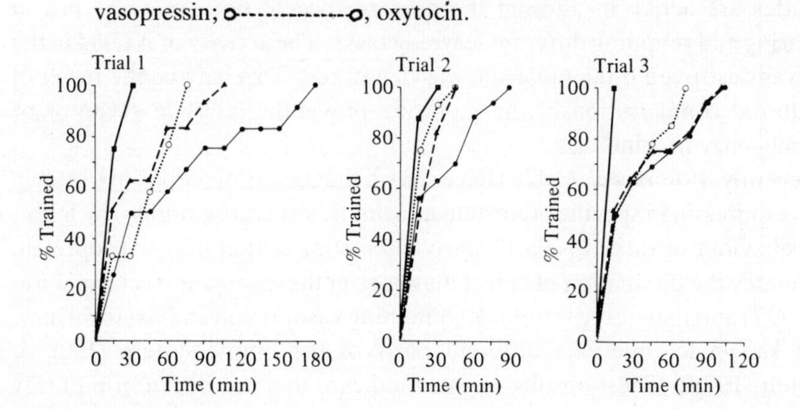

Although the fish received their only injection of a particular peptide 1 h before trial 1, they still performed significantly better in trial 2 than did control fish. Only the fish administered with ACTH showed significantly increased performance in trial 3. Whether the fish had a specifically enhanced retention of information, or whether they only performed better in trial 2 because they had performed so well in trial 1, is difficult to ascertain. Unlike the enhancement of acquisition of the learning task produced by ACTH, intracranial injections of denatured ACTH significantly inhibited the response of the fish to the float in trial 1. In trial 2 the fish still performed less effectively than the control fish, while in trial 3 their performance was only marginally better than the controls but still significantly worse than fish receiving ACTH. The normal structural integrity of the peptide is therefore required for ACTH to improve the learning and memory performance of fish.

Discussion

The above results show that injections of ACTH, lysine-vasopressin or oxytocin, enhance the acquisition and increase the retention of an adaptive motor response in fish. Previous work on the effects of pituitary peptides on memory and learning has been conducted primarily in rats, using only conditioned active and passive avoidance responses; although vasopressin, ACTH and MSH facilitate the consolidation and retrieval of memory processes in rats (Bohus *et al.*, 1973; De Wied *et al.*, 1974), oxytocin attentuates the responses (Bohus *et al.*, 1978). All the peptides we tested, including oxytocin, increased acquisition and retention of the adaptive motor response in the fish. In the fish ACTH had a longer effect than vasopressin; in rats the pattern of changes found is the reverse (De Wied *et al.*, 1974). Because of these differences between rats and fish it is difficult to evaluate whether the peptides act in a similar fashion in the two animals. Nevertheless, pituitary peptides are active in altering the performance of both rats and fish in acquiring and retaining different learning tasks. The activity of ACTH in the fish was destroyed if the molecule was denatured. This may be the result of an altered configuration of the molecule preventing peptide–receptor or peptide–enzyme binding.

Recently, Bohus *et al.* (1978) have shown that microinjections of oxytocin and vasopressin in specific brainstem and limbic structures modify the learning behaviour of rats. It is particularly interesting that arginine-vasopressin modulates the dissipation of catecholamines in these same areas (Tanaka *et al.*, 1977) and that nerve terminals containing vasopressin and oxytocin have been shown to innervate different parts of the limbic system (Elde & Hokfelt, 1979). These results seem to indicate that the different pituitary peptides may have a physiological role in controlling or modifying behavioural responses.

Pituitary peptides are, therefore, active in both mammals and fish in altering adaptive behaviour, though there are differences in the response of these animals to the peptides and further work is required to determine whether the effects of the peptides are due to similar mechanisms in the different species.

References

Barchas, J. D., Akil, A., Elliot, G. R., Holman, R. B. & Watson, S. S. (1978). Behavioural neurochemistry: neuroregulators and behavioural states. *Science*, **200**, 964.

Barker, J. L. (1975). Peptides: roles in neuronal excitability. *Physiological Review*, **56**, 435.

Bohus, B., Gispen, W. H. & De Wied, D. (1973). Effect of lysine-vasopressin and $ACTH_{4,10}$ on conditioned avoidance behaviour of hypophysectomized rats. *Neuroendocrinology*, **11**, 137.

Bohus, B., Urban, I., Van Wimersma Griedanus, Tj.B. & De Wied, D. (1978). Opposite effects of oxytocin and vasopressin on avoidance behaviour and hippocampal theta rhythm in the rat. *Neuropharmacology*, **17**, 237.

De Wied, D., Bohus, B. & Van Wimersma Griedanus, Tj.B. (1974). The hypothalamo-neurohypophyseal system and the preservation of conditioned avoidance behaviour in rats. *Progress in Brain Research*, **41**, 417.

Elde, R. & Hokfelt, T. (1979). Localization of hypophysiotropic peptides and other biologically active peptides within the brain. *Annual Review of Physiology*, **41**, 587.

Greven, H. M. & De Wied, D. (1975). The influence of peptides derived from corticotropin (ACTH) on performance: structure–activity studies. *Progress in Brain Research*, **39**, 429.

Kaplan, B. B., Dyer, J. C. & Sirlin, J. L. (1973). Macromolecules and behaviour: effects of behavioural training on transfer R.N.A.s of goldfish brain. *Brain Research*, **56**, 239.

Shashoua, V. E. (1968). RNA changes in goldfish brain during learning, *Nature, London*, **217**, 238.

Shashoua, V. E. (1976). Identification of specific changes in the pattern of brain protein synthesis after training. *Science*, **193**, 1264.

Tanaka, M., De Kloct, E. R., De Wied, D. & Versteag, D. A. G. (1977). Arginine vasopressin affects catecholamine metabolism in specific brain nuclei. *Life Sciences*, **20**, 1799.

Van Wimersma Griedanus, Tj.B. & De Wied, D. (1975). The role of vasopressin in memory processes. *Progress in Brain Research*, **42**, 187.

Van Wimersma Griedanus, Tj.B., Dogterom, J. & De Wied, D. (1975). Intraventricular administration of anti-vasopressin serum inhibits memory consolidation in rats. *Life Sciences*, **16**, 637.

Woolf, C. J., Willies, G. H., Hepburn, H. R. & Rosendorff, C. (1974). Time dependence of a neurochemical correlate of a learning task: a non-disruptive approach to memory consolidation, *Experientia*, **30**, 760.

INDEX